Pathways to Modern Chemical Physics

Salvatore Califano

Pathways to Modern Chemical Physics

Springer

Salvatore Califano
Laboratorio Europeo di Spettroscopie non
Universitá di Firenze
Firenze
Italy

ISBN 978-3-642-28179-2 ISBN 978-3-642-28180-8 (eBook)
DOI 10.1007/978-3-642-28180-8
Springer Heidelberg New York Dordrecht London

Library of Congress Control Number: 2012940202

Printed on acid-free paper

Springer is part of Springer Science+Business Media (www.springer.com)

Preface

This book originates from the suggestion made by several colleagues to extract certain sections from a two-volume book that I recently published in Italian with the Bollati-Boringhieri publishing house. The sections concerned deal with recent developments in chemical physics and the intention was to implement them with additional material in order to produce a book in English, explicitly dealing with the progress of chemical physics, in particular that realized in the last two centuries. As a professor of chemical physics, I felt encouraged to fill this gap by producing a book that could offer to new generations of chemistry students a testimony of the commitments, hopes, and dreams that my generation has experienced throughout this fascinating adventure.

Although chemistry has its roots in alchemy, or even earlier in the old Sumerian, Babylonian, and Egyptian cultures, chemical physics became an independent discipline only in the second half of the eighteenth century. At this time the efforts of several scientists interested in developing the basic theoretical aspects of chemistry and their relationships with physics gave rise to the birth of this new discipline and to the creation of the first chairs and journals of chemical physics. A reasonable official birth date of chemical physics seems to be the year 1887, when Ostwald founded the *Zeitschrift für physikalische Chemie*. A journal which was clearly inspired by chemical physics, the *Annalen der Physik und der Physikalischen Chemie*, had in fact already been founded in 1819 by Ludwig Wilhelm Gilbert (1769–1824), but it changed its name and interests after only few years, and in 1824 it transformed into the *Annalen der Physik und Chemie* edited by J.G. Poggengorff.

The connection between chemistry and physics, however, goes back much further, as testified by the *Sceptical Chymist* of Robert Boyle in 1661, bearing the subtitle *Chymico-physical Doubts and Paradoxes* and by a lecture held at the St. Petersburg University by the Russian chemist Mikhail Lomonosov in 1752, entitled "A Course in True Physical Chemistry". In the eighteenth century, specialization was not yet tightly structured and ties between behavior and composition of matter were such that it was often difficult to classify the contributions of different researchers in current terms. For instance, the mathematician and physicist Pierre Simon Laplace contributed, together with Lavoisier, to water synthesis, and at the beginning of the nineteenth century another mathematician, André-Marie Ampère, developed a hypothesis parallel to that of Avogadro. Newton, a physicist, was interested in alchemy, while Faraday, a chemist, had provided fundamental contributions to electromagnetism.

By the nineteenth century, however, chemical physics was already a mature science and developed along several separate directions: thermodynamics, kinetics, electrochemistry, molecular structure determination, and colligative properties of the solutions. The premises to this type of organization of the discipline had already existed since the eighteenth century: the affinity theory, born with Boerhaave, Macquer, Geoffroy, and Berthollet, assumed the form of a proto-theory of the chemical bond. The caloric theory and the study of phase transitions of Lavoisier paved the way to thermodynamics. The series of Berzelius created the basis on which the theory of electrolytic dissociation could be developed. Atomic spectroscopy originated from the works of Fraunhofer, Kirchhoff, and Bunsen, while Stanislao Cannizzaro and Hermann Kopp introduced the concepts of dissociation and chemical equilibrium from which chemical kinetics originated. With the discovery of the electron, the affinity concept found its correct interpretation and was transformed into the modern chemical bond theory that was then completely assimilated into the developing quantum mechanics. In the evolution of the different branches of chemistry, lines of thought, new methodologies, and different problems and solutions cumulated and amalgamated in time, giving rise to a science that perfectly blends the theoretical structure and quantitative exactness of physics with the necessary systematization and classification typical of the natural sciences. As a consequence of this complex and articulated situation, chemical physics represents today a natural bridge connecting very different scientific disciplines such as physics, molecular biology, geology, mineralogy, and even astrophysics.

Public opinion sees chemistry as essentially an experimental laboratory practice and a chemist as an alchemist or, at best, a dowdily clad gentleman of the nineteenth century mixing chemicals in a test tube in an old-fashioned laboratory. However, in fact chemistry possesses a very important theoretical structure, elaborated essentially in the last two centuries, that has contributed to the development of all kinds of scientific knowledge and to the formulation of the theories that now constitute the supporting framework of modern science.

The vision that the public has of chemistry results from a secular diffidence and an old form of humanism, bound to the concept of an unremitting separation between the two cultures, leading to two incompatible directions of thought and consequent mistrust of scientific conquests. The fear was that they could orient the development of modern society toward Orwellian scientific power, able to condition the freedom and the humanity of future societies. In contrast to this opinion, this book represents an attempt to convince the reader that chemistry, and in particular the more theoretical structure represented by chemical physics, is an integral part of the general culture of mankind and that, more than any other branch of science, had a profound influence on the growth of culture in modern societies. The development of chemistry, and particularly that realized in the last two centuries, coincides with the development of civilization, owing to the strong influence that chemistry has on most of the scientific, technological, and even social growth of mankind. The birth and growth of modern industrial structure and the important social transformations that arise from it, are indeed strictly connected with theoretical developments in chemistry and reflect its historical evolution, all its conceptual transformations and contradictions, and all its constant enrichments and extraordinary achievements.

The history of modern chemical physics is, in many respects, the history of the great theoretical and technological achievements that humankind has realized in developing its control of nature and in realizing its freedom from hunger, sicknesses, poverty, and sorrow. It is essentially the history of the contributions that ranks of chemists have made to improve the knowledge of the structure of our world and of the fascinating intellectual adventure represented by their participation in the development of natural philosophy. It is also the story of how the human mind has succeeded in penetrating the secrets of nature and of how knowledge and the ideas of ancient times have flowed together in the structure of modern culture, making it possible to modify a hostile nature, adapting it to the needs of modern society in continuous improvement.

It is actually inherent to the nature of chemical physics to give more space to the ideas that have produced variations of the paradigms supporting the theoretical structure of modern chemistry rather than to results obtained in practical applications. This book is therefore more a history of ideas than an account of results, and for this reason the historical development of chemical physics is presented in terms of problems, although still paying the necessary attention to the illustration of the heroes in this wonderful adventure. Svante Arrhenius, Nobel laureate for his contributions to modern electrochemistry, gave a very enthusiastic evaluation of the importance of chemical physics in scientific culture. He said:

> The theoretical side of physical chemistry is and will probably remain the dominant one; it is by this peculiarity that it has exerted such a great influence upon the neighboring sciences, pure and applied, and on this ground physical chemistry may be regarded as an excellent school of exact reasoning for all students of the natural sciences.

This book is organized into eight main chapters, each separated in turn into several sections presenting the fundamental directions in which chemical physics has evolved since its birth. The first three chapters deal with the history of the growth of classical chemical physics, particularly thermodynamics, chemical equilibrium, and electrochemistry. Classical thermodynamics describes energy transformations in terms of macroscopic variables and has always been the battle horse of theoreticians, owing to its lucid and logically perfect treatment of the transformations of any physical system. The development of classical thermodynamics is supplemented by that of statistical thermodynamics, relative to irreversible and out-of-equilibrium processes and to the interpretation of the microscopic interactions between individual particles or quantum-mechanical states. Chapter 2 treats another classical problem, that of chemical equilibrium, presenting a detailed treatment of the modern approach to the problem in terms of catalysis and including a description of the most advanced techniques utilized in the study of catalytic processes. Chapter 3 covers the strict connection between matter and electricity, the different theories of ionic processes in solution, the modern version of ionic transport in solution, and the study of colligative properties. Chapter 4 is the first in the part of the book dedicated to advanced research problems typical of modern chemical physics, including rotational, vibrational, nuclear, and electronic resonance spectroscopy, as well as x-rays and neutron diffraction. Chapters 5 and 6 are dedicated to the history of the discovery of the electron and nuclear structure, the development of models of the atom, theories of the chemical bond in the framework of the old quantum theory, and the related problems of natural and artificial radioactive substances. Chapter 7 then covers the birth of quantum mechanics and its extension to the study of the quantum behavior of molecules with particular reference to the various methods of approximation that have allowed the quantum treatment of large molecular systems. Finally, Chap. 8 illustrates how all these theoretical aspects have converged in the study of the mechanisms of chemical reactions, i.e., in offering to the organic chemists a theoretical justification of their work on synthesis.

Florence, Italy
9-April 2012

Salvatore Califano

Acknowledgments

In concluding this brief presentation of the book I wish to make clear how much I feel indebted to all friends and colleagues that have helped me with their advice and suggestions to improve the text. First of all I would like to express my gratitude for the seminal discussions on the history of science with Paolo Rossi over years of fraternal friendship. I also wish to acknowledge the help and suggestions from my friends and colleagues Riccardo Cheli, Gianni Cardini, Giovanni Giacometti, Roberto Moccia, Roberto Righini, and Vincenzo Schettino. Special thanks are due to Elizabeth Hawkins for her valuable help in the organization and presentation of the book and for her assistance in improving the English text.

Finally I must express my gratitude to my wife Mimma for the respect and love for culture that she has developed in me over so many years of our life together. I hope that she will accept this book as an excuse for the long evenings in which I have foresaken her company for that of the ghosts of the great personages that have contributed to the construction of the marvelous building blocks of chemistry over the centuries.

Salvatore Califano

Contents

1 **Chemistry and Energy** 1
1.1 Thermodynamics and Heat Theory 1
1.2 Thermal Engines 3
1.3 Entropy and Free Energy 6
1.4 Thermochemistry 9
1.5 Statistical Thermodynamics 13
1.6 Non-equilibrium Thermodynamics 15
References 23

2 **Kinetics and Chemical Equilibrium** 27
2.1 Affinity and Reactivity 27
2.2 Chemical Equilibrium 28
2.3 The Quantum Mechanical Approach to Chemical Kinetics 34
2.4 Catalysis 40
References 52

3 **Matter and Electricity** 59
3.1 The Association of Matter and Electricity 59
3.2 Solutions 70
3.3 The Debye–Hückel Theory 78
3.4 Acids and Bases 81
References 83

4 **Chemical Physics Structural Techniques** 87
4.1 Development of Physical Techniques in Chemistry 87
4.2 X-Ray Diffraction 88
4.3 Neutron Diffraction 91
4.4 Vibrational Spectroscopy 92
4.5 Rotational Spectroscopy 96
4.6 Nuclear Magnetic Resonance (NMR) 100
4.7 Electron Spin Resonance Spectroscopy EPR 103
References 105

5 The Electron and Atomic Structure 109
5.1 Birth of the Electron 109
5.2 Models of the Atom 112
5.3 The Old Quantum Theory 117
5.4 The Electronic Theories of the Chemical Bond 122
5.5 The Aufbau Principle 128
5.6 Electron Spin 130
References 132

6 Radioactivity 135
6.1 The Revival of Inorganic Chemistry 135
6.2 The Curie Couple 136
6.3 Hunting for New Radio Elements 143
6.4 Transmutation of the Elements 145
6.5 Completion of the Periodic Table 152
6.6 Transuranium Elements 158
References 163

7 From Quantum Mechanics to Quantum Chemistry 167
7.1 Planck's Quanta 167
7.2 Quantum Mechanics 172
7.3 Quantum Chemistry 183
7.4 Interatomic and Intermolecular Forces 187
7.5 The Valence Bond Theory 190
7.6 The Molecular Orbital Theory 199
7.7 The English School of Quantum Chemistry 208
7.8 The Density Functional Theory 211
7.9 Molecular Dynamics Simulation 213
References 216

8 The Mechanisms of Chemical Reactions 223
8.1 Reaction Mechanisms and Residual Affinities 223
8.2 The Robinson–Ingold Controversy 229
8.3 Short Living Molecules: Carbanions and Radicals 237
8.4 Theoretical Organic Chemistry 248
8.5 Organic Photochemistry 254
References 264

Index 271

Chapter 1
Chemistry and Energy

If someone points out to you that your pet theory of the universe is in disagreement with Maxwell's equations – then so much the worse for Maxwell's equations. If it is found to be contradicted by observation – well, these experimentalists do bungle things sometimes. But if your theory is found to be against the second law of thermodynamics I can give you no hope; there is nothing for it but to collapse in deepest humiliation. (Arthur Eddington, The Nature of the Physical World)

1.1 Thermodynamics and Heat Theory

The connection between chemistry and physics played a decisive role when the use of energy became central to a world in rapid transformation from an agricultural to an industrial structure. The control of the transformation of thermal energy into mechanical energy, initiated in England with the industrial revolution, acquired a much deeper importance with the industrial mechanization which profoundly modified the organization of the work, as well as the economy and the social structure of developed countries.

The only important source of energy available at the time came from wood and fossil fuels. It was thus natural that the attention of chemists, physicists and engineers would be focused on thermal energy, "the heat". Up until the first half of the nineteenth century, heat was considered as a weightless fluid called *caloric*. The term was created by Lavoisier, who imagined that it was contained in matter, but was able to escape in combustion processes where it flowed from hot to cold bodies. The idea that heat was connected to other forms of energy had been, however, already put forward by Davy, who in 1799 had shown that two pieces of ice, rubbed against one another, produced enough heat to melt (Davy 1799). This experiment offered significant evidence against the caloric theory, but at that time was not considered seriously in the scientific community. The connection between heat and mechanical energy was instead well understood by the American, English naturalized, Benjamin Thompson, later Count of

S. Califano, *Pathways to Modern Chemical Physics*,
DOI 10.1007/978-3-642-28180-8_1,

Rumford, who realized the equivalence of work and heat while assisting the boring of cannons at the arsenal in Munich, Bavaria.

Benjamin Thompson (1753–1814) was one of the most disquieting personages in the history of science, at the same time a soldier, an inventor, a spy and a physicist as well as a good politician (Brown 2001). Born in Massachusetts, at the age of 19 he married a rich widow of *Rumford* (today *Concord*, capital of *New Hampshire*) who he soon abandoned. While serving in the American army, he became a spy for the English intelligence service. Once manifest, he escaped to London where, thanks to the protection of Lord George Germain, secretary of State, he was appointed in 1780 Under-Secretary of State for the colonies. In the meantime, he also became a scientist and a gunpowder expert (Thompson 1781).

A restless spirit, mixture of genius and charlatanry, of selfishness and altruism, Thompson left England in 1785, moving to Bavaria where he spent 11 years reaching the highest ranks in the Government, up to the position of minister of war. In Bavaria he proved his great ability for organization: he reorganized the army, established workhouses for the poor people, improved the education of the working class, developed the industry, diffused the cultivation of the potato and eliminated the bands of thieves that ravaged the country. Among others in 1789, he transformed his property into a great urban public park of Munich, known still today as the "English Garten". When he left, the Munich municipality erected a statue of him in front of the Maximilian University and gave his name to a street in the city.

Owing to his expertise in explosives, Thompson became interested in several problems connected to heat conduction (Thompson 1786). He also invented a calorimeter, studied different improvements for chimneys and fireplaces and even new techniques for food cooking. However, it was when manufacturing Bavarian armaments that in 1798 Thompson made his most important discovery. While watching the manufacture of cannon barrels by boring solid blocks of metal with large drills, he realized that the heat released in the process was by no means proportional to the amount of chip produced, as required by the caloric theory. On the contrary the heat was continuously released as the drilling went on and continued to be released for a long time even when the boring was stopped. He thus realized that the friction due to the mechanical boring procedure gave rise to heat and concluded that heat was itself a form of energy. He published his finding in 1798, in the paper *Enquiry concerning the Source of Heat which is excited by Friction* (Thompson 1798).

For his political and scientific merits the Great Elector of Bavaria made him in 1790 a Count of the Holy Roman Empire with the title of Count of Rumford, from the name of the American township where he had got married. He went back to England and founded in 1799 in collaboration with Sir Joseph Banks, the *Royal Institution of Great Britain*. This institution in a short time became the most important cultural center of the country and hosted the greatest British scientists and intellectuals. In 1802 he moved to Paris where he met the widow of Lavoisier who he married in 1804, after the death of his American wife. The marriage, however, was not a happy one; at times it was even stormy. The couple soon separated and Thompson spent the rest of his life far from Paris, in the villa that Lavoisier had built in Auteil (Brown 2001).

1.2 Thermal Engines

At the beginning of the nineteenth century the transformation of thermal energy into mechanical energy required the construction of efficient thermal engines as well as the understanding of the theoretical and practical problems connected to their activity.

Thermal engines able to perform mechanical work were invented in England at the beginning of the industrial revolution. The first of these, invented in 1695 by the engineer Thomas Savery (1650–1715) to pump water from the bottom of coal mines, was a kind of Denis Papin's *steam digester,* a type of pressure cooker in which steam under pressure forced the water to rise in a pipe and to reach the ground surface (Ceccarelli 2007). The first steam engine with a piston, built in 1712 by the English blacksmith Thomas Newcomen (1663–1729), used instead the atmospheric pressure to work. It functioned through the introduction of vapour in a cylinder containing a piston in which the steam, first heated and then cooled, created a depression that moved the piston. The engine, pompously called "motor" and operated for the first time in 1705, had very low efficiency and required long time intervals between two successive steps (Rolt and Allen 1998). A steam apparatus truly deserving the name "engine", was only developed much later by the Scott, James Watt, a pupil of Joseph Black, who worked at Glasgow University as a builder of scientific instruments. In 1763, repairing a Newcomen engine, Watt realized that most of the heat was lost to first warm up and then cool down the cylinder to condensate the steam. After 3 years of experiments and meditations he had the idea of condensing the steam in a separate vessel. In this way it was no longer necessary to first warm up and then cool down the cylinder and the engine became much faster, safer and more efficient. The piston engine became a motor of broad application only in 1781, when Watt developed a mechanical system able to transform the straight motion of the piston into a rotation. Watt also introduced the term *horse-power* to measure the engine power, i.e. the ratio between the work made and the time required to do it (Marsden 2002).

Despite Watt's invention and the following improvements, a physical explanation of the steam engine action was still unknown. Its theoretical analysis was first undertaken about 40 years later by the Parisian Nicolas Léonard Sadi Carnot (1796–1832), *cadet* of the École Polytechnique and later engineer in the French army. In a manuscript from 1822 to 1823, never published and discovered in handwritten form only in 1966, Carnot tried to quantify the work produced by a kilogram of steam. In a 120 pages of text (Carnot 1824) published at his own expenses in 1824, he faced the problem that there was an occurrence of an upper limit to the work produced by a heat source. The book, written in a simple and clear form with a minimum number of equations, required little knowledge of algebra and arithmetic. The most important part of the text concerned the description of an ideal system, which was utilized to clarify the general principles of thermal engines independently from their form and practical application. Carnot, educated according to the knowledge of his time, believed that the caloric fluid was able to flow from a warm to a cold body. Using the classical concepts of fluid mechanics, he thought that since the work of hydraulic engines depends on the variation of the water level, the work done by a thermal engine should be due to the transfer of caloric fluid from a warm to a cold source and thus should depend only on the temperature jump and not on the nature of the fluid (Carnot 1824):

> La production de la puissance motrice est donc due, dans les machines à vapeur, non à une consommation réelle du calorique, mais à son transport d'un corps chaud à un corps froid, c'est à dire à son rétablissement d'équilibre, équilibre supposé rompu par quelque cause que ce soit, par une action chimique, telle que la combustion, ou par toute autre. Nous verrons bientôt que ce principe est applicable à toute machine mise en mouvement par la chaleur........La puissance motrice de la chaleur est indépendante des agents mis en ouvre pour la réaliser; sa quantité est fixée uniquement par les températures des corps entre lesquels se fait en dernier résultat le transport du calorique.

Carnot also introduced the concept of reversibility of an ideal engine, assuming that a thermal machine could work cyclically transforming heat into work and work into heat. He realized, however, probably upon suggestion of his father, that this cyclic action, known as the Carnot cycle, was purely ideal, a portion of the caloric being always lost in the cycle, and therefore suggested that any real engine was always condemned to a lower efficiency than the theoretical one.

Carnot text did not receive much attention from the scientific community for this work and remained practically ignored until rediscovered by Émile Clapeyron, who resuming Carnot's ideas, redeemed them from the old concept of caloric. Benoît Paul Émile Clapeyron (1799–1864) was also Parisian and a student at the École Polytechnique. After 10 years spent directing building construction and teaching mathematics in St. Petersburg, he returned in 1830 to France to participate to the development of the French railways system. In 1834, Clapeyron became involved in the study of the driving force. He gave mathematical form to Carnot's ideas and integrated them with a graphic representation of Carnot's cycle in the form of a pressure/volume diagram (Clapeyron 1843). In 1843 he also developed Carnot's idea of reversible transformations and in 1844, was appointed professor at the École des Ponts et Chaussées, elaborated the famous *Clausius-Clapeyron* equation

$$\frac{\partial \mathrm{P}}{\partial \mathrm{T}} = \frac{\mathrm{L}}{\mathrm{T}\Delta \mathrm{V}}$$

connecting the pressure variations as a function of temperature T with the volume variation ΔV and with the latent heat L that represents the heat released or absorbed in a transition of state at constant temperature.

The equivalence between mechanical work and heat was definitively assessed in the period 1840–1848 by Dalton's pupil *James Prescott Joule* (1818–1889), a brewer and physics amateur at Manchester. Joule conducted a series of experiments and measured the ratio of heat/work. In one of his first papers (Joule 1841) Joules measures the heat developed in a conductor by the action of the electric current. He found the relationship now known as *Joule's law*, which states that the quantity of heat Q developed in a time *t* by an electrical current flowing through a circuit is proportional to his resistance R and to the square of the current intensity I

$$Q = R I^2 t$$

Joule conducted several experiments to study the transformation of mechanical work into heat, spinning with the aid of a free falling weight in a system of paddle wheels in an insulated barrel of water. The work done by the spinning paddles was transformed into heat that warmed up the water. In 1843 he presented to a meeting of the *British Association for the Advancement of Science* a manuscript on the transformation of mechanical work into heat. This was received without enthusiasm and refused publication by several scientific journals. It was eventually published by the Philosophical Magazine with the title *The Mechanical Equivalent of Heat* (Joule 1843).

Joule's experiments proved without doubt that to perform work or to add heat were two equivalent ways of increasing the energy of a physical system. In 1847, Joule delivered in the reading room of *St. Ann's Church* in Manchester, a lecture entitled *Matter, Living Force, and Heat,* which was consequently published between 5 May and 12 May in the Manchester Curier, presenting a comprehensive and popular speech his version of the energy conservation principle, the first principle of thermodynamics, which states that when work is produced a quantity of heat proportional to the work done is consumed:

> You see therefore, that living force can be converted into heat, and that heat may be converted into living force, or its equivalent attraction through space. All three, therefore – namely heat, living force and attraction through space (to which I might also add light, were it consistent with the scope of the present lecture) – are mutually convertible into one another.

In 1847 Joule was still treating heat in the tradition of the *vis-viva* (Brush 2003), and in fact in his speech said:

> This force possessed by moving bodies is termed by mechanical philosophers vis viva or living force. ... wherever living force is apparently destroyed an equivalent is produced which in process of time can be reconverted into living force. This equivalent is heat.

Almost at the same time the equivalence between work and heat was realized also by the German physician *Julius Robert von Mayer* (1814–1878). During a long journey to the Dutch East Indies, Mayer discovered that the sailors venous blood was more red at the tropics than in Europe and deduced that this was due to a greater concentration of oxygen in the blood, since in hot countries there is a lower consumption of oxygen to keep the body warm. He concluded therefore that the muscular work, as well as all other possible forms of energy could be converted one into another.

Mayer also realized that the heat necessary to keep the temperature of the body constant is balanced by the oxidation of a given amount of food and therefore that work and energy are not created but transformed into one another. In 1841 he sent an essay entitled *Erhaltungssatz der Kraft* to the *Annalen der Physik.* The essay postulates a law of conservation of force and was not accepted by the editor because it contained too many mistakes in basic physics. In 1842, Mayer, following the advice of Johann Gottlieb Nörremberg, professor at Tübingen, in a new paper

(Mayer 1842) presented his evaluation of the mechanical equivalent of heat, which was obtained measuring the heat produced by the compression of air. His incapability to express himself in an acceptable language of physics made it difficult for him to convince the scientific community of the correctness of his ideas, so that many contemporary physicists, including even von Helmholtz and Joule, opposed to his conclusions. Since his contribution to physics was not recognized and the merit of the first principle was attributed to Joule, with whom he developed a bitter dispute over priority, he suffered in 1848 of a serious depressive crises made even worse by the death in rapid succession of his two children. This drove him to attempt suicide, an event that committed him to a mental health clinic. In the meantime, however, his fame had grown and the recognition of his work started to increase, although too late. In 1859 he received the title of doctor *ad honorem* by the philosophical faculty of the University of Tübingen; in 1862 John Tyndall made a public revival of his works in a lecture at the Royal Institution and in 1867 he was awarded the nobility title of *von Mayer*. In 1867 he published the book *Die Mechanik der Warme* (Mayer 1867) presenting in complete form the mechanical theory of heat.

As much unrecognized, both in his native country and in the rest of Europe, was also the contribution of the Danish engineer *Ludwig August Colding* (1815–1888) who too developed the principle of energy conservation, independently from Joule and Mayer. In 1843 he presented to the Real Danish Academy his thesis on the "forces" in which he maintained that the quantity of heat evolved is proportional to the *lost moving forces*. Colding, inspector of roads and bridges in Copenhagen, had a broad scientific knowledge ranging from meteorology, oceanography, hydrology, to fluid mechanics and thermodynamics. He was also one of the founders in 1872 of the Danish institute of meteorology. His religious ideas, his friendship with the physicist Hans Christian Ørsted and his interest in the *Naturphilosophie*, influenced him to develop the *principle of imperishability of the forces*, establishing a concept equivalent to that of the transformation of work into heat (Colding 1871). Unfortunately, Colding did not evaluate, however, the mechanical equivalent of heat, as done by Joule in the same year.

1.3 Entropy and Free Energy

The ideas of Carnot and Clapeyron had a profound influence on the German mathematical physicist Clausius, who worked out in original form the heat theory giving birth to a new science, thermodynamics.

Rudolf Julius Emanuel Clausius (1822–1888), professor of physics in Berlin, at the ETH Zürich, in Würzburg and finally in Bonn, analyzed with extreme precision the problem of the transformation of heat into work, starting from Carnot's principle and from the successive developments of Clapeyron. In 1850 he formulated exactly the first principle of thermodynamics, stating that *in a physical system without energy exchanges with the external world, the total energy is conserved*. He proposed also the second principle of thermodynamics stating that *heat cannot flow from a cold to a warm body without the expense of energy* (Clausius 1850). In this paper, Clausius observed that Carnot's theorem concerning the equivalence of heat Q and work W in the form Q = W, was incorrect since it missed the evaluation of the amount of heat lost in the process and thus not transformable into work. In the period 1857–1865 Clausius published his famous works on the kinetic theory of gases, including in the theoretical treatment the translational, rotational and vibrational motions of the molecules (Clausius 1862) and defining the concept of mean free path. In 1865 he introduced also one of the fundamental concepts of classical thermodynamics, that of entropy (Clausius 1865), as the ratio $\Delta S = \Delta Q/T$ between the quantity ΔQ of heat transferred in a process from a warm to a cold source and therefore not utilized to make work and the temperature T at which the process takes place. Clausius selected the word *entropy* due to its resemblance to that of energy, since according to him the two physical concepts were strictly connected:

> ...So schlage ich vor die Größe S nach dem griechischen Worte ἡ τροπή, die Verwandlung, die Entropie des Körpers zu nennen. Das Wort Entropie habe ich absichtlich dem Worte Energie möglichst ähnlich gebildet, denn die beiden Größen, welche durch diese Worte benannt worden sollen, sind ihren physikalischen Bedeutungen nach einander so nahe verwandt, daß eine gewisse Gleichartigkeit in der Benennung mir zweckmäßig zu seyn scheint.[1]

In the same paper (Clausius 1865) Clausius defined also an irreversible transformation showing that in such a transformation the entropy always increases. He formulated the two principles of thermodynamics in this way:

1. *Die Energie der Welt ist konstant.* (The energy of the Universe is constant)
2. *Die Entropie der Welt strebt einem Maximum zu.* (The entropy of the Universe tends to a maximum)

Clausius' concept of entropy originated from the idea that part of the energy used to perform a work is lost by dissipation and therefore is not utilizable. This concept was already included in a 1803 paper on the efficiency of thermal engines written by Lazare Carnot, father of Sadi Carnot and establishing that in the motion of the parts of any engine there are always losses of *useful work* (Carnot 1803), namely of energy.

Clausius also developed the concept of total energy H of a physical system, today universally known as enthalpy. The term enthalpy, derived from the Greek ενθαλπος (enthalpos, to introduce heat), was invented at the beginning of the twentieth century by the Dutch physicist Heike Kamerlingh Onnes (1853–1926) at the first meeting of the Institute of Refrigeration in Paris in 1908. He also clarified his connection with the internal energy U in the form $H = U + PV$, where P and V are the volume and the pressure of the system, respectively.

---

[1] I propose to name the quantity S the entropy of the system, after the Greek word [τροπη trope], the transformation. I have deliberately chosen the word entropy to be as similar as possible to the word energy: the two quantities to be named by these words are so closely related in physical significance that a certain similarity in their names appears to be appropriate.

The equivalence between work and heat was discussed also by the German *Hermann Ludwig Ferdinand von Helmholtz* (1821–1894) professor of physiology in Königsberg, Bonn and Heidelberg, and after, professor of physics in Berlin from 1871. From 1888 onwards he was also director of the *Physikalisch-Technische Reichsanstalt* of Charlottenburg. In 1847 Helmholtz published a treatise (Helmholtz 1847) on energy conservation, supported by a very broad philosophical culture, built essentially by reading Kant, who was great friend of his father.

Helmholtz introduced in thermodynamics the concept of potential energy which he called *tension force*. He also formulated, independently from Joule, the energy conservation principle by studying the muscle metabolism. Resuming Carnot's and Clapeyron's work, he postulated the existence of a correspondence between all forms of work, mechanical, thermal, electric and magnetic, considering all of them as a manifestation of a single physical quantity, the energy.

The same conclusions, even if in a completely different way, were reached by the English mathematician and physicist *William Thomson* (1824–1907), Baron of Kelvin, one of the greatest physicists of the nineteenth century. Thomson produced important works on mathematical analysis, electricity and thermodynamics and strongly contributed to the unification of modern physics.

In 1847 Thomson, during the annual meeting of the *British Association for the Advancement of Science* at Oxford, had the opportunity to listen to a speech of Joule attacking the caloric theory and the thermal engines of Carnot and Clapeyron and insisting on the convertibility of mechanical work into heat.

From that moment on, Thomson started to become interested in Carnot and Clapeyron's work. His thermodynamic research drove him to propose in 1848 the absolute temperature scale (Thomson 1848) in which a unit of heat, transferred from a body at a temperature T_1 on the new scale to another body at the temperature $T_2 = T_1 - 1$, produces always the same amount of work, independently from the value of T:

> The characteristic property of the scale which I now propose is, that all degrees have the same value; that is, that a unit of heat descending from a body *A* at the temperature $T°$ of this scale, to a body *B* at the temperature $(T - 1)°$, would give out the same mechanical effect, whatever be the number *T*.

On this basis he defined the zero point of the absolute temperature scale as that at which it would be impossible to transfer heat from one body to another.

From 1849 to 1851 he elaborated his dynamic heat theory, reconciling Carnot's ideas with those of Joule. Starting from the fact that the volume of water increases on freezing, he concluded that his melting point should decrease with increasing pressure, since otherwise it would give rise to a *perpetuum mobile*. Finally in 1851 he proposed his own version of the second principle of thermodynamics (Thomson 1851) equivalent to that formulated by Clausius:

> It is impossible, by means of inanimate material agency, to derive mechanical effect from any portion of matter by cooling it below the temperature of the coldest of the surrounding objects.

The title of baron, awarded him by the English Government in 1892 for his great scientific merits, referred to the name of the Kelvin River running close to his university at Glasgow.

1.4 Thermochemistry

In the second half of the nineteenth century, several chemists dedicated themselves to the accurate measurement of the quantity of heat produced in chemical reactions. Already in 1782, Lavoisier and Laplace used a calorimeter to measure the latent and specific heat of several substances. The true founder of calorimetry was, however, Germain Ivanovich Hess (1802–1850). Hess was born in Geneva but emigrated as a child to Russia where he studied medicine at the Dorpat (today Tartu) university and where he obtained the Ph.D. in 1825 with a thesis entitled *Studies of chemical composition and healing properties of Russian mineral waters*. After a stimulating meeting with Berzelius, he moved to Stockholm to collaborate with him. Back in Russia and settled as physician at Irkutsk, he oriented his activity in 1830 towards chemical research, becoming after a short while professor at the St Petersburg institute of technology. Influenced by Berzelius' ideas, he decided to evaluate chemical affinity by measuring the amount of heat developed in neutralization reactions of an acid with several bases and was able to show that the amount of heat produced is always the same, independently of the nature of the bases. The paper that made him famous (Hess 1840) contains the basic principle of thermochemistry, and is known as Hess's law. The law states that the amount of heat produced or absorbed in a reaction is independent of the path followed and of the number of intermediate steps. For instance, if the three hydrogen atoms of phosphoric acid H_3PO_4 are substituted one at a time with sodium atoms in three successive steps of neutralization, the amount of heat produced in the global reaction is equal to the sum of the reaction heats of the three sequential steps.

Hess's law is nothing else than a different formulation of the energy conservation principle that extends its validity from purely physical processes like transitions of state, to chemical processes such as reactions. Hess also found that when two neutral salt solutions are mixed together, the final solution remains neutral and no heat is produced. He generalized these results with the term thermo-neutrality and this law is valid for salts of strong acids and bases which completely dissociated in solution (Hess 1840). Berthelot discovered later several exceptions to this law such as in the case of weak acids and bases which not completely dissociated.

The Irishman Thomas Andrews (1813–1885), professor of chemistry at Belfast Queen's College, famous for his discovery of the critical temperature of gases, was also interested in the measurement of neutralization heats. Andrews measured in the period 1841–1848, the neutralization heat of a base with several acids and showed that the amount of heat produced was the same in all cases. Andrews discovered also that when a metal displaces another metal in a solution of its salts, the thermal effects are independent of the nature of the acid and found that Hess's law of the constancy of the sum of partial heats applies also to salts of metals in solution.

The research on the neutralization of acids and bases of Hess and Andrews were complemented by a series of accurate measurements made by the French chemists Pierre Antoine Favre (1813–1880) and Johann Theobald Silbermann (1806–1859) who succeeded in proving that these heats were the sum of two constant terms, one dependent on the acid and the other on the bases (Favre and Silbermann 1853).

Thermochemical data became an even more significant experimental support to the affinity theory with the research of the French chemist, science historian and politician Marcellin Pierre Eugène Berthelot (1827–1907). Born in Paris, assistant of Balard at the College de France in 1851, author in 1854 of a Ph.D. thesis *Sur les combinaisons de la glycérine avec les acides*, he became in 1859 professor of organic chemistry at the École Supérieure de Pharmacie and in 1865 at the Collège de France. He was also involved in social and political activities. First member and later perpetual secretary of the Académie, he became general inspector of higher education in 1876, life senator in 1881, minister of public instruction in 1886 and of foreign affairs in 1895–1896.

Berthelot was a convinced adversary of the *vis vitalis* theory, strongly supported instead by Berzelius, who maintained that the formation of organic substances was controlled by interactions different from those occurring in the inorganic world (Berthelot 1860). Berthelot proved instead with the synthesis of several hydrocarbons, natural fats and sugars that organic compounds obey the same laws that control the formation of the inorganic ones. His opposition to the vitalistic approach, and his belief that the organic world was controlled by the same mechanical laws operating in the universe, gave rise to his interests in thermochemistry and calorimetry. Indeed he produced an enormous number of experiments and two books on these arguments, *Essai de Mécanique Chimique fondée sur la thermochimie* of 1879 and *Thermochimie* of 1897.

Berthelot was also convinced that reactions producing heat (exothermic reactions) are spontaneous, whereas those absorbing heat (endothermic) are not. Berthelot's idea, even if plausible, was, however, not totally correct since there are both spontaneous reactions that are not exothermic, as well as reactions that proceed spontaneously absorbing heat from the external world.

His research on heats of reactions drove him to study the theory of explosives which he later described in the book *Sur la force de la poudre et des matières explosives* (Berthelot 1872). Thanks to his competencies in explosives, Berthelot had an important role for France during the siege of Paris in 1870–1871.

Berthelot was also a science historian. He published several essays and books on the history of science, chemistry and alchemy: *Collection des anciens alchimistes grecs* (translation of several Greek, Sirian and Arab writings of alchemy) of 1888; *Les Origines de l'alchimie* (1885); *Introduction à l'étude de la chimie des anciens et du moyen âge* (1889). In *La Chimie au moyen âge* (1893) he wrote (Berthelot 1893):

> La chimie est née d'hier: il y a cent ans à peine qu' elle a pris la forme d' une science moderne. Cependant les progrès rapides qu' elle a faits depuis ont concouru, plus peut-être que ceux d' aucune autre science, à transformer l' industrie et la civilisation matérielle, et à donner à la race humaine sa puissance chaque jour croissante sur la nature. C' est assez dire quel intérêt présente l' histoire des commencements de la chimie. Or ceux-ci ont un caractère tout spécial: la chimie n' est pas une science primitive, comme la géométrie ou l' astronomie; elles' est constituée sur les débris d' une formation scientifique antérieure; formation demi-chimérique et demi-positive, fondée elle-même sur le trésor lentement amassé des découvertes pratiques de la métallurgie, de la médecine, de l' industrie et de l' économie domestique.

In parallel to Berthelot's activities, an important group of thermo-chemical research was carried out by the Dane Hans Peter Jørgen Julius Thomsen (1826–1909) who was born and lived in Copenhagen where he taught chemistry at the polytechnic and then at the university from 1866 to 1891. In the period 1869–1882 he took a large number of measurements of the heat produced or absorbed in reactions of combustion and of oxidation-reduction of acids with bases. His works were collected into the four-volume treatise *Thermochemische Untersuchungen* (Thomsen 1882–1886).

Thomsen and Berthelot were both convinced that reaction heat measurements supplied a direct evaluation of chemical affinity. They independently formulated the principle of maximum work which specifies that in all chemical reactions the process that will occur is the one giving rise to the greatest production of heat.

This principle is certainly applicable to a large number of reactions, but is far from being a true general law of thermochemistry, since it ignores the contribution of entropy to chemical processes. The occurrence of reversible reactions, producing heat in one direction and absorbing it in the other, were soon utilized by several critics, especially by Helmholtz, to invalidate the principle. In 1882 Helmholtz proved in his paper *Die Thermodynamik Chemischer Vorgänge* (The Thermodynamics of Chemical Operation) that affinity should be measured not by the heat evolved, but by the maximum work (free energy) produced in condition of reversibility (Helmholtz 1882–1895). These criticisms convinced both Berthelot and Thomsen to refuse their principle. Later Nernst proved that the Thomsen-Berthelot principle is completely valid only close to the zero of the absolute temperature.

In 1906 thermochemistry joined theoretical thermodynamics thanks to the work of the German chemist Walther Hermann Nernst (1864–1941), 1920 Nobel laureate in chemistry for the formulation of the third principle of thermodynamics, based just on thermochemistry research.

Walther Nernst, after having attended the universities of Zürich, Berlin, Graz and finally of Würzburg where he received his Ph.D. in 1887 with a thesis in electrochemistry (Nernst 1887), went to work with Ostwald at the university of Leipzig, where he had the possibility to meet scientists of the level of Arrhenius and van't Hoff. In this environment of the highest scientific quality, he started his research, producing in a short time fundamental contributions to electrochemistry, solution theory, thermodynamics, solid state physics and photochemistry. He collected these contributions in his famous 1893 treatise *Theoretische Chemie vom Standpunkte der Avogadro'schen Regel und der Thermodynamik* (Nernst 1893).

In 1895, Nernst published in collaboration with Arthur Schoenflies (1853–1928) a mathematics textbook for chemists (Nernst et al. 1895) and a book of theoretical chemistry, largely utilized in German universities.

In 1888 he formulated the concept of a solubility product in a saturated solution of electrolytes (Nernst 1888) as a product of the concentration of the ions present in solution. He derived also his famous equation (Nernst 1889) connecting the chemical equilibrium in solution with the driving force, which gives rise to the potential of a galvanic cell, defines the redox potential at equilibrium of a half cell and in 1893 developed a theory of the dissociation of ionic compounds in water. In 1897 Nernst invented a lamp which used instead of a glowing tungsten filament, a ceramic rod heated to incandescence, a system more efficient than in existing lamps. A source of infrared radiation well known to infrared spectroscopists as the "Nernst Glower" comes from the lamp.

In 1894 Nernst founded the Göttingen institute of chemical physics and electrochemistry. From 1905 he became professor, first of chemistry, and then of physics at the University of Berlin, and in 1924 he became director of the *Physikalisch-Chemisches Institut*. In Berlin Nernst became interested in calorimetry, and with a group of students he carried out an important project of reaction heats and of specific heat measurements of solids at very low temperatures. Nernst realized that by lowering the temperature the enthalpy ΔH becomes more and more alike to the constant pressure free Energy $\Delta G = \Delta H - T\Delta S$ and that free energy variations become smaller and smaller. If the free energy ΔG and the enthalpy ΔH tend to equalize, the entropy ΔS must necessarily tend to zero. Nernst thus deduced that, as the temperature approached little by little the absolute zero, chemical reactions would occur with smaller and smaller entropy variations and therefore that the maximum work obtainable in a chemical process could be evaluated by measuring the amount of heat developed in the process (namely the enthalpy) at temperatures close to absolute zero. Nernst established in this way the third principle of thermodynamics, asserting that the entropy of a system at equilibrium tends to zero when the temperature tends to absolute zero (−273°C). Nernst presented the paper on 23 December 1905 at a meeting of the *Königliche Gesellschaft der Wissenschaften zu Göttingen* and again on 20 December 1906 at the meeting in Berlin of the *Königlich Preussische Akademie der Wissenschaften* (Nernst 1906).

In 1909 Max Planck improved Nernst definition of the III principle, pointing out that a $T = 0$ the entropy of any body is equal to zero and that in the same way also the thermodynamic coefficients vanish. In the same 1909 the Greek mathematician Constantin Carathéodory (1873–1950), who at the time taught at the university of Hannover, published a completely axiomatic formulation of thermodynamics that used only mechanical concepts in the framework of the mathematical *theory of differential forms* (Carathéodory 1909).

Nernst collected in a volume all experimental researches of thermodynamics developed at the Berlin *Physikalisch-Chemisches Institut*, to prove the validity of the III principle (Nernst 1918). The III principle is perfectly valid in the classical formulation of Nernst and Planck. For quantum systems at $T = 0$ the entropy is not zero since the vibrational energy is never zero and thus the definition becomes an approximation.

The validity of the third principle of thermodynamics and the idea that it was a fundamental rule of nature were largely proved by the researches of the American physical chemist William Francis Giauque (1895–1982) professor at Berkeley who with his students and coworkers performed a large number of researches at very low temperatures (Giauque et al. 1933, 1935). In particular Giauque and his student Duncan P. MacDougall, by investigating the effect of magnetic fields for the evaluation of the entropy of paramagnetic systems, invented in 1933 (Giauque and MacDougall 1936) the adiabatic demagnetization technique that allows to reach temperatures very close to the absolute zero.

1.5 Statistical Thermodynamics

Classical thermodynamics is a formidable phenomenological theory that considers energy transformations in macroscopic systems without specifying the nature and the mechanism of the transformation. Thanks to the work of Ludwig Boltzmann and Josiah Willard Gibbs (Cercignani 1998) a different formulation, the statistical thermodynamics based on the microscopic knowledge of the systems, developed at the end of the nineteenth century.

Statistical thermodynamics originated from the development of the kinetic theory of gases (Cercignani 1998) due to the Austrian physicist Ludwig Boltzmann (1844–1906) who, on the basis of previous ideas of Maxwell, applied statistical criteria to the dynamics of the particles of a gas obeying the laws of classical mechanics.

Boltzmann started from the idea that heat and temperature are two different macroscopic manifestations of the motion of the particles, atoms or molecules, composing the gas (Boltzmann 1871a, b). This approach allows to better interpret thermodynamic quantities like energy and entropy, eliminating the metaphysical meaning of their mathematical definition and reducing them to pure mechanical processes. Boltzmann in 1872 published one of his most important papers, *Weitere Studien* (Boltzmann 1872), that included two famous results, known as the Boltzmann equation and the H theorem (Tolman 1938). In this paper he discussed the importance of the statistical approach in the framework of the kinetic theory of gases (Gallavotti 1994), starting from the idea that the number of particles in a gas is so large and their motion so fast, that only average values can be observed. Starting from these ideas he produced a new demonstration of the second principle of thermodynamics on purely statistical and probabilistic grounds (Cohen 1996).

In Boltzmann treatment the entropy S of an isolated physical system at equilibrium is defined as the natural logarithm of the number W of distinct microscopic states available to the system for each given energy value.

The famous Boltzmann equation

$$S = k \ln W$$

that statistically defines the entropy function, is engraved on Boltzmann's grave, in the *Zentralfriedhof* cemetery of Vienna. The k constant connects the macro and the microscopic world, establishing the absolute thermodynamic temperature scale in terms of the energy of the particles: $E = kT$.

In 1877 Boltzmann used statistical concepts to interpret the concept of entropy, showing that entropy can be considered as a measure of the disorder of the physical systems and that the second principle of thermodynamics expresses the tendency of isolated systems to evolve towards the situation of maximum possible disorder, corresponding to the maximum of entropy (Boltzmann 1877).

Boltzmann's papers were written with a prolix and obscure style that made their reading very difficult even to high level physicists, so that many of his contemporaries did not accept his ideas. An implacable adversary of Boltzmann was Ostwald who, despite of being his friend and colleague, violently attacked the use of statistical criteria to discuss problems of classical dynamics. He was contrary to the introduction in a macroscopic theory such as thermodynamics of the microscopic concept of atom in which he did not believe. These attacks continued for a long time, driving Boltzmann to a serious depression that pushed him to commit suicide. In vacation to Duino, close to Trieste, he profited of the fact that his wife and daughter were gone to the seashore to hang himself. A long time was necessary before Boltzmann's work could be understood and appreciated.

Today statistical thermodynamics is correctly formalized by the works of Josiah Willard Gibbs (1839–1903), professor of mathematical physics at the university of Yale, one of the greatest American chemical-physicists, who was the first to use the term statistical mechanics and made accessible to a large audience Boltzmann's and Maxwell's ideas. In 1876, starting from the results of Clausius and Helmholtz, Gibbs published a series of papers for a total of 330 pages, in which he defined the free energy ΔG of a thermodynamic system

$$\Delta G = \Delta H - T\Delta S$$

as the energy completely available to perform work, removing from the total energy ΔH (enthalpy) the energy losses $T\Delta S$ defined by Clausius relationship.

Gibbs' work (Gibbs 1875–1876) was published in the period 1875–1878 in several sections of an obscure local journal, the *Transactions of the Connecticut Academy*, and became known only after Ostvald and Le Chatelier had translated it in German and French, respectively. In these papers Gibbs applied the thermodynamic idea to the interpretation of several physical-chemical processes, connecting in an homogeneous interpretation facts that at the time seemed unrelated and inexplicable. For a dynamical system of N identical particles, Gibbs defined as *canonical ensemble* the body of all states of the system having different choices of coordinates and momenta proposing the *distribution function* $e^{-E/kT}d\tau$, a function of coordinates, momenta and time which gives the number of particles per unit volume in the phase space. Starting from the idea that the free energy depends on the concentration of the chemical species present in the system, he introduced also a new thermodynamic quantity, the *chemical potential*, defined as the variation of the free energy due to an infinitesimal variation of the concentration of a chemical species present (Gibbs 1902). The concept of chemical potential was soon found to be of great importance for the developments of chemical kinetics, when it was clear its function of driving force in chemical reactions.

Using the definition of chemical potential, a variation ΔG of Gibbs free energy assumes the form

$$\Delta G = \sum_{i} \mu_i \Delta c_i$$

where c_i represents the concentration of the ith chemical species and μ_i the corresponding chemical potential.

The mathematical formalism developed by Gibbs in several papers (Gibbs 1891a, b–1893a, b) was utilized by Gilbert Levis and Merle Randal in the United States and by Edward Armand Guggenheim (1901–1970) in England for the evaluation of the free energy of a large number of chemical compounds. The book by Lewis and Randall *Thermodynamics and the Free Energy of Chemical Substances* of 1923 and that by Guggenheim, *Modern Thermodynamics by the Methods of Willard Gibbs* of 1933 are fundamental classics of modern chemical-physics. Before the publication of these two textbooks, the most known text of thermodynamics, especially in Germany was the 1912 treatise *Lehrbuch der Thermochemie und Thermodynamik* written by Otto Sackur, that, once translated into English in 1917 by the American naturalized Scotsman George Ernest Gibson (1884–1959), became the official textbook of thermodynamics in American universities until 1923, when it was replaced by that of Lewis and Randall.

In 1911 the German chemical physicist Otto Sackur (1880–1914), after a post-doctor period spent at London with William Ramsay and then at Berlin with Walter Nernst, published in the Annalen der Physik (Sackur 1911) a paper in which he introduced quantum mechanical concepts in the statistical calculation of the entropy of an ideal monoatomic gas, connecting Nernst third principle of thermo-dynamics with Planck's h constant. Almost at the same time the young Dutch theoretician Hugo Tetrode, born in a well-established middle-class family of Amsterdam, at the age of only 17 published in 1912 on the Annalen der Physik two papers on the same argument (Tetrode 1912).

1.6 Non-equilibrium Thermodynamics

At the beginning of the twentieth century classical thermodynamics was perfectly formalized and allowed to correctly handle reversible processes, namely transitions from one equilibrium state of a system to another, still preserving the equilibrium conditions. Real physical processes are, however, never truly reversible and therefore equilibrium thermodynamics can be utilized only in an ideal way, even if it allows to ascertain whether a process can occur spontaneously or not.

In the case of systems out of equilibrium the equations of classical thermodynamics represent therefore only inequalities. For instance at the melting temperature the free energy G_s of a solid body is perfectly equivalent to that G_l of the liquid, $G_l = G_s$, whereas at lower temperatures this relationship becomes the inequality $G_l > G_s$. These inequalities aid thus only to show the direction in which the system evolves spontaneously. Out of equilibrium processes can be, however, kept in a stationary situation of equilibrium, by supplying energy to the system as in the case of a liquid kept at the boiling temperature by a continuous supply of heat.

Even if it was clear to everybody that irreversible processes were by and large the greatest majority of the real ones, the formalism of classical thermodynamics remained for long time limited to equilibrium systems and the dynamics of physical processes continued to be described in terms of succession of quasi equilibrium states. In these equilibrium states it was not indispensable to evaluate in details entropy variations, being largely sufficient to know that the entropy would in any case reach its maximum possible value. Irreversible processes leading to entropy increases were therefore considered more or less as a tedious complication, since they degraded the energy preventing that thermal engines could reach their optimal efficiency.

In the second half of the nineteenth century, however, isolated attempts started to appear to extend the thermodynamic formalism valid for equilibrium systems to those out of equilibrium, namely to dynamical processes subject to external forces that give rise to fluxes of energy or matter, producing always an increase of entropy.

The first who considered out of equilibrium processes in thermodynamic terms was Lord Kelvin in 1852 (Thomson 1852), through a presentation in parallel to Clausius of his own definition of the second principle of thermodynamic which included thermal or mechanical energy dissipation processes. In Lord Kelvin's formulation when heat is produced in an irreversible process (for instance friction), the dissipation of energy make it impossible to restore the initial conditions. In the same way if heat is diffused by conduction, an amount of energy is lost and it becomes impossible to go back to the initial situation.

In 1854 Lord Kelvin discussed two well known out of equilibrium processes, connected to the thermoelectric potential of a thermocouple, the Seebeck and Peltier effects (Thomson 1854) in which charges and thermal energy fluxes are indissolubly connected. In 1873 also Lord Rayleigh (J.W. Strutt) started to be interested in out of equilibrium fluid systems. By broadening Carnot's and Clausius approaches, he introduced in the dynamics of viscous fluids the concept of dissipation function, as the quantity controlling the rate of mechanical to thermal energy transformation due to the presence of frictional forces proportional to the velocity (Rayleigh 1873). In 1878 von Helmoltz (1878) extended Lord Kelvin's treatment to the coupling between concentration gradients and fluxes of electric current in electrolytes and formulated a corresponding reciprocal relation.

Lord Rayleigh treatment considered only mechanical processes and did not take into account temperature variations as responsible of fluxes in physical systems. The generalization of the concept of dissipation function to temperature gradients turned out to be of great importance for the successive developments of the thermodynamics of out of equilibrium systems.

In this growth of new ideas it became clear that local entropy variations were the true significant variables for non-equilibrium processes. In the direction of fluxes and of entropy changes developed thus the works of the Polish theoretical physicist Władysław Natanson (1864–1937) (Natanson 1896), who influenced the ideas of the French philosopher, physicist and mathematician Pierre Maurice Marie Duhem (1861–1916). Duhem extended the concept of thermodynamic potential to other domains of physics (Duhem 1886, 1887), convinced that a completely general version of thermodynamics would furnish a theory able to incorporate the whole physics and chemistry (Duhem 1911). In the same direction of thought progressed also the researches of the Viennese physicist Gustav Jaumann (1863–1924), leader of the Viennese school of statistical thermodynamics bound to the names of Ludwig Boltzmann, Josef Loschmidt, and Josef Stefan (Jaumann 1911), as well as those of the Austro-American chemical-physicist Alfred James Lotka (1880–1949) who tried to interpret in terms of out of equilibrium thermodynamics even the Darwinian theory (Lotka 1922).

Important contributions to the theory of irreversibility were also due to the American physicist Carl Eckart (1902–1973) who, after having done significant work to formalize quantum mechanics (Eckart 1940), derived the laws of diffusion from the principles of irreversible processes and computed entropy variations in the dynamics of viscous fluids.

The thermodynamics of irreversible processes became, however, a true independent branch of chemical physics only after the second world war, when the scientific community started to understand the importance of the works of the Norwegian Lars Onsager and of the group of physicists and chemists who, especially in Belgium and in the Netherlands, contributed to its sensational development along the same line of thoughts. Lars Onsager one of the greatest theoretical talents of the twentieth century, developed in facts at a very high level the thermodynamics of non-equilibrium processes for systems in situations out of equilibrium. For his research he obtained in 1968 the Nobel Prize for chemistry. In 1920 Lars Onsager (1903–1976) had joined the Norges Tekniske Hogskole of Trondheim to study engineering. The study of mathematics equipped him with an extraordinary theoretical background that allowed him to face without difficulties problems of high conceptual level. This excellent scientific preparation was integrated by a solid general culture and a profound knowledge of the classics and of the Norwegian culture, including northern sagas. He loved to declaim them to friends and relatives in original version or even to translate them into English.

After 5 years of studies at Trondheim, once graduated, he moved to Zurich to work until 1928 in Peter Debye's laboratory. He introduced an important improvement in the Debye-Hückel theory of strong electrolytes, development known after as the *Onsager limiting law*. Onsager's contribution to electrochemistry and his collaboration with Debye during the Zurich years are discussed in the following section on electrochemistry, while here we shall essentially deal with his contributions to thermodynamics.

In 1928 Onsager emigrated to the United States as associate professor at the John Hopkins University where he got interested in the mass action law and where he was charged of a chemistry course for first year students. Onsager's lectures of basic chemistry come out to be a disaster for his incapacity to communicate with persons with a mental level and a knowledge of mathematics lower than his own. He carried on his shoulders this difficulty of communication with common mortals for his full life. Next year, left John Hopkins, Onsager was engaged at Brown University where the director of the chemistry department, Charles A. Kraus, was an experimentalist far from his high theoretical level, but with a good nose to identify smart fellows. At Brown Onsager developed actually his fundamental work that was later honored with the Nobel Prize. Even in his new university Onsager had problems with his colleagues mostly because his research themes seemed absolutely incomprehensible to the faculty members. The teaching at Brown was also not more productive than at Johns Hopkins, his lectures of statistical mechanics being officially named by the students' lectures of *sadistical mechanics*. Nevertheless once in a while a rare student able to follow him did show up. During one of these lectures in which he filled the blackboard with complex equations, a young student stood up pointing out that Onsager had made a small error. This student was Raymond Matthew Fuoss (1905–1987), who after made his Ph.D. thesis under his supervision and became later one of his best coworkers at Yale.

Onsager stayed at Brown until 1933 when, due to the economy depression that hit the United States, it became difficult to pay him a salary for lectures that nobody understood and he had the possibility, thanks to a research contract, to join the chemistry department at the Yale university where he remained for the rest of his life and retired in 1972.

Onsager started in 1931 his interest in irreversible processes dealing with equilibrium systems in stationary conditions in which several extensive properties can fluctuate at the same time (Onsager 1931a, b). A stationary process out of equilibrium is clearly the electrolytic one, in which an electric current forces the ions to migrates towards opposite poles. Onsager had already considered since the Zurich times the influence of the electric field on ion's motion in an electrolytic solution and in 1931 decided to face the problem in a completely general form taking into account the combined effect of diffusion and electric conduction.

He established that the force K_i acting on the ith ion with charge e_i is given by the relation

$$K_i = -\nabla\mu_i - e_i\nabla\varphi_i$$

where $\nabla\mu_i$ and $\nabla\varphi_i$ are the gradients of the chemical and electrical potential, respectively. These gradients represent the thermodynamic forces that control the fluxes of extensive quantities in the system. In Onsager's theory of ionic transport in electrolytes, the ionic flux J_i was controlled by a linear combination of all forces acting on the ions

$$J_i = \sum_j L_{ij}K_j$$

Onsager, using the criterion of microscopic reversibility and considerations of statistical mechanics was able to show that the coefficients L_{ij} are elements of a symmetric positively definite matrix, such that

$$L_{ij} = L_{ji}$$

These relations are called *Onsager's reciprocal relations* (Onsager 1931a, b). In the development of his theory Onsager was stimulated by the experimental results of Claus Nissen Riiber (1867–1936) on the tautomeric kinetics of galactose (Riiber et al. 1929). Riiber, using chemical-physical techniques (interferometry, dilatometry and rotatory power) had shown that at least three different optical isomers of galactose exist in solution and supposed that each of these isomers might transform into any one of the others.

Riiber's problem was complicated by the fact that in solution both the cyclic equilibrium

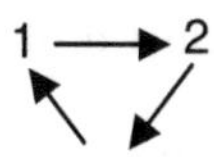

and the ensemble of separated equilibriums $1 \rightleftarrows 2$, $1 \rightleftarrows 3$ and $2 \rightleftarrows 3$ could in principle exist. In analyzing Riiber's results, Onsager started from the idea that the probability of the $1 \rightarrow 2$ process was identical to that of the $2 \rightarrow 1$ inverse process and so on for all other ones. He further proved that the cyclic process, although mathematically correct, was physically unreasonable and therefore that the balance of the three separated equilibriums was the only one able to correctly reproduce Riiber's data.

In the study of out of equilibrium processes it is necessary to know the probability of the temporal succession of intermediate non-equilibrium states. Onsager realized that this probability is determined by the principle of minimum energy dissipation introduced by Helmholtz in 1868 (Helmholtz 1868) and successively discussed by Lord Rayleigh in 1873 (Rayleigh 1873). The minimum energy dissipation leads to the dynamical superposition of casual processes of Markovnian type such that in each instant the system has no memory of his past.

Onsager used also with great ability the principle of microscopic reversibility formulated by Richard C. Tolman (1881–1948) in 1924 and published after in his treatise of statistical mechanics of 1938, which establishes that in a reversible reaction the direct reaction mechanism is exactly the inverse of that in the opposite direction. As a consequence the series of intermediate transition states in the direct reaction is the mirror image of that of the inverse reaction. As Onsager wrote in the Nobel lecture given at Stockholm in 1968:

> I looked for a way to apply the condition of microscopic reversibility to transport processes, and after a while I found a handle on the problem: the natural fluctuations in the distribution of molecules and energy due to the random thermal motion. According to a principle formulated by Boltzmann, the nature of thermal (and chemical) equilibrium is statistical, and the statistics of spontaneous deviation is determined by the associated changes of the

> thermodynamic master function-that is, the entropy. Here was a firm connection with the thermodynamics, and we connect with the laws of transport as soon as we may assume that a spontaneous deviation from the equilibrium decays according to the same laws as one that has been produced artificially.

Onsager generalized the results obtained for the ionic transport in an electrolytic solution to the general case of the transport of any extensive property of the system such as volume, electric charge, concentration of chemical species etc., in stationary non equilibrium conditions.

At equilibrium both the free energy and the entropy are constant. An irreversible proccss instead continuously dissipates energy and creates entropy. The amount of dissipated energy is given by Clausius relationship and must correspond to the energy flux dJ multiplied by the generalized force X responsible of the process:

$$\mathrm{TdS} = \mathrm{XdJ}$$

For example in the case of an electric current circulating in a conductor, the amount of dissipated heat is the product of the current dJ times the electromotive force X.

In the most general case, using the Kamerlingh Onnes definition of enthalpy, the entropy variation assumes the form

$$\mathrm{dS} = \frac{1}{\mathrm{T}}\mathrm{dU} + \frac{\mathrm{p}}{\mathrm{T}}\mathrm{dV} - \sum_{\mathrm{i}=1}^{\mathrm{k}} \frac{\mu_{\mathrm{i}}}{\mathrm{T}}\mathrm{dc}_i$$

expressed as a function of the intensive quantities temperature T, pressure p and chemical potentials μ_i as well as of the differentials of the extensive quantities energy U, volume V and concentrations c_i. If instead the extensive quantities U, V and c_i are assumed as variables, the corresponding thermodynamic forces are then the gradients of 1/T, p/T and μ_i/T, respectively. These thermodynamic forces are those controlling the fluxes of the extensive quantities. The connection between the fluxes J_i and the intensive variables was treated by Onsager using the Taylor series development, in the assumption that the thermodynamic forces vary slowly with time and that the fluxes are weak.

Onsager announced in 1929 to have obtained the reciprocal relations but published them in official form only in 1931. In 1933 he left Brown for a position of contract researcher at Yale; in 1934 became associate professor and in that occasion, in the total embarrassment of the faculty it came out that he had never written a Ph.D. thesis, title considered essential to teach at an American university, especially in a top one like Yale. Onsager decided then to write a thesis that he prepared in a very short time with the title *Solutions of the Mathieu Equation of Period 4π and Certain Related Functions,* title horrifying and incomprehensible for the member of the chemistry department but also for those of physics who declared to be unable to judge it. Happily in the department of mathematics there was an expert of modern algebras who accepted with enthusiasm the thesis that he judged very favorably. The chemistry department was then forced, to cut a poor figure, to close an eye and to grant him the Ph.D. title in 1935.

At Yale Onsager continued to produce works of very high level, as for instance the solution of the famous problem of the two-dimensional Ising model (Onsager 1942), an infinite planar lattice of very small magnets (spins), that had been considered by several high level physicists such as Hans Bethe, Rudolf Peierls, Hendrik Kramers and Gregory Wannier. Onsager had very few students and coworkers contributing to his researches on the thermodynamics of irreversible processes; among them the Israeli theoretical physicist Bruria Kaufman (1918–2010), Stefan Machlup (1927–2008) and above all Ray Fuoss (1905–1987) with whom he wrote in 1932 an important paper on irreversible processes in electrolytes (Onsager & Fuoss 1932).

The results of a paper published in 1953 with Stefan Machlup (Onsager et al. 1953) concerning the thermodynamic behavior and the spontaneous fluctuations of a many particles ensemble subjected to a variety of forces and the calculation of the probability of a given succession of non equilibrium states is named in the recent literature as *Onsager-Machlup-Laplace approximation*. In 1945 the Dutch theoretical physicist Hendrik Brugt Gerhard Casimir (1909–2000) re-formulated the principle of microscopic reversibility, extending Onsager reciprocal relations to a broader class of processes than that considered in the original paper of Onsager (Casimir 1945).

The study of the thermodynamics of irreversible processes progressed in the second half of the century thanks to the works of Sommerfeld's pupil Josef Meixner (1908–1994) (Meixner 1965), of Sybren Ruurds de Groot (1916–) and of Pavel Mazur (1922–2001) (de Groot and Mazur 1962) and in the field of biology of Aharon Katzir Katchalsky (1914–1972) (Katchalsky 1976; Katchalsky and Curran Peter 1965).

The development of the thermodynamics of irreversible processes was essentially due to the works of the theoretical physicist Josef Meixner, professor at the *Rheinisch-Westfälische Technische Hochschule Aix-la-Chapelle*. Meixner developed from 1941 onwards, a phenomenological theory of irreversible processes that included the application of Onsager's reciprocal relations to a consistent number of physical processes (Meixner 1965, 1973).

Particularly significant was also the contribution of the Russian-Belgian naturalized, Ilya Prigogine who developed the concept of dissipative structures and established their role in irreversible processes. Ilya Prigogine (1917–2003) was born in Moscow into a family that left soviet Russia in 1921 and emigrated first to Germany and then in 1929 to Belgium, where Ilya entered the free university of Brussels and started his scientific career. In 1959 he became director of the Solvay institute in Brussels, and after, professor at the University of Texas in Austin, where in 1967 he founded a centre for the study of quantum complex systems. As a member of the famous Brussels thermodynamics school, he developed in the period 1950–1970 an important series of research on the thermodynamics of irreversible processes. His book *Introduction to Thermodynamics of Irreversible Processes* (Prigogine 1955) is a true classic in the field, written with great clarity and competence.

Especially important was his research on the formation of non-equilibrium structures and on the self-organization of out of equilibrium systems. He contributed to the development of the theory of dissipative structures, i.e. of the coherent space-time structures responsible of the exchange of matter and energy in open systems. In particular, he considered living bodies as examples of highly organized systems in which irreversible phenomena play an essential role. For non equilibrium stationary states near equilibrium, he developed the theorem of minimum entropy production, correlating the stability of thermodynamic equilibrium states with the stability of biological systems. For this research he was awarded the Nobel Prize for chemistry in 1977. Central to the scientific activity of Prigogine was also the analysis of the concept of time and of the constructive role of the arrow of time in irreversible processes, incompatible with time reversible theories such as Newtonian and quantum dynamics.

His philosophical conceptions of time, dominated by a profound adhesion to the ideas of Henry Bergson, gave rise to considerable hostility in the community of experts. After he was awarded the Nobel Prize, he started to write books of divulgation of scientific-social problems and of his own ideas, among which the best known are *La fin des certitudes* (Prigogine 1996) and *La nouvelle alliance*, written in collaboration with his pupil Isabelle Stengers (Prigogine and Stengers 1979).

Many acknowledge that Prigogine's activity spans a wide range of expertise, from the natural to social sciences. For this reason and because of his activity as a writer, his name is better known to the public today than the names of Lars Onsager and Ludwig Boltzmann, the two giants who created the great structure of statistical thermodynamics.

In conclusion, both classical and non equilibrium thermodynamics are highly conceptual theoretical tools that, although devised to describe the time evolution of ideal macroscopic systems, offer a perfect interpretation of real processes in the physical world, simply by using abstract mathematics. Thermodynamics describes perfectly transformations of physical systems in which heat is transferred and transformed into other forms of energy without considering the intermediate stages of the whole process, as well as the mechanisms involved in the transformation. The beauty of this thermodynamic approach, which has had a great impact on our affluence and civilization, is essentially related to the fact that it makes full use of the powerful, although still mysterious power of mathematics to build an abstract world that correctly reproduces the real one. A famous theoretical physicist Arnold Sommerfeld, the true father of theoretical physics in Germany before the Second World War, described in a nice and humorous way the importance of thermodynamics:

> Thermodynamics is a funny subject. The first time you go through it, you don't understand it at all. The second time you go through it, you think you understand it, except for one or two small points. The third time you go through it, you know you don't understand it, but by that time you are so used to it, it doesn't bother you anymore.

References

Berthelot M (1860) Sur la fermentation glucosique du sucre de canne. Compt Rend 50:980–984

Berthelot M (1872) Sur la force de la poudre et des matières explosives. Gauthier-Villars, Paris

Berthelot M (1893) La Chimie au moyen âge. Imprimerie nationale, Paris

Boltzmann L (1871a) Über das Wärmegleichgewicht zwischen mehratomigen Gasmolekülen. Wien Sitzungsberichte der Kgl. Akademie der Wissenschaften in Wien 63:397–418

Boltzmann L (1871b) Einige allgemeine Sätze über Wärmegleichgewicht. Wien Bericht 63:679–711

Boltzmann L (1872) Weitere Studien über das Wärmegleichgewicht unter Gasmolekülen. Wien Bericht 66:275–370

Boltzmann L (1877) Über die Beziehung zwischen dem zweiten Hauptsatze der mechanische Wärmetheorie und der Warscheinlichkeitsrechnung respektive den Sätzen über des Wärmegleichgewicht. Wien Bericht 76:373–435

Brown GI (2001) Count Rumford: the extraordinary life of a scientific genius-scientist, soldier, statesman, spy. Sutton Publishing

Brush SG (2003) The kinetic theory of gases. World Scientific Publishing Co., River Edge, NJ.

Carathéodory C (1909) Untersuchungen über die Grundlagen der Thermodynamik. Math Ann 67:355–386

Carnot L (1803) Principes fondamentaux de l'équilibre et du mouvement. Deterville, Paris

Carnot S (1824) Réflexions sur la puissance motrice du feu et sur les machines propres à développer cette puissance. Bachelier, Paris

Casimir HBG (1945) On Onsager's principle of microscopic reversibility. Rev Mod Phys 17:343–350

Ceccarelli M (ed) (2007) Distinguished figures in mechanism and machine science: their contributions and legacies series: history of mechanism and machine science, vol 1. Springer, Dordrecht

Cercignani C (1998) Ludwig Boltzmann, the man who trusted atoms. Oxford University Press, Oxford

Clapeyron BPE (1834) *Puissance motrice de la chaleur*. Journal de l'École Royale Polytechnique, Vingt-troisième cahier, Tome XIV, 153–190

Clapeyron BPE (1843) Puissance motrice de la chaleur. J l'École Roy Polytechnique Vingt-troisième cah XIV:153–190, published later in German in the Poggendorff's Annalen der Physik, LIX, 446, 566, 1843

Clausius RJE (1850) Über die bewegende Kraft der Wärme und die Gesetze, welche sich daraus ableiten lassen. Ann Phys Chem 79:368–397, 500–524

Clausius RJE (1862) Über die Wärmeleitung gasförmiger Körper. Ann Phys 115:1–57

Clausius RJE (1865) Über verschiedene für die Anwendung bequeme Formen der. Hauptgleichung der mechanischen Wärmetheorie. Ann Phys 125:353–400

Cohen EGD (1996) Boltzmann and statistical mechanics. In: Ludwig B (ed) Boltzmann's legacy 150 years after his birth, vol 131, Atti dei Convegni Lincei. Accademia Nazionale dei Lincei, Rome, pp 9–23

Colding LA (1871) On the universal powers of nature and their mutual dependence. Philos Mag 42:1–20

Davy H (1799) An essay on heat, light, and the combinations of light. In: Beddoes T (ed) Contributions to physical and medical knowledge, principally from the West of England. Biggs & Cottle, Bristol

de Groot SR, Mazur P (1962) Nonequilibrium thermodynamics. North Holland, Amsterdam, Riedition, Dover, New York, 1984

Duhem P (1886) Le potentiel thermodynamique et ses applications à la mécanique chimique et á la théorie des phénoménes électriques. A. Hermann, Paris

Duhem P (1887) Étude sur les travaux thermodynamiques de J. Willard Gibbs. Bull Sci Math 11:122–159

Duhem P (1911) Traité de l'énergétique. Gauthier-Villars, Paris
Eckart C (1940) The thermodynamics of irreversible processes. I. The simple fluid. Phys Rev 58:267
Favre PA, Silbermann JT (1853) Recherches sur les quantités de chaleur degagées dans les actions chimiques et moléculaires. Ann Chim Phys 38:494
Gallavotti G (1994) Ergodicity, ensembles, irreversibility in Boltzmann and beyond. J Stat Phys 78:1571–1589
Giauque WF (1953) Determination of thermodynamic temperatures near 0°K without introducing heat below 1°K. Phys Rev 92:1339–1345
Giauque WF, MacDougall DP (1933) Attainment of temperatures below 1° absolute by demagnetization of $Gd_2(SO_4)_3 \cdot 8H_2O$. Phys Rev 43:768
Giauque WF, MacDougall DP (1935) The thermodynamic temperature scale in the region below 1° absolute. Phys Rev 47:885–886
Gibbs W (1875–1876) On the equilibrium of heterogeneous substances. T Conn Acad III:108–248
Gibbs JW (1891a) On the rôle of quaternions in the algebra of vectors. Nature 43:511–513
Gibbs JW (1891b) Quaternions and the Ausdehnungslehre. Nature 44:79–82
Gibbs JW (1893a) Quaternions and the algebra of vectors. Nature 47:463–463
Gibbs JW (1893b) Quaternions and vector analysis. Nature 48:364–367
Gibbs W (1902) Elementary principles in statistical mechanics. Yale University Press, New Haven
Harcourt AV, Esson W (1867) On the laws of connexion between the conditions of a chemical change and its amount. Philos Trans 157:117
Hess GH (1840) Thermochemische Untersuchungen. Ann Phys 50:385
Jaumann G (1911) Geschlossenes system physikalischer und chemischer differential-gesetze. Sitzber Akad Wiss Wien 120:385–530
Joule JP (1841) On the heat evolved by metallic conductors of electricity. Philos Mag 19:260
Joule JP (1843) The mechanical equivalent of heat. Philos Mag 23:263
Katchalsky AK (1976) Biophysics and other topics. Academic, London
Katchalsky AK, Curran Peter F (1965) Nonequilibrium thermodynamics in biophysics. Harvard University Press, Cambridge, MA
Thomson W (Lord Kelvin) (1848) On an absolute thermometric scale founded on Carnot's theory of the motive power of heat, and calculated from Regnault's observations. Trans Roy Soc Edinburgh Math Phys Pap 1:100–106
Lotka A (1922) Contribution to the energetics of evolution. Proc Nat Acad Sci 8:1467
MacDougall DP, Giauque WF (1936) The production of temperatures below 1°A. The heat capacities of water, gadolinium nitrobenzene sulfonate heptahydrate and gadolinium anthraquinone sulfonate. J Am Chem Soc 58:1032–1037
Marsden B (2002) Watt's perfect engine. Columbia University Press, New York
Mayer JR (1842) Bemerkungen über die Kräfte der unbelebten Natur. Liebig's Ann Chem Pharm 42:234
Mayer JR (1867) Die Mechanik der Warme. Verlag der J. G. Cotta'schen Buchhandlung, Stuttgart
Meixner J (ed) (1965) Statistical Mechanics of Equilibrium and Non-equilibrium. In: Proceedings of the 1964 international symposium on statistical mechanics and thermodynamics, North-Holland Publishing Co., Germany
Meixner J (1973) The entropy problem in thermodynamics of processes. Rheol Acta 12:465–467
Natanson W (1896) Uber die Gezetze nicht umkehrbarer Vorgange. Z Phys Chem 21:193–200
Nernst W (1887) Über die electromotorischen Kräfte, welche durch den Magnetismus in von einem Wärmestrome durchflossenen Metallplatten geweckt werden. Inauguraldissertation, Universität Würzburg, Ann Phys Chem N.F. 31:760–789
Nernst HW (1888) Zur Kinetik der in Losung befindlichen Korper I Theorie der Diffusion. Z Phys Chem 2:613
Nernst HW (1889) Die elektromotorische Wirksamkeit der Jonen. Z Phys Chem 4:129
Nernst W (1893) Theoretische Chemie vom Standpunkte der Avogadro'schen Regel und der Thermodynamik. Enke, Stuttgart

Nernst W, Schönflies A (1895) *Einfürung in die mathematische Behandlung der Naturwissenschaften. Kurzgefasstes Lehrbuch der Differential- und Integralrechnung mit besonderer Berücksichtigung der Chemie*. München
Nernst W (1906) Über die Berechnung chemischer Gleichgewichte aus thermischen Messungen. Nachr Kgl Ges Wiss Gött 1:1–40
Nernst W (1918) Zur Anwendung des Einsteinschen photochemischen Äquivalentgesetzes. I Z Elektrochem 24:335–336
Nernst W, Schönflies A (1895) Einführung in die mathematische Behandlung der Naturwissenschaften. Kurzgefasstes Lehrbuch der Differential- und Integralrechnung mit besonderer Berücksichtigung der Chemie. Dr. E. Wolff, München
Onsager L (1931a) Reciprocal relations in irreversible processes, I. Phys Rev 37:405–426
Onsager L (1931b) Reciprocal relations in irreversible processes, II. Phys Rev 38:2265–2279
Onsager L (1942) Crystal statistics. I. A two-dimensional model with an order-disorder transition. Phys Rev 65:117–149
Onsager L (1968) The motion of ions: principles and concepts, nobel lecture. Norstedt, Stockholm; Science (1969) 166:1359–1364
Onsager L, Fuoss RM (1932) Irreversible processes in electrolytes diffusion, conductance, and viscous flow in arbitrary mixtures of strong electrolytes. J Phys Chem 36:2689
Onsager L, Machlup S (1953) Fluctuations and irreversible processes. II. Phys Rev 91:1512–1515
Prigogine I (1955) Introduction to thermodynamics of irreversible processes. Wiley, New York
Prigogine I (1996) La fin des certitudes. Ed. Odile Jacob, Paris
Prigogine I, Stengers I (1979) La Nouvelle alliance. Gallimar, Paris
Riiber CN, Minsaas J, Lyche RT (1929) On the modifications of galactose. J Chem Soc 2173
Rolt LTC, Allen JS (1998) The steam engine of Thomas Newcomen. Landmark Publishing, Ashbourne Derbs
Thompson B (Count Rumford) (1781) New experiments with gun-powder. Philos Trans Roy Soc London 71:229–328
Thompson B (Count Rumford) (1786) New experiments upon heat. Philos Trans Roy Soc 76:273
Thompson B (Count Rumford) (1798) An experimental enquiry concerning the source of the heat which is excited by friction. Philos Trans Roy Soc 88:80–102
Sackur O (1911) Zur kinetischen Begründung des Nernstschen Wärmetheorems. Ann Phys 339:455–468
Tetrode H (1912) Die chemische Konstante der Gase und das elementare Wirkungsquantum. Ann Phys 343:255–256, 344, 434–442
Thomsen J (1882–1886) Thermochemische Untersuchungen, vol 4. J.A. Barth, Leipzig
Thomson W (1851) On the dynamical theory of heat; with numerical results deduced from Mr. Joule's equivalent of a thermal unit and M. Regnault's observations on steam. Trans Roy Soc Edinburgh Math Phys Papers 1:175–183
Thomson W (1852) On a universal tendency in nature to the dissipation of mechanical energy. Philos Mag 4:304–306
Thomson W (1854) On a mechanical theory of thermo-electric currents. In: Proceedings of the Royal Society, Edinburgh, pp 91–98
Tolman RC (1938) The principles of statistical mechanics. Oxford University Press, London, pp 134–140
von Helmholtz H (1847) Über die Erhaltung der Kraft. Ed. G. Reimer, Berlin
von Helmholtz H (1868) On discontinuous movements of fluids. Philos Mag 36:337–346
von Helmholtz H (1878) Über galvanische Ströme, verursacht durch Concentrationsunterschiede; Folgeren aus der mechanischen Wärmetheorie. Ann Phys Chem 3:201–216
von Helmholtz H (1882–1895) Wissenschaftliche Abhandlungen von Hermann von Helmholtz, vol 3. J.A. Barth, Leipzig, **2**, 958–978
Strutt see Rayleigh (John William Strutt) (1873) Some general theorems relating to vibrations. Proc Math Soc London 4:357–368
Rayleigh (John William Strutt) (1873) Some general theorems relating to vibrations. Proc Math Soc London 4:357–368

Chapter 2
Kinetics and Chemical Equilibrium

The natural sciences are sometimes said to have no concern with values, nor to seek morality and goodness, and therefore belong to an inferior order of things. Counter-claims are made that they are the only living and dynamic studies... Both contentions are wrong. Language, Literature and Philosophy express, reflect and contemplate the world. But it is a world in which men will never be content to stay at rest, and so these disciplines cannot be cut off from the great searching into the nature of things without being deprived of life-blood. (Sir Cyril Norman Hinshelwood)

2.1 Affinity and Reactivity

The concept of the chemical bond, a concept which forms the basis of modern chemistry, had its roots in the idea of chemical affinity, inherited from the alchemist's vision of the interaction between objects as corresponding to the one between human beings. This idea occupied the minds of many leading chemists of the past in their attempts to rationalize the reactivity and behavior of chemical compounds, as well as their changes of physical state with temperature and pressure.

In particular Claude Louis Berthollet published, between 1800 and 1803, a new theory of affinity, forcing a re-evaluation of previous ideas on the subject and a basic revision of the nature of chemical reactions. He focused attention on the multiplicity of forces causing chemical reactions, in contrast to the specificity of the attractive forces between substances which represented the essence of the so-called elective affinity of the previous chemists and alchemists.

According to Berthollet, affinities were a manifestation of a universal attraction, all particles exerting an attraction towards all others, trying to bring them together in chemical combination. Hence, combinations between particles in variable proportions were likely to occur. This last assertion was inconsistent with Dalton's atomic theory and with Proust's principle of definite proportions and gave rise to a serious polemic that ended with the triumph of Proust's theory.

S. Califano, *Pathways to Modern Chemical Physics*,
DOI 10.1007/978-3-642-28180-8_2,

The concept of affinity survived for a good part of the nineteenth century and preserved his mysterious meaning, bound to an anthropomorphic vision of matter which associated bonds between atoms to attractions between human beings. Only through the development of classical physical theories such as thermodynamics and kinetics, associated with the study of thermal processes, were new horizons revealed in the understanding of the tendency of molecules to react.

Once the chemical community realized that statistical and thermodynamic factors govern the behaviour of chemical reactions, the idea started to develop that the same factors could also affect their reaction rates and that a measure of these rates could offer a direct evaluation of chemical affinity, as already suggested in 1777 by Karl Friedrich Wenzel (1740–1793). This idea then gave rise to one of the most important chapters in the newly emerging chemical physics, (Wenzel 1777) chemical kinetics, namely the study of the factors controlling reaction rates and responsible for the establishment of chemical equilibria in reversible reactions.

2.2 Chemical Equilibrium

The physico-chemical aspects of time evolution in chemical reactions were addressed for the first time in 1850 by Ludwig Ferdinand Wilhelmy (1812–1864) in the framework of a polarimetric investigation of the cane sugar inversion catalyzed by inorganic acids. Wilhelmy, who had a solid mathematical background, showed that the initial reaction rate was proportional to the sugar concentration and proposed (Wilhelmy 1850) a differential equation to describe its decrease in time. Wilhelmy's results remained practically ignored for a long time until they became known, thanks to Wilhelm Ostwald who was inspired by them to establish in 1884 an analytical method to measure the strength of acids from their ability to catalyze the sugar inversion. An important consequence of Wilhelmy's work was that the inversion rate is an exponential function of the inverse of the absolute temperature. Similar results were reached by the English chemist Augustus George Vernon Harcourt (1834–1919) in the study of the acid catalyzed clock reactions between iodide and hydrogen peroxide and between oxalic acid and potassium permanganate. The mathematician William Esson (1838–1916) interpreted then Harcourt's kinetics data in terms of differential equations not too different from those used today (Harcourt and Esson 1867).

Of great interest for the understanding of the kinetics of equilibrium reactions was a group of papers published by Marcellin Berthelot and Pean St. Gilles (Berthelot and St. Gilles 1862) relative to esterification reactions of the type $RCOOH + R'OH \rightleftarrows RCOOR' + H_2O$. They studied the effects of temperature and of reactant's concentration on the direct reaction rate, concluding that it is proportional to the product of the concentration of the acid and of the alcohol.

A mechanistic attempt to explain the transformations occurring in a chemical reaction in equilibrium conditions was first made by the Austrian physicist Leopold

Pfaundler von Hadermur (1839–1920), professor at the University of Innsbruck, who treated the reactions in the framework of the kinetic theory of gases in terms of molecular collisions (Pfaundler von Hadermur 1867). Pfaundler assumed that at a given temperature the same number of molecules was formed and decomposed by collisions. This explanation implied that not all molecules had the same amount of translational and internal energy, since only a limited number of collisions was effective to give rise to the reaction, either by dissociating or by forming molecules. In Pfaundler's theory chemical affinity was thus defined in purely kinetic and statistical terms as involving all possible internal motions of the molecules. Later Berthelot took up, even if only qualitatively, in his *Essai de mécanique Chimique*, the ideas expressed by Pfaundler.

A more complete and general treatment of equilibrium displacements in chemical reactions was developed in the same period by two Norwegian scholars bound since their childhood by a fraternal friendship, Peter Waage (1833–1900) professor of chemistry at the University of Christiania and his brother in law Cato Maximilian Guldberg (1836–1902), a mathematician at the same university. Guldberg and Waage, whose collaboration started by studying Berthollet's affinity theory, realized that the concept of mass was not easily utilizable in a solution and that it was by far more convenient to use "active masses", i.e. concentrations. They started from the idea that for a generic equation $\alpha A + \beta B \rightleftarrows \gamma C + \delta D$, the direct reaction rate at equilibrium must be equal to that of the inverse reaction and that the two rates are proportional to the product of the concentrations of the reactants according to the reactions

$$v_{dir} = k_{dir}[A]^{\alpha}[B]^{\beta} \quad v_{inv} = k_{inv}[C]^{\gamma}[D]^{\delta},$$

where square parentheses indicate concentrations. By equalizing the two reaction rates, $v_{dir} = v_{inv}$, they obtained the equation known to all first year students of chemistry as the *mass action law*:

$$\frac{[A]^{\alpha}[B]^{\beta}}{[C]^{\gamma}[D]^{\delta}} = K$$

In 1862 Guldberg and Waage presented a short note to the Norwegian Academy of sciences with their first report on the problem. The note, entitled *Studier over Affiniteten*, was published in 1863 in Norwegian (Waage et al. 1864), a language unknown to European chemists, and remained therefore totally ignored. The same fate occurred to a second more detailed note, *Étude sur les affinités chimiques* (Guldberg et al. 1867), published this time in French in 1867. It was only in 1877 that Ostwald, once read this second paper confirmed its validity with a series of experiments. In 1877 van't Hoff, who also ignored Guldberg and Waage's paper, obtained the same law using the kinetic theory (van't Hoff 1877). At this point the two Norwegian friends decided to publish a new version of their work, this time in German (Guldberg et al. 1879), resuming in the discussion the ideas of Pfaundler. Eventually van't Hoff quoted their contribution, automatically acknowledging their priority.

In the late nineteenth century the chemist who best understood the importance of the new physical-chemical concepts in the study of chemical reactions was Jacobus van't Hoff (1852–1911) who had already achieved considerable fame in 1875 after the German translation of his 1874 paper that formulated the idea of the tetrahedral carbon atom, creating the basis of the new field of stereochemistry.

In 1870 van't Hoff, influenced by a paper of Wislicenus on the tri-dimensional representation of the isomers of lactic acid, developed the idea that the optical activity of organic compounds was due to the asymmetry of the carbon atom. Van't Hoff published a 13 pages essay with an unreadable Dutch title (van't Hoff 1874) in which he proposed to represent in three dimensions the structure of organic molecules, assuming that the four valences of the carbon atom were directed towards the vertices of a tetrahedron. A direct consequence of this idea was that when the carbon atom is asymmetric, i.e. when the four valences are saturated by four different atomic groups as in the case of lactic acid, two optically active isomers must exist, one being the mirror image of the other.

Almost at the same time The French Alsatian chemist Joseph Achille Le Bel (1847–1930) published an article (Le Bel 1874) based on Pasteur's hypothesis of the molecular asymmetry in which he reached the same conclusions of van't Hoff.

Van't Hoff's paper was translated in French and published in 1875 with the title La chimie dans l'espace (van't Hoff 1875) and after in German as Die Lagerung der Atome im Raume (van't Hoff 1877). An enlarged version of this paper, Dix années dans l'histoire d'une théorie, was later published at Rotterdam by van't Hoff in 1887 and at Paris in 1892 with the title Stéréochimie.

The German translation strongly contributed to the diffusion in Europe of van't Hoff's ideas and gave origin to a violent and ferocious polemics with Hermann Kolbe, the most important German chemist of the time, who attacked van't Hoff with a paper full of terrible insults published on the Journal für Praktische Chemie (Kolbe 1887). Van't Hoff theory was strongly appreciated by many important chemists such as Wislicenus and Wurtz, and gained his final consecration thank to the decisive support of Friedrich Wilhelm Ostwald, of Hans Heinrich Landolt, author of a fundamental book on the optical rotatory power of organic molecules (Landolt 1898), of Johann Friedrich Adolf Von Baeyer and of Emil Fischer. Finally in 1913 The two Braggs (see Chap. 4) determined experimentally the tetrahedral structure of the carbon in a diamond crystal.

Jacobus van't Hoff, after graduation at the polytechnic of Delft in 1871, decided to become a scientist, studying first mathematics at Delft for 1 year and then attending for about 6 months Kekulé's chemical laboratory in Bonn and for 1 year that of Wurtz at Paris. In 1874 he went back to the Netherlands where he obtained the Ph.D. at Utrecht with a thesis on the cyanoacetic and malonic acids entitled *Bijdrage tot de Kennis van Cyaanazijnzuren en Malonzuur*. In 1876 he was accepted as an assistant at the veterinary college of Utrecht and the year after moved to the same position of the University of Amsterdam. Only in 1878, when he was already known all over Europe for his theory of the stereochemistry of the carbon atom, he was promoted to the position of professor of chemistry, mineralogy and geology, a position that he maintained for 18 years until he accepted a move to Berlin as honorary professor and member of the Real Academy of Prussia.

A romantic dreamer, lover of music and poetry, van't Hoff was a convinced supporter of the importance of fantasy in scientific research. In his inaugural lecture *Verbeeldingskracht in de Wetenschap* (the power of imagination in science) that he hold at the University of Amsterdam, he defended the role of imagination in scientific investigation (van't Hoff 1878), presenting a series of examples showing how several famous scientists had developed the ability to visualize the properties of imaginary objects without having seen them.

Trained as an organic chemist, he was one of the first to become interested in chemical physics, thanks to his excellent preparation in mathematics and physics. As he pointed out in his two-volume treatise, *Ansichten über die organische Chemie* (1878–1881), discoveries made in his studies of the spatial organization of the molecules persuaded him that the chemical properties of organic molecules depend essentially on their physical structure.

Van't Hoff entered by rights the chemical physics community with the book *Études de Dynamique chimique* (1884), in which he examined the problem of identifying the conditions that control the equilibrium of reversible reactions (van't Hoff 1884).

Reversible reactions had already studied by Berthollet in the framework of his research on affinity. In his books *Recherches sur les Lois de Affinité* (1801) and *Essai de statique chimique* (1803) he reached the conclusion that chemical reactions do not always proceed up to the end, but often reach an equilibrium situation that depends on the amount of reactants involved (Berthollet 1801). Also, Williamson had studied reversible reactions during his research on ethers (Williamson 1850) concluding that in a reaction of the type $A + B \rightleftarrows C + D$, the equilibrium is not static but dynamic. He concluded that the reactions takes place at the same rate in both directions, giving the impression of having stopped. The idea that equilibrium is reached when the rates of the two opposite reactions equalize, was also clearly stated almost at the same time by the Italian Faustino Malaguti (Malaguti 1853).

The principal interests of van't Hoff in the 1880s were concentrated on chemical reaction rates as a measure of chemical affinity. In 1884 he laid the mathematical basis of chemical kinetics, starting from the idea that each stoichiometric reaction is the sum of a series of elementary events in which the molecules enter in direct contact one with the other in order to react. The probability that the molecules react is therefore greater, the greater the concentration of the reacting species. In the simple case of a monomolecular reaction in which the reaction rate depends on the concentration c of a single chemical species, he formulated the differential equation

$$v = -\frac{dc}{dt} = kc$$

where k is a constant that he named *rate constant* since it represents the concentration decrease per unit time for an unitary concentration ($c = 1$). On the basis of the number of molecules taking part in the elementary reaction event, van't Hoff classified the reactions in mono- and bi-molecular, showing that more complex reactions which apparently seem to involve the contemporary collision of several molecules, occur in reality through a succession of mono or bi-molecular processes.

As with all chemists of his time, van't Hoff was interested in measuring the affinity which, although considered responsible for chemical reactions, continued to be not easily quantifiable and translatable in measurable quantities. An attempt to find a numerical value for a quantity connected to the affinity had been made before, in 1844, by Eilhard Mitscherlich (1794–1863) who discovered isomorphism and who, working with Berzelius at the Karolinska Institutet in Stockholm, became interested in the affinity problem. Mitscherlich attempted to measure the attractive force of water of hydration in hydrate salts by controlling the decrease of the vapour tension in these crystals. By introducing crystals of the Glauber salt ($Na_2SO_4{\cdot}10H_2O$) in the empty space of a barometer, he observed that the mercury column would drop by 5.45 mm with respect to pure water that instead produced a decrease of 8.72 mm. From these data he deduced that the difference of 3.27 mm measured the affinity of sodium sulphate for hydration water, corresponding to a force of about 12 g/cm^2 (Mitscherlich 1844).

When van't Hoff read Mitscherlich's paper, he was surprised by the very low value of the hydration force, convinced that even the weakest chemical forces should have been much stronger. As he stated later, the botanist Hugo de Vries, whom he met in a street in Amsterdam, had told him that his colleague, the pharmacist and botanist Wilhelm Pfeffer (1845–1920), had observed a dependence of the osmotic pressure on the temperature, finding for each degree of temperature lowering a decrease of the osmotic pressure of about 1/270.

The osmotic pressure is the hydrostatic pressure present in a solution in contact with the pure solvent through a semi-permeable membrane which allows only the solvent molecules to get across. The osmotic pressure, discovered in 1748 (Nollet 1748) by Jean Antoine Nollet (1700–1770), was at this time used essentially by botanists, being considered responsible of the motion of water in plants. Even if some chemists, like Thomas Graham and Justus von Liebig, had shown some interest in the problem essentially in relationship to their biological researches, osmotic pressure remained of limited interest for chemists. Van't Hoff instead immediately realized the importance of osmotic pressure measurements for the evaluation of chemical affinity and decided to use a cell of the type invented by Pfeffer to test his ideas.

Pfeffer (1887) had constructed a cell for osmotic pressure measurements in which a wall of porous ceramic holds a semi-permeable membrane of copper ferrocyanide separating the solution from the pure solvent. Using a diluted solution of cane sugar in water, van't Hoff was able to show that the attraction between water and sugar was about 100 times larger than that measured by Mitscherlich between the sulphate and the hydration water. From these measurements he further deduced a much more important and general result, i.e. that in diluted solutions the particles of the solute behave like the particles of an ideal gas and therefore that the osmotic pressure is proportional to the absolute temperature as the pressure in the gas phase.

In 1886 van't Hoff published a new text in French entitled *L'Équilibre chimique dans l'État dilué gazeux ou dissous* that presented his own ideas on the chemical physics of diluted solutions, establishing a perfect parallelism between ideal gases and diluted solutions in the form of the equation $\pi = \text{í}cRT$ where π is the osmotic pressure, R the universal gas constant, T the absolute temperature, c the concentration and í an empirical corrective coefficient. This equation, practically identical to the equation of state of the gases, $PV = nRT$, was found to be perfectly valid (van't Hoff 1887) for diluted solutions of organic compounds such as sugar with í = 1. Van't Hoff, determined the numerical value of í using different physical methods such as the increase of the boiling point or the decrease of the melting point. He realized that for solutions of inorganic acids, bases or salts í assumed instead a value close to 2. Van't Hoff did not offer a valid explanation for the meaning of his empirical constant í in these cases. Only in 1887 Arrhenius, informed by van't Hoff, suggested (Arrhenius 1887, p. 631) that a value of í of the order of 2 was nothing else than a measure of the fact that the electrolytes were dissociated in solution and therefore that the number of particles of the solute was almost the double. Van't Hoff accepted without hesitation Arrhenius's explanation and re-interpreted his data in terms of ionic dissociation in a paper that he published the same year with the enthusiastic adhesion of Ostwald (van't Hoff 1887).

The analogy between diluted solutions and ideal gases was very useful in extending the second principle of thermodynamics to solutions (van't Hoff 1894). One of the important questions of chemical kinetics concerned in fact at that time the influence of temperature on reversible reactions. In 1884 van't Hoff developed the fundamental thermodynamic relationships controlling the displacement from equilibrium of a reaction with temperature (van't Hoff 1884). In his *Etudes de Dynamique Chimique* he explained the exponential dependence of the reaction rate from the inverse temperature assuming that the equilibrium constants obey the exponential relationship

$$k = Ae^{-\Delta G/RT}$$

where ΔG is the free energy variation (units of J/mole).

Although this equation was proposed by van't Hoff in 1884, it is universally known as the Arrhenius equation, since Svante Arrhenius was the first to offer in 1889 its physical interpretation (Arrhenius 1889). Van't Hoff actually had not made any hypothesis on the physical meaning of the pre-exponential factor A. Arrhenius instead suggested that, in order that a reaction could take place, the reacting molecules had to possess an energy greater than a limiting value that he called *activation energy* E_a. At the temperature T the fraction of molecules possessing a kinetic energy larger than E_a is given by the statistical distribution law of Boltzmann and is proportional to the factor $e^{-Ea/RT}$. In the Arrhenius equation the fraction of free energy available to give rise to the reaction is thus only the one superior to the value of E_a. Therefore the previous equation must be rewritten in the form

$$k = Ae^{-E_a/RT}$$

Van't Hoff used this equation to obtain a relationship that allowed the determination of the standard enthalpy variation ΔH° of a reactive process, avoiding the recourse to calorimetric measurements. From the Arrhenius equation, by calculation of the logarithm of the k constant one obtains,

$$\ln k = -\frac{\Delta G}{RT} = -\frac{\Delta H - T\Delta S}{RT}$$

and assuming that the temperature variation of the enthalpy ΔH ($\Delta H = \Delta H^0$) and of the entropy ΔS is negligible in a restricted temperature interval, one gets

$$\left(\frac{\partial(\ln \mathrm{k})}{\partial \mathrm{T}}\right)_{\mathrm{P}} = \frac{\Delta \mathrm{H}^0}{\mathrm{RT}^2}$$

known as van't Hoff isochore since obtained the first time for a constant volume situation. Thus van't Hoff reached the conclusion that in a reversible reaction a shift of the equilibrium tends always to compensate the temperature variation. If the temperature is lowered the equilibrium shifts in the direction that produces heat, whereas a temperature increase gives rise to the opposite effect (van't Hoff 1898). This conclusion is in reality a particular case of the more general principle formulated in 1885 by the French chemist Le Chatelier that states that *each system tends to counteract any change imposed from the exterior by minimizing his effect.* Henry Le Chatelier (1850–1936), son of a Parisian engineer who contributed to the creation of the French aluminum industry, was an applied chemist specializing in the production of cement who, after having formulated the principle that bears his name (Le Chatelier 1884), translated in 1899 the thermodynamic works of Gibbs in French to prove the validity of his principle.

2.3 The Quantum Mechanical Approach to Chemical Kinetics

The twentieth century witnessed an explosion of research into kinetics, due to its importance in the study of chemical reactivity. Central to the understanding of kinetic processes were the classical researches of Sir Cyril Hinshelwood (1897–1967) who studied in depth the effect of pressure on the kinetics of reaction in the gas phase (Stubbs et al. 1951). In particular he discussed the competition between effective collisions that supply the excess energy (activation processes) to trigger the reaction and inhibiting collisions in which an activated molecule loses his excess energy transferring it to another molecule or transforming it into kinetic energy. Sir Cyril Hinshelwood was awarded the 1956 chemistry Nobel Prize for his research into kinetics (Thompson et al. 1929) and for his fundamental contributions to the theory of chain reactions (Gibson and Hinshelwood 1928). The chain reaction's theory had been already started by Nernst who in 1918 conceived the mechanism of chain reactions to explain the high quantum yield of the photochemical reaction of chlorine with hydrogen (Nernst 1918a, b) according to the scheme

$$\mathrm{Cl_2} + h\nu = 2\,\mathrm{Cl}^* \qquad \mathrm{Cl}^* + \mathrm{H_2} = \mathrm{HCl} + \mathrm{H}$$
$$\mathrm{H} + \mathrm{Cl_2} = \mathrm{HCl} + \mathrm{Cl} \qquad \mathrm{Cl} + \mathrm{Cl} + \mathrm{M} = \mathrm{Cl_2} + \mathrm{M},$$

where M represents any object, for instance the walls of the reaction vessel, which eliminate the excess energy developed in the reaction. Nernst's theory was further developed by his successor at Berlin, Max Ernst August Bodenstein (1871–1942), who studied in detail gas phase reactions (Bodenstein 1913), in particular those of hydrogen with chlorine (Bodenstein and Dux 1913) and iodine and developed the quasi-stationary approximation which neglects the time variation of the concentration of intermediate species (Bodenstein 1922), assuming that the concentration variation of any highly reactive intermediate species is negligible in comparison to the speed of its formation and decay. The kinetic equation of the intermediate species is then substituted by an algebraic equation used to eliminate the corresponding concentration from the ensemble of the kinetic equations.

Hinshelwood studied several apparently simple elementary reactions as the one between molecules of hydrogen and oxygen to produce water, and showed their complexity. Hinshelwood was a key representative of the great scientific tradition of Oxford University. As a scientist with a strong classical and philosophical background and a passionate reader of Dante (he spoke perfectly Italian and) he created a style of humanistic-scientific writing of great elegance and courtliness through the refinement of his cultural formation. His books, *The Kinetics of Chemical Change* (1926), *The Structure of Physical Chemistry* (1951) and *The Chemical Kinetics of the Bacterial Cell*, continue to be famous and unsurpassed.

At the end of the 1930s Hinshelwood oriented his research to the study of bacterial growth, a complex problem that he faced using the same theoretical tools of his research in kinetics (Hinshelwood 1952). In this field he defined the total integration principle to characterize the ability of bacterial cells to adapt their enzymatic balance to the external environment (Hinshelwood 1953).

Fundamental for the understanding of chain reactions were also the contributions of the Russian Nikolay Semyonov (1896–1986), one of the greatest experts of combustion reactions and of explosive processes. Semyonov, who studied at St. Petersburg, was professor at the polytechnic institute from 1928 until 1931 when he went to direct the Chemical Physics Institute of the Russian Academy of Sciences. In the 1930s Semyonov founded the mathematical theory of auto-ignition reactions (Semyonov 1928) that allows the prediction of the auto-ignition temperature of mixtures from the knowledge of the rate and heat dissipation constants (Semyonov 1940). During the same period, he developed, with colleagues, the general theory of flame propagation processes in gases. These researches led to his interest in the chain reactions for which he developed, in parallel with Hinshelwood, a general theory of the reaction kinetics and of the branching processes and a detailed analysis of the mechanisms of chain ignition (Semyonov 1929) as well as of those leading to the end termination of reactive chains. His treatise *Chain Reactions*, written in Russian in 1934 and translated into English the next year (Semyonov 1934), represented a milestone in the study of chain reactions. For his research activity he shared with Hinshelwood the chemistry Nobel Prize in 1956.

The link between classical kinetics, in particular the van't Hoff and Arrhenius theories, and the rising quantum mechanics was realized by Fritz London, Henry Eyring and Michael Polanyi soon after the publication of the seminal paper of Heitler and London on the calculation of the quantum energy of the hydrogen molecule (Heitler and London 1927).

Henry Eyring (1901–1981), was born and grew up in a Mormon community, and remained intimately bound to the religion all his life. He studied mining engineering, metallurgy and chemistry at the University of Arizona and obtained a Ph.D. degree in chemistry at Berkeley in 1927. After that he spent, thanks to a State fellowship, 1 year (1929–1930) at the Kaiser Wilhelm Institute of Berlin, where he met one of the most interesting personalities in Berlin society at the time, the Hungarian Michael Polanyi (1891–1976), chemist, philosopher and economist. Polanyi strongly influenced Eyring's scientific research showing him the road to the quantum mechanical treatment of kinetic processes. Michael Polanyi had started in 1919 to investigate gas phase reactions and had developed a theory which remained rather in the shadow (Polanyi 1932), of the works of Hinshelwood. When Eyring reached Berlin as a post-doctoral researcher with Polanyi, the idea that chemical reactions occurred through the displacement of the reactive system along a path on a multi-dimensional potential energy surface was already spreading. This idea was essentially based on a paper published in 1928 by Fritz London (London 1928a) in a collection of contributions of several authors to celebrate the 60 birthday of Sommerfeld (Sommerfeld Festschrift). In this paper, London, extending the theory developed with Heitler for the hydrogen molecule, discussed in detail in Chap. 4, presented an approximate calculation of the potential energy of a system made of one atom and one molecule of hydrogen (London 1928a, b, 1929), computed in terms of exchange and Coulomb energy between couples of atoms, leading to the formation of an unstable H_3 molecule.

Starting from London's idea, Eyring and Polanyi developed in 1931 their transition state theory, and succeeded in performing the quantum mechanical calculation of the potential energy surface for the collinear reaction

$$H_c + H_a—H_b \rightarrow H_3 \rightarrow H_c—H_a + H_b$$

between one hydrogen atom and one molecule of hydrogen (Eyring and Polanyi 1931), with formation of an unstable transition state, the H_3 ($H_c—H_a—H_b$) molecule (activated complex).

A three-dimensional potential energy surface of this reaction is schematically drawn in the figure below as a function of the relative position of all atoms involved and shows the energy path followed by the reagents in their evolution towards the final products. During the reaction they migrate from one to another energy minimum, climbing a barrier h which shows the relative height of the intermediate stage to be crossed to reach the final products. The height and the position of the barrier h are completely defined by the level of accuracy of the quantum calculation.

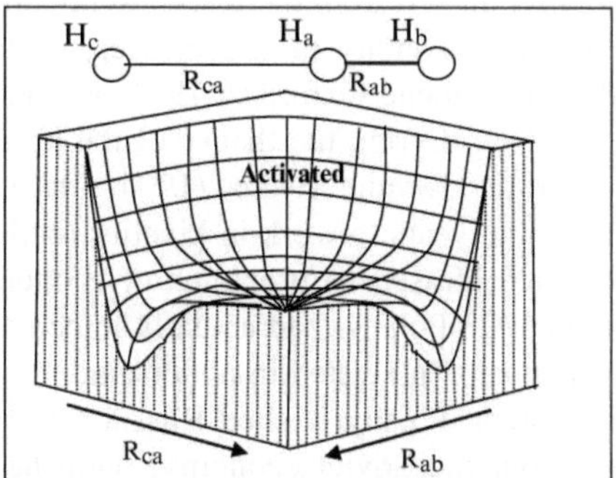

.Schematic representation of the potential energy surface of the reaction: $H_c + H_a—H_b \rightarrow H_c—H_a + H_b$

The complete quantum mechanical calculation of the variation of the electronic energy in a chemical reaction was not feasible in the 1930s, since it requires to compute the energy for several tens of thousands of possible relative positions of the atoms as well as their representation with several variables functions, operations that require the use of powerful electronic computers available only at the end of the twentieth century. Nevertheless Eyring's approximate method of calculation, applied to the reaction of hydrogen with halogens, proved to be effective in predicting whether a reaction occurred through bimolecular collisions or through a mechanism at the atomic level. In 1955 the Japanese Shin Sato of the Technology Institute of Tokyo (Sato 1955) extended the calculations of potential energy surfaces to other reactions between an atom and a diatomic molecule.

At the end of 1930 Eyring went back to the United States where he continued to study the potential energy surfaces of elementary reactions (Eyring 1935). He developed the theory of the absolute reaction rates and of the activated complex and established an explicit expression for the pre-exponential Arrhenius factor in the form

$$k = \frac{k_B T}{h} e^{-\Delta G^*/RT}$$

where ΔG^* represents the activation energy, and where k_B and h are the Boltzmann and the Planck constants, respectively. Eyring became one of the fathers of the modern kinetic theory, author of several books and of more than 450 scientific papers.

The transition state theory was, however, not accepted with enthusiasm by the scientific community, since Eyring and Polanyi's papers were not easy to read and in addition introduced the concept of the activated complex, an object that for the time being was purely hypothetical, did not possess a true physical reality and for many chemists did look as a kind of "Deus ex machina". Several years had to pass before ultrafast laser technology could supply true spectroscopic evidence for the existence of short living molecules as intermediate stages of chemical reactions.

In 1933 Michael Polanyi moved from the Kaiser Wilhelm Institute for Physical Chemistry in Berlin to the University of Manchester, where he stayed until he became a Fellow at Merton College Oxford in 1959. At Manchester, after having developed independently from Eyring the theory of the absolute reaction rates in collaboration with Meredith Gwynne Evans (1904–1952) (Evans and Polanyi 1935), he continued for some years to work in kinetics until he abandoned definitively the field and devoted himself to problems of economics, politics and the philosophy of science. In 1928 Polanyi had actually joined a group including Leo Szilard and John von Neumann interested in understanding with Jacob Marschak, the Russian economic system. Polanyi visited Russia several times and in 1935, wrote a critical paper about the Soviet economy, published by the *Manchester School of Social and Economic Sciences.*

The research on reaction dynamics was, however, continued by his son John Polanyi, inventor of the infrared luminescence technique. John Polanyi was born in Germany but educated in England and emigrated in 1952 to Canada where he first worked at the National Council then moved in 1956 to the University of Toronto where he became professor in 1974. John Polanyi studied in depth the distribution of energy in the excited quantum levels of the reaction products, measuring their infrared emission in simple exothermic reactions (Cashion et al. 1958). Over the course of several years, he developed the technique of infrared emission for several reactive processes and succeeded in building a clear picture of the energy distribution in the vibro-rotational levels of the reaction products (Polanyi 1963, 1967). Of extreme interest was the experimental evidence of the existence of transition states that he collected by means of crossed molecular beams experiments (Polanyi et al. 1995). In 1986 John Polanyi obtained the Nobel Prize for chemistry together with Yuan Lee and Dudley Herschbach, authors of fundamental research in reactive systems by means of the molecular beams technique.

Dudley Robert Herschbach (1932–. . .), obtained a Ph.D. in chemical physics at Harvard in 1956 under the supervision of Bright Wilson Jr. with a thesis in microwave spectroscopy and in 1959 moved to the University of California at Berkeley, where he started a research project on reaction kinetics using molecular beam experiments, that made him known all over the world as a specialist in this new field (Herschbach et al. 1956). Returning to Harvard in 1963 as a professor, he started the study of a series of bimolecular reactions produced by the collision of alkali atoms with molecules (Herschbach 1966) and analyzed in detail the reactive collision as a function of several parameters (Miller et al. 1967) including incidence angles, particle speed, type of electronic state and so on. Amongst his most interesting experiments were those concerning the collisions between a deuterium atom and halogen molecules (McDonald et al. 1972), those relative to diffusion processes by reactive collisions (Herschbach 1973) as well as the multiple collision processes (King et al. 1973).

The study of reactive collisions in crossed molecular beams presents a large number of experimental difficulties that were resolved thanks to the cleverness and experimental ability of the Chinese Yuan Tseh Lee (1936–), born and educated in Taiwan, who in 1962 emigrated to Berkeley to obtain a Ph.D. in 1965. There he started the study of the reactions between ions and molecules by means of molecular beam experiments. In 1967 Lee moved to Herschbach laboratory at Harvard where he completed his scientific preparation (Lee et al. 1969) and started a brilliant career that brought him first to Chicago in 1968 and then again to Berkeley in 1974. Fundamental to the development of his research in reactive dynamics and for the identification of transition states was the technique of crossed beam experiments (Chen et al. 1973), that he developed to a high degree of sophistication (Zhang et al. 1997). In 1994, after having received the award of the Nobel Prize, he decided to go back to his native Taiwan where he dedicated himself to the social and scientific developments of the population, incurring, however, the hostility of students because of the excessive bureaucratization of his reforms and for the authoritarianism of his decisions.

The quantum theory of the dynamics of the reactive processes was extended in the period 1956–1965 by Rudolph Arthur Marcus (1923–) to oxidation-reduction reactions involving electron transfer processes from reactive centers of complex molecules in proteins, in semiconductors and in electrochemical aggregates in solution (Marcus 1956, 1965). Rudolph Arthur Marcus (1923–) born in Montreal in Quebec, Canada, was educated at McGill University where he graduated in 1956 in chemistry under the supervision of Carl A. Winkler, specialist in the study of the rates of chemical reactions who had been a student of Sir Cyril Hinshelwood at Oxford. As a young student at McGill he had the opportunity to attend the group of Edgar William Richard Steacie (1900–1962) an international figure in the study of free-radical reactions. In 1949 he moved to the United States where he worked first as post-doctoral student with Oscar K. Rice at the University of North Carolina at Chapel Hill, then at the Polytechnic Institute of Brooklyn, until in 1978 he accepted an offer from the California Institute of Technology to come there as Arthur Amos Noyes Professor of Chemistry.

At CalTech he started to work on the problem of reaction rates and, although still very young, was able to formulate a particular case of a theory of monomolecular gas phase reactions known as the Rice-Ramsperger-Kassel theory that he published in 1951 and in a more general formulation in 1952 (Marcus 1952).

Marcus and his group completed and extended the kinetic theory of monomolecular reactions and of intramolecular dynamics (Marcus 1952) started in 1921 by Frederick Alexander Lindemann (1886–1957), Viscount of Cherwell, who succeeded in explaining the first order kinetic of several monomolecular reactions, assuming that these were controlled by bimolecular collision processes followed by monomolecular dissociation events that define the determining stage of the reaction speed.

Lindemann's theory was developed by Sir Cyril Hinshelwood assimilating the molecular degrees of freedom to harmonic oscillators and using statistical methods to evaluate the collisional activation probability of the molecule (Levine and

Bernstein 1987). The theory was subsequently improved by Oskar Knefler Rice (1903–1978) and Herman Carl Ramsperger (1896–1932) (Rice and Ramsperger 1927, 1928). In the Rice-Ramsperger theory (RR theory) the dissociation rate of an excited molecule is treated as a function of its energetic content. In this theoretical approach the dissociation rate is assumed proportional to the number of internal degrees of freedom of the molecule and the dissociation is interpreted as due to the localization of a critical amount of energy in a given internal degree of freedom. The distribution of the excitation energy among the different molecular degrees of freedom had been independently discussed also by Louis S. Kassel (Kassel 1928) and by Marcus, so that the theory finally took the name of RRKM (Rice, Ramsperger, Kassel, Marcus) theory. In 1952 Rudolph Marcus (Marcus 1952) introduced in the framework of Eyring's transition state theory the idea that the molecular excitation energy could be separable into fixed and variable contributions and that only these latter, able to shift among the different degrees of freedom, were responsible for the reaction.

One of the most significant results of Marcus's electron transfer theory concerns the parabolic relationship between the driving force and the reaction rate. As the driving force increases, the reaction rate first increases, then reaches a maximum and finally decreases. The occurrence of an "inverted region" for the process was for some time seen with skepticism by the chemical community until it was experimentally documented (Miller et al. 1984; Deisenhofer et al. 1984).

2.4 Catalysis

Information concerning catalytic reactions was extremely limited before the beginning of the nineteenth century, when kinetic theory started to be an independent branch of chemistry, although humans had applied the principles for thousands of years in fermentation processes.

The first known catalytic processes involved reactions in solution catalyzed by the addition of small amounts of acid (homogeneous catalysis). This occurred in 1548 when the German physician and botanist Valerius Cordus (1515–1544) used oil of vitriol to transform ethyl alcohol in the corresponding ether, probably following an old recipe imported from Middle East by Portuguese explorers (Cordus 1548). At the end of the eighteenth century, acid catalysis started to play an important role in chemical kinetics, when several chemists, even without a rational explanation for it, realized that the addition of small amounts of an acid or of a basis, could modify and speed up reactions.

Antoine Augustin Parmentier realized for instance in 1781 that addition of acetic acid accelerated the transformation of potato flour into a sweet substance by means of cream of tartar and in 1811 the Russian Sigismund Konstantin Kirchhoff (1764–1833) found that the hydrolysis of starch in glucose was made faster by adding few drops of sulphuric acid (Kirchhoff 1811). Afterwards Ludwig Wilhelmy (1812–1864) found that the inversion process of cane sugar was made easier by inorganic acids (Wilhelmy 1850) and Augustus George Vernon Harcourt (1834–1919) discovered the importance of acid catalysis in clock reactions (Shorter 1980).

At the beginning of the nineteenth century the interpretation of chemical processes was still dominated by the affinity concept, but slowly new results started to appear, suggesting that different mechanisms could contribute to orient the course of a reaction. In 1833 Anselme Payen (1795–1871) and Jean-François Persoz (1805–1868) attributed the starch transformation discovered by Kirchhoff to the action of a particular biological substance, that they called *diastase*, and proved that at 100°C it loosed its activity (Payen and Persoz 1833). The starch fermentation was studied also by Johann Wolfgang Döbereiner who in 1822 found that starch transformed into alcohol only after having been converted in sugar (Döbereiner 1822). In 1877, the German physiologist Wilhelm Kühne (1837–1900), pupil of Claude Bernard and Virkow, isolated tripsine from gastric juice (Kühne 1877) and coined the world *ενζυμον*, enzyme, from the Greek, *εν* in and *ζυμον* ferment, to describe cellular fermentation:

> Um Missverständnissen vorzubeugen und lästige Umschreibungen zu vermeiden schlägt Vortragender vor, die ungeformten oder nicht organisirten Fermente, deren Wirkung ohne Anwesenheit von Organismen und ausserhalb derselben erfolgen kann, als Enzyme zu bezeichnen.[1]

Later the term *enzyme* was used only for non living systems and substituted by the word *ferment* for living organisms. Enzymes are essentially globular proteins and just from protein research derived the studies of enzymatic catalysis. At the beginning of the nineteenth century also metals were identified as reaction accelerators and in particular platinum gained a significant position in the catalysis world as a prima donna in heterogeneous catalysis.

In 1813 Louis Jacques Thenard (1777–1857) discovered that ammonia decomposed in nitrogen and hydrogen by flowing on red-hot metal. Subsequently, in collaboration with Pierre Dulong (Dulong and Thenard 1823a, b), later to be internationally known for the specific law of solids, Thenard showed that the ability to decompose ammonia decreased in the series copper, silver, gold and platinum. They proposed that this was due to electric charges on the metal. Furthermore, in 1817 Humphry Davy proved that platinum accelerated several organic reactions without being altered (Davy 1817) and in 1820 his cousin Edmund Davy (1785–1857) prepared a platinum sponge able to absorb large amounts of gas and realized that in the presence of finely divided platinum, alcohol vapors were transformed in acetic acid (Davy 1820). He showed also that boiling a mixture of platinum sulphate with ether and alcohol, one obtains a finely divided precipitate of platinum capable of absorbing

[1] In order to obviate misunderstandings and avoid cumbersome periphrases, the author, a lecturer, suggests to designate as "enzymes" the unformed or not organized ferments, whose action can occur without the presence of organisms and outside of the same.

hydrogen and possessing great catalytic properties, up to the point of becoming red-hot in the presence of a mixture of city gas and oxygen.

After some years Justus Liebig took up the idea and succeeded in preparing a platinum sponge able to absorb up to 250 times its volume of oxygen (Liebig 1829). In 1823 Johann Wolfgang Döbereiner discovered also the catalytic virtues of platinum, realizing that finely divided platinum in contact with hydrogen becomes red-hot while the hydrogen burns (Döbereiner 1823). In 1824 the Italian physicist and physician Ambrogio Fusinieri (1773–1853) from Vicenza re-interpreted Döbereiner's experiment, (Fusinieri 1824) putting forward the hypothesis that a solid layer of gas was adsorbed and continuously rebuilt on the platinum surface as the gas was consumed in the combustion reaction. The catalytic activity of platinum was studied also by William Henry, the author of the law bearing his name on the dissolution of gases in liquids, who discovered the inverse effect, namely that ethylene would stop the action of platinum on the mixture hydrogen–oxygen (Henry 1824). Soon after, Michael Faraday showed the ability of platinum to recombine hydrogen and oxygen obtained from the electrolysis of water (Faraday 1834).

A first attempt to interpret the mechanism of catalysis was made by Berzelius who in a report to the Swedish Academy of Sciences of 1835, published in 1836 (Berzelius 1836), reviewed a large number of results on both homogeneous and heterogeneous catalytic reactions and proposed the existence of a "new catalytic force", acting on the matter. He also coined the word catalysis, combining together the Greek words κατά (down) and λύσις (solution, loosening). According to Berzelius, a catalyst was a substance able to start a reaction without taking part in it and thus without being consumed. In 1839 Justus von Liebig, tied to the theory of organic radicals, postulated that the difference among the chemical properties of the elements forming a radical, could give rise to the break of their attraction, the breaking being favored by the temperature, by water, but also by a third body, the catalyst, which did not take part in the reaction (Liebig 1839).

A different interpretation of the catalytic mechanism was instead developed in 1845 by Julius Mayer who, in the frame of his thermodynamic researches, had devoted himself to the study of photosynthetic processes able to convert light into chemical energy. Mayer put forward the idea that the catalyst was able to release large amounts of "sleeping energy" that could allow the reaction to break out. In 1876, Mayer maintained that catalysis was a particular case of the more general concept of primer (Mayer 1876), a kind of chemical trigger able to start the reaction. The idea that the catalyst, without interacting with the reagents, could speed up the reaction producing intermediate products capable of opening new and faster paths to the reacting molecules, was developed by Christian Friedrich Schönbein (1799–1868), the discoverer of ozone and of gun cotton, who asserted that a reaction is not a single process, but occurs through a time ordered series of intermediate events (Schönbein 1848).

The possibility that a catalyst could release energy to facilitate the reaction, did show up again after some years in the catalysis theory proposed by the German Friedrich Stohmann (1832–1897) on the basis of his research into the combustion heat of foodstuffs. Stohmann thought that catalysis was a process in which the energy released by the catalyst was transformed in motions of the atoms of the reacting molecules which reorganized themselves, giving rise to a more stable system by emission of energy (Stohmann 1894). Ostwald, however, did not agree with Stohmann and dedicated a large portion of a review of Stohmann's paper published on the *Zeitschrift für physikalische Chemie* (Ostwald 1894) to his own vision of the problem, asserting that the catalyst did not alter the reaction mechanism, but simply accelerated its kinetics, lowering the energy barrier necessary to prime the reaction. He maintained that several unstable systems exist such that their transformation into stable systems is so slow as to let them appear stable; the catalyst does nothing else than speed up the reaction which in any case would have occurred without its presence although at a much slower rate. In particular for gas phase reactions, Ostwald suggested that the metal catalytic effect was due to pure physical processes of adsorption in which the gases entered the cavities of the porous metals where their close contact, combined with local heating processes favored the reaction. In 1901 Ostwald synthetically formulated his own definition of catalysis (Ostwald 1902):

> Ein Katalysator ist jeder Stoff, der, ohne im Endprodukt einer chemischen Reaktion zu erscheinen, ihre Geschwindigkeit verändert.[2]

The theory of catalysis was the principal argument in favor of the award of the Nobel Prize to Ostwald in 1909 (Ostwald 1902) The Nobel lecture delivered by Ostwald at the Royal Academy of Sweden (Ostwald 1910) represented the definitive consecration of catalysis as one of the fundamental branches of chemical physics. In a short time it gained enormous influence on the development of the chemical industry.

Ostwald patented in 1902 the catalytic process for the production of nitric acid from ammonia. The procedure had been already developed and patented by *Charles Frédéric Kuhlmann* (1803–1881) more than 60 years before, but owing to the high price of ammonia was still considered as a pure academic exercise. The later date of 1908 is often reported for Ostwald's patent starting time, probably because of the unavoidable bureaucratic delays to make it operative or perhaps because only at that time the Fritz Haber ammonia synthesis (Haber and Le Rossignol 1910) was really effectual and the ammonia price became acceptable. Ammonia production in Germany was still controlled by BASF (*Badische Anilin und Soda Fabrik*) directed by Alwin Mittasch (1869–1953), pupil of Ostwald, who also maintained in 1930 that a catalysis was a particular case of chemical primer. The concept of primer was fashionable in the 1930s in Germany when the National Socialist regime encouraged the rhetoric of integration between nature and science for its autarkic programs. Several authors of the Nazi regime such as Walter Greiling (1900–1986), Karl Aloys Schenzinger (1886–1962) and Anton Zischka (1904–1997), used the concepts of primer and catalysis in books of Nazi propaganda.

[2] A catalyst is a substance that alters a chemical reaction rate without being part of the final products.

Ostwald's process coupled to the Haber ammonia synthesis, led to large scale production of fertilizers and explosives, giving new breath to Germany during the first world war, when the guano import from Chile was blocked.

Support for Ostwald's theory of catalysis as a pure physical process came from the researches of the French chemist Jacques Duclaux (1877–1978), an expert of the chemistry of colloidal substances who, after the first world war, started a research project on the catalytic activity of colloids, based on the idea that catalysis was due to adsorption processes. Of the same idea was Henry Moissan, discoverer of fluorine who, studying in collaboration with Charles Moureu (1863–1929) the interaction of acetylene with finely divided metals, nickel, iron and cobalt (Moissan and Moureu 1896), was convinced that acetylene was adsorbed in the metal pores where by effect of heat a pyrolytic reaction took place, originating a mixture of carbon, benzene and hydrogen. Next year Paul Sabatier (1854–1941), professor at the University of Toulouse and his co-worker, the abbot Jean Baptiste Senderens (1856–1937), proved with a similar experiment on ethylene that Moissan's conclusions were wrong since ethylene in contact with the metal burns deposing carbon, but produces ethane and not hydrogen. From this they deduced that the action of nickel was to catalyze the attack to the ethylenic double bond, in other words to give origin to a true chemical reaction (Sabatier and Senderens 1897). On the basis of this result Sabatier and Senderens concluded that the catalysis was not, as maintained by Ostwald, a purely physical process, but that finely divided metals were able to absorb large amounts of gas and that this property was very specific and confirmed a selective action of a pure chemical nature.

With this experiment and subsequent research (Sabatier 1897), Paul Sabatier made a fundamental contribution to heterogeneous catalysis, developing the catalytic hydrogenation technique by which ethylene was transformed in ethane using finely divided nickel as catalyst.

The catalytic hydrogenation method of Sabatier-Senderens initiated, in the period 1906–1919, a series of important papers in collaboration not only with Senderens but also with his students Alfonse Mailhe, Marcel Murat, Leo Espil and Georges Gaudion, published mostly in the *Comptes Rendus* de l'Académie des Sciences and in the *Bulletin de la Société Chimique de France*. In 1902 Sabatier and Senderens demonstrated also that by flowing a mixture of CO and H_2 in the ratio 1:3 in volumes on finely divided nickel at 250° one gets, with a 100% yield, methane and water

$$CO + 3H_2 \rightleftarrows CH_4 + H_2O \quad CO_2 + 4H_2 \rightleftarrows CH_4 + 2H_2O$$

and that at higher temperature the same kind of reaction occurs also with a mixture of carbonic anhydride and hydrogen in the ratio 1:4. This type of reaction became known as the "Sabatier process" (Sabatier and Senderens 1902a, b).

For his contributions to catalysis and in particular for the catalytic hydrogenation, illustrated in great detail in his famous 1913 treatise *La Catalyse en Chimie Organique*, Sabatier was awarded the Nobel Prize in 1912 together with Victor Grignard.

Sabatier was a very reserved man, very tied to Toulouse and indifferent to success to the point that he never agreed to move to Paris to the chair left vacant in 1907 by the death of Moissan. Nevertheless, when he became famous and particularly after the Nobel award, he incurred without understanding why, the resentment of Abbot Senderens, who felt denied official recognition for his contribution. As a consequence of their misunderstanding, their relationships which had lasted more than 30 years eventually deteriorated.

Later research on heterogeneous catalysis allowed several other elements to catalyze even very complex reactions, although elements belonging to group eight, in particular platinum, were the most prominent elements used for a good part of the century.

The understanding of the elementary mechanisms of heterogeneous catalysis took a great step forward in 1916 when Irvin Langmuir developed a theory of the chemisorptions of gases on metallic supports which rapidly became the starting point of the modern theories of heterogeneous catalysis (Langmuir 1916). According to Langmuir's model, a gas was adsorbed on a metallic surface thanks to residual valences, closely resembling those introduced by Thiele, Flürscheim and Lapworth in the discussion of molecules with conjugated bonds. In this way true compounds of variable composition were formed between the gas and the metal and these favored the reaction (Langmuir 1916, 1917, 1918).

Langmuir's theory offered a simple mechanism of attack of the gas molecules on the catalyst surface of the type

$$A_g + S \rightleftarrows AS$$

where A_g is a gas molecule and S an adsorption site. To simplify the mathematical treatment, Langmuir assumed that the metallic surface was uniform with adsorption sites S all equivalents, that there was a single adsorption mechanism, that the adsorbed molecules did not react together and that they formed at most a single monolayer.

With these simplifications the theory allowed to connect at constant temperature the monolayer fraction θ of adsorbed gas molecules with the pressure P, according to the relationship

$$\theta = \frac{\alpha \cdot P}{1 + \alpha \cdot P}$$

where α was a constant characteristic of the bond energy between the gas and the substrate, representing the ratio between the equilibrium constants of the direct and inverse reactions. To a first approximation α was inversely proportional to temperature. The curves of θ as a function of P are known as Langmuir's adsorption isotherms.

In 1938, since the hypothesis of a single gas monolayer adsorbed on the metal surface was too limitative except at very low pressures, the Hungarian Stephen Brunauer (1903–1986) in collaboration with Paul Hugh Emmett (1900–1985) and with the future father of the atomic bomb Edward Teller (1908–2003) (Brunauer et al. 1938) modified Langmuir's adsorption theory, developing a new one called BET from the initials of their names, taking into account the possibility of formation of multilayers of gaseous molecules A_g physiadsorbed at the sites S of the catalyst according to the reactions

$$A_g + S \rightleftarrows AS$$
$$A_g + AS \rightleftarrows A_2S$$
$$\ldots\ldots\ldots\ldots\ldots\ldots\ldots\ldots$$
$$A_g + A_mS \rightleftarrows A_{m+1}S$$

In the analysis of the experimental data Langmuir's isotherms which predicts the occurrence of chemical bonds due to residual valences between the gas molecules and the atoms of the metallic surface, show in general a better agreement especially at low pressure and for chemisorptions processes, whereas the BET isotherms explain better the physisorption on non micro-porous materials in which the molecules are bound to the solid surface by weak van der Waals forces.

The first hypotheses on the mechanism involved in heterogeneous catalysis arose from the research of Adalbert and Ladislas Farkas, two Hungarian brothers who in 1927 went to the Fritz Haber Institute in Germany where both worked on molecular hydrogen and in particular on the ortho-para transformation catalyzed by tungsten. In 1933 the Farkas brothers were forced to flee Germany to escape Nazi persecution and being unable to go to the United States or England, emigrated to Israel. Later Adalbert went to the United States where he obtained academic positions, while Ladislas continued to work in the chemical industry.

After a seminal paper by Adalbert (Farkas 1931) and another by Ladislas in collaboration with Hans Sachsse (1906–1992) on the ortho-para transformation induced by the presence of paramagnetic (Farkas and Farkas 1934) and oxygen molecules, (Farkas and Sachsse 1933) the Farkas (Farkas and Farkas 1934) and in collaboration with Paul Karl Maria Harteck (1902–1985) (Farkas et al. 1934) proposed an interesting mechanism for the ortho-para transformation process following a previous treatment (Wigner 1933) by the famous physicist Eugene Paul Wigner (1902–1995), another high level Hungarian emigré. The theory assumed that changes from ortho- to para-hydrogen occur because the inhomogeneous magnetic field in the vicinity of a paramagnetic ion or molecule acts as a perturbation. In this way during a collision between a hydrogen molecule and the paramagnetic ion on the metal surface, the otherwise forbidden ortho-para transition can occur.

The reaction mechanism proposed by the Farkas brothers for the ortho-para transformation process opened the road to intense research activity on the elementary mechanisms of the catalytic processes. After a few years the transformation was also studied by one of the best English experts of catalysis, the chemist Eric Rideal starting from the idea, borrowed from the BET isotherms theory that on the solid surface multilayer's of adsorbed molecules could be deposited.

Eric Keightley Rideal (1890–1974) after a period of postdoctoral training in Germany from 1910 to 1913, first at Aachen and then at Bonn, acquired considerable experience in heterogeneous catalysis working with Hugh S. Taylor (1890–1974) at the Haber-Bosch process for the ammonia synthesis from hydrogen and nitrogen. In 1930 he became professor at Cambridge where he founded a world-renowned known laboratory of colloid chemistry. For the conversion reaction of para to ortho molecular hydrogen Rideal proposed the mechanism bearing his name that assumes that a para H_2 molecule, weekly physiadsorbed in the second layer on the solid surface (M), reacts with an hydrogen atom strongly chemisorbed in the underlying layer (Rideal 1939) exchanging an H atom according to the scheme

$$\mathrm{pH{-}H + M{-}H \rightarrow H{-}M + oH{-}H}$$

The same type of mechanism known as the Eley-Rideal, was proposed successively by Daniel Douglas Eley and by Rideal himself for reactions between two molecules, one adsorbed on the catalyst and the other coming from the gas phase (Eley and Rideal 1940, 1941). Eley who had started his research activity with Michael Polanyi (Eley and Polanyi 1935) entered later Rideal's research group and eventually became editor of the famous series *Advances in Catalysis* (Academic Press, New York).

A more efficient mechanism, based on Langmuir's adsorption theory and known as the Langmuir-Hinshelwood mechanism, was developed by Sir Cyril Norman Hinshelwood (Hinshelwood 1940) starting from the idea that two molecules A and B, both adsorbed on the catalyst surface could give rise to a bimolecular reaction according to the scheme (Gadsby et al. 1946)

$$\mathrm{A + S \leftrightarrow AS}$$
$$\mathrm{B + S \leftrightarrow AS}$$
$$\mathrm{AS + BS \leftrightarrow AB + 2S}$$

In the first half of the twenty-first century the industrial applications of catalysis had a substantial boost from the work of the Muscovite Vladimir Nikolaevich Ipatieff (1867–1952) who as a young man took up a military career studying at the famous Russian military academy *Mikhailovskaia artilleriiskaia akademiia*. After his graduation in chemistry at the University of St. Petersburg in 1907 he reached the highest grades in the Czar's Nicholas II army up to the position of general. After the October revolution Ipatieff remained in the Soviet Union where in 1927 he founded the high pressure institute and where he was even awarded the prestigious Lenin prize.

In 1931 to escape Stalin's purges he took advantage of a trip abroad to escape to the United States where he became research director at the Universal Oil Products at Des Plaines as well as professor of chemistry at the Northwestern University. In the same year a young Polish chemist, Herman Pines (1902–1996), joined his group and became his closest coworker up to his death. Ipatieff and Pines set right in the 1930s the catalytic technique of addition of a paraffin molecule to the double bond of an olefin in the presence of a strong acid, operation considered impossible until then since paraffins were considered chemically inert molecules as stated by their Latin name *parum affinis*. In 1932 they proved that isobutane reacts at −35°C with olefins and this alkylation reaction opened the road to the synthesis of gasoline with high octane number. The synthesis of iso-octane from butene and isobutane that they patented in 1938 allowed to produce high quality fuel that played a decisive role for the Royal Air Force victory during the Aerial Battle of Britain in 1941.

In 1928 Johannes Nicolaus Brønsted (Brønsted 1928) reconsidered the problem of homogeneous acid and basic catalysis, correlating the free energy variation to the activation energy and showing that strongly exothermic reactions possess a low activation energy. A mathematical form of this concept was produced in 1936 by Brønsted's pupil Ronald Percy Bell (1907–1996) (Bell 1935, 1936) and after him by Meredith G. Evans and Michael Polanyi, when this latter joined the University of Manchester. These authors tackled the problem in terms of the potential energy surface (Evans and Polanyi 1937) and in order to simplify the numerical calculations approximated the form of the potential energy curve with pieces of linear segments, producing a linear relationship between the activation energy ε_a and the free energy ΔG in the form $\varepsilon_a = k_1 + k_2 \Delta G$ (Evans and Polanyi 1938) where the constants k_1 and $k_2 > 0$ depend on the inclination of the segments of curve. The linear relation between the free energy ΔG and the activation energy is known as the *Bell-Evans-Polanyi principle*. A more accurate representation, due to Rudolph Arthur Marcus (Marcus 1968), approximated the potential energy curve with two parabolas centered at the two energy minima and added a quadratic term to the previous equation.

In the period 1930–1960, research was mostly oriented to individuate the role played by the electronic structure of metals and semiconductors in catalytic processes. In this way the electronic theory of catalysis developed as a valid combination of solid state physics and adsorption theories. The electronic theory of catalysis was essentially a phenomenological one and had a considerable development in Russia after the pioneering research of the Ukrainian Lev Vladimirovich Pisarzhevsky (1871–1938), the first who correlated in 1916 the catalytic activity of solids with their electronic structure (Pisarzhevsky 1955). The contribution of the Russian school to heterogeneous catalysis is well documented in the book of Georgii K. Boreskov, "Heterogeneous Catalysis" (Boreskov 2003) and reached its heights with the institution in 1958 of the Soviet Institute of catalysis dedicated today to its founder and first director Boreskov, a great supporter of the importance of the chemical nature of heterogeneous catalysis. In 1953 he formulated the rule of the approximate perseverance of the catalytic capacity of substance with the same chemical composition. Even if unable to explain the intimate mechanisms of catalysis, the electronic theory opened the road to the applications of the concepts of transition state and of collective surface effects later incorporated in the quantum theory of catalysis.

The concept of surface collective phenomena was introduced in the catalysis theory by Dennis Albert Dowden (1950) and after by Karl Hauffe (1950) and Fyodor F. Vol'kenshteĭn (Vol'kenshteĭn 1960), who contributed a large amount of experimental research to correlate catalytic activity to the chemical structure of semiconductors and alloys used in heterogeneous catalysis.

With the coming of quantum mechanics and of the physics of surfaces it became clear that on the metallic surfaces were present sites particularly active in catalysis. A fundamental contribution to this problem was given by the Berliner Georg-Maria Schwab (1899–1984) who started his interest in catalysis working at the catalytic decomposition of methane and ammonia (Schwab and Pietsch 1926) in collaboration with Erich Pietsch (1902–1979) who later would become one of the principal figures of scientific divulgation in Germany as editor of the *Gmelin Handbuch der anorganischen Chemie*. In 1929 they developed together the "adlineation" theory describing the reaction along the line between two surface phases, that assigns particular catalytic efficiency to structures with high density of adsorption (Schwab and Pietsch 1929). Schwab discussed in detail the catalytic activity of lattice defects and dislocations at the surface of solid catalysts that expose to adsorption different crystal faces giving rise to an increase of the density of adsorbed gas molecules on particularly active sites. In 1929 in collaboration with Erika Cremer (1900–1996) a German pioneer in gas-chromatography, he introduced in his kinetic approach to catalysis the concept of compensation effect that he named the "theta rule", referring to the thermal preparation used to fix the thermodynamic distribution of catalytic centers (Cremer 1929). In 1930 Schwab and Hermann Schultes discussed also the importance of finely divided metal catalysts dispersed on a support of metal oxides, argument that Schwab reconsidered in 1950 (Schwab 1950). He wrote in 1931 his comprehensive treatise *Katalyse vom Standpunkt der Kemischen Kinetik* (Verlag J. Springer, Berlin 1931) translated in English in 1937 as *Catalysis from the Standpoint of Chemical Kinetics,* (D Van Nostrand Co 1937) and acted as editor of the *Handbuch der Katalyse* (Springer Verlag, Wien) from 1940 to 1957. In 1939 he emigrated to Greece as director of the chemical physics department of the Nikolaos Kanellopoulos Institute in the Piraeus, until in 1950 he went back to Germany as director of the Institute of Chemical Physics of the University of Munich where he directed a broad research project on catalysis.

With the development of new experimental techniques such as LEED (*Low Energy Electron Diffraction*), UPS (*Ultraviolet photoelectron Spectroscopy*) and STM (*Scanning tunneling Microscope*) which revolutionized the study of surfaces in the second half of the twentieth century, the understanding of the elementary mechanisms of catalysis made a significant quality jump.

The LEED technique is a sophisticated application of the discovery of electron diffraction made in 1927 by Clinton Joseph Davisson (1881–1958) and his assistant Lester Halbert Germer (1896–1971) (Davisson and Germer 1927) following the hypothesis postulated in 1924 by Louis de Broglie. One month after George Paget Thomson, son of J.J. Thomson with his coworker Alexander Reid, published their electron diffraction obtained with electron kinetic energy thousand times higher than that used by Davisson and Germer (Thomson and Reid 1927). These two experiments opened a new era for the application of electron diffraction to the study of matter.

Electron diffraction did not, however, become a popular tool for the study of surfaces until the early 1960s, owing to the difficulty of correctly monitoring the directions and intensities of the diffracted beams with the available vacuum techniques and detection methods. In the early 1960s ultra high vacuum became widely available and better detection methods were devised and thus the LEED technique experienced an intensive diffusion in catalytic laboratories, thanks also to the realization of new techniques for the reconstruction of clean metal surfaces. Its operability as a true research instrument in chemical physics for the determination of bond angles and bond lengths of metal surfaces and of molecular systems adsorbed on them, became a true reality only when a new dynamical electron diffraction theory which took into account the possibility of multiple scattering was realized.

Another important physical technique for the study of catalytic surfaces is the Scanning tunneling Microscopy developed in 1981 by the German Gerd Binnig (1947–) and the Swiss Heinrich Rohrer (1933–) at the IBM research laboratories in Zurich (Binnig et al. 1986). The tunnelling microscope technique is based on the tunnel effect realized by letting a small conducting tip to approach very closely a surface so that electrons can tunnel through the vacuum when a voltage difference is applied among them. With this kind of microscope a resolution of Ångströms can be easily obtained, well suited to image individual atoms and to explore the presence of impurities attached to a surface. An important advantage of this technique with respect to LEED is that it can be applied not only in ultra high vacuum but also in air, or in a liquid and even at high temperatures.

The intelligent application of these techniques characterized the research of the German Gerhard Ertl and of the Hungarian Gabor Somorjai, the two figures who have contributed most to the development of modern catalysis theories.

Gerhard Ertl (1936-) was born in Stuttgart where he graduated in 1961 after a post doctoral period in Paris first and then at the Münich Technische Universität where he completed his Ph.D. thesis in 1965. In his research he faced the study of the surface structure of metallic catalysts and semiconductors and developed avant-garde techniques allowing observation of atoms or molecules absorbed on very clean surfaces, with active sites completely free from adsorbed atmospheric oxygen or nitrogen molecules.

In 1974, by coupling LEED measurements of electron diffraction to desorption techniques and to theoretical models, he succeeded in explaining how hydrogen is tight to palladium (Conrad, Ertl, Latta 1974) and nickel (Christmann et al. 1974) metallic surfaces. In 1977 he worked out a theory of the molecular mechanisms in the famous Haber Bosch ammonia synthesis, concentrating his attention on the process that he considered the determining step for the reaction rate constants, namely the dissociation of the N_2 molecules into single atoms of nitrogen. Ertl first showed that nitrogen atoms derived from the breaking of the $N \equiv N$ triple bond are present on the iron catalyst surface (Bozso et al. 1977) and proposed a model of the iron-nitrogen bond. He proved also that the activation energy is different for different crystallographic planes of the iron, pointing out the greater efficiency of the (111), (110) e (100) planes and clarified the details of the reaction mechanism (Ertl and Huber 1980).

Later Ertl studied the non linear mechanisms involved in the classical oxidation reaction of carbon monoxide CO on platinum and palladium catalysts (Skottke et al.

1987). He identified at the microscopic level the causes of the non linear behaviour of the reaction and the conditions that give rise to oscillating reactions (Cox et al. 1985) and contributed to the theory underlying the production of catalytic exhaust pipes for cars and of combustion cells. He investigated also the formation of molecular aggregates on the surface of tiny crystallites of ice in the stratosphere, responsible for reactions that damage the ozone layer protecting the Earth from solar radiation. For his researches Gerhard Ertl received in 2007 the Nobel Prize for chemistry.

Of equal importance is the research and under many aspects equivalent are the researches of the Hungarian Gabor Arpad Somorjai (1935–), who as a child escaped the holocaust thanks to the help of the Swedish diplomat Raoul Wallenberg who in 1944 procured for him and his mother false Swedish passports. Somorjai studied chemical engineering at Budapest but in 1956, having taken part in the Hungarian revolution, escaped to the United States where he obtained the Ph.D. in 1960 (Somorjai 1960) at the University of California, Berkeley, where he still works as a teacher and a researcher since 1964, after a period at the IBM research laboratories at Yorktown Heights, New York. At Berkeley Somorjai discovered that inhomogeneous catalytic reactions occur where there are defects on the catalyst surface: when these defects break down, new bonds are formed between the atoms, with creation of new organic compounds.

Since 1965 Somorjai has directed a research project concerning the study of the molecular bases of heterogeneous catalysis by characterizing the structure of the crystal surfaces and determining the nature of the bonds with adsorbed molecules.

In these researches Somorjai utilized single crystals of transition metals in a large number of catalytic reactions both in high pressure cells and under high vacuum, applying advanced techniques to control in *situ* the catalyst surface during the reactions. Since 1994 he has been interested in nanostructures (Somorjai and Park 2008) produced either by lithography through bombardment with electrons, or by epitaxial growth (Habas et al. 2007), as well as in photo lithography and in colloidal science, studying among others the hardness and the resistance to friction of the catalysts using either atomic force microscopy (AFM or SFM) or generation of overtones by non linear optics.

In the 1990s Somorjai started an important collaboration with his colleague at Berkeley Yuen-Ron Shen (1935–), one of the greatest experts of non-linear optics. Together they developed the technique of overtone generation by frequency sum or difference (Cremer et al. 1996) which Somorjai applied to the study of catalytic reactions on surfaces (Somorjai and McCrea 2000).

Somorjai, from his interest in the Haber-Bosch process, reached different conclusions from those of Ertl as far as the slow stage of the reaction is concerned. According to Somorjai the slow stage is the one in which the ammonia molecules are detached from the catalyst. Somorjai derived this idea from the knowledge of the fact that in the industrial synthetic process several molecules of NH_3 are produced at the same time and these partially fill the catalyst surface preventing the adsorption of new nitrogen molecules (Spencer et al. 1982).

The tradition of heterogeneous catalysis studies, established at Cambridge by Rideal and Eley, was successfully continued by the Welshman John Meurig Thomas who complemented his efforts in the elucidation of catalysis with an equally powerful interest in general culture and in the popularization of science. His balance of interests was shown through his position as director of the Royal Institution, a role that identified him as the natural heir of his scientific hero, Michael Faraday. He celebrated this in 1991 in a beautiful book entitled *Michael Faraday and the Royal Institution: The Genius of Man and Place,* (Taylor and Francis 1991)

Born in the small mining town of Llanelli in South Wales, he graduated in 1958 from the University of Wales, Swansea under the supervision of Keble Sykes, completing his Ph.D. work through a collaboration with the Queen Mary College, University of London. After a year he joined the Department of Chemistry at the University of Wales, Bangor, where he started his academic career until in 1969 he became professor of chemistry at the University College of Wales, Aberystwyth. In 1978, he was appointed Head of the Department of Physical Chemistry at the University of Cambridge and in 1986 succeeded George Porter as Director of the Royal Institution of Great Britain. In his research, Thomas has used different chemical-physical techniques in particular synchrotron radiation, high-resolution X ray diffraction and spectroscopy to investigate the nature and the location of active sites of solid catalysts under operating conditions (Wright et al. 1986) and to devise new mesoporous (Raja et al. 2003), micro-porous, and molecular sieve catalysts. He has also contributed to the development of industrial applications of heterogeneous catalysis, inventing the solvent-free (Thomas et al. 2001) catalytic synthesis of ethyl acetate, which has reached today a production of 200,000 t/year in the UK (Ballantine et al. 1984). He has also devised a single-step, solvent-free process for the production of caprolactam, the raw material for nylon-6 (Thomas et al. 2005).

References

Arrhenius S (1887) Über die Dissociation der im Wasser gelösten Stoffe. Z Phys Chem 1:631

Arrhenius S (1889) Über die Reaktionsgeschwindigkeit bei der Inversion von Rohzucker durch Säuren. Z Phys Chem 4:226–248

Ballantine JA, Purnell JH, Thomas JM (1984) Sheet silicates: broad spectrum catalysts for organic synthesis. J Mol Catal 27:157–167

Bell RP (1935) The theory of reactions involving proton transfers. Trans Faraday Soc 31:875

Bell RP (1936) The theory of reactions involving proton transfers. Proc R Soc Lond 154A:414

Berthelot M, St. Gilles P (1862) De la formation et de la décomposition des éthers. Ann Phys Chim 65:385

Berthollet CL (1801) Recherches sur les lois de l'affinité. Baudouin, an IX, Paris

Berzelius JJ (1836) Quelques Idées sur une nouvelle Force agissant dans les Corps Organiques. Ann Chim Phys 61:146

Binnig G, Rohrer H (1986) Scanning tunneling microscopy. IBM J Res Dev 30:4

Bodenstein M (1913) Eine theorie der photochemischen reaktionsgeschwindigkeiten. Z Phys Chem 85:329–397

Bodenstein M (1922) Chemische Kinetik. Erg Exakt Naturwiss 1:197–209

Bodenstein M, Dux W (1913) Photochemische Kinetik des Chlor-knallgases. Z Phys Chem 85:297

Boreskov GK (2003) Heterogeneous catalysis. traduzione inglese di K.I. Zamaraev e A. V. Kzasin, Nova Science Publisher, Hauppauge

Bozso F, Ertl G, Grunze M, Weiss M (1977) Interaction of nitrogen with iron surfaces. 1. Fe(100) and Fe(111). J Catal 49:18–41

Brønsted JN (1928) Acid and basic catalysis. Chem Rev 5:231–338

Brunauer S, Emmett PH, Teller E (1938) Adsorption of gases in multi molecular layers. J Am Chem Soc 60:309

Cashion K, Polanyi JC (1958) Infrared chemiluminescence from the gaseous reaction atomic H plus Cl_2. J Chem Phys 29:455

Chen CH, Siska PE, Lee YT (1973) Intermolecular potentials from crossed beam differential elastic scattering measurements, VIII. J Chem Phys 59:601–610

Christmann K, Schober O, Ertl G, Neumann M (1974) Adsorption of hydrogen on nickel single crystal surfaces. J Chem Phys 60:4528

Conrad H, Ertl G, Latta EE (1974) Adsorption of hydrogen on palladium single-crystal surfaces. Surf Sci 41:435–446

Cordus V (1548) Pharmacorum Conficiendorum Ratio. Jacobus Gazellus, Paris

Cox MP, Ertl G, Imbihl R (1985) Spatial self-organization of surface-structure during an oscillating catalytic reaction. Phys Rev Lett 54:1725

Cremer E, Schwab GM (1929) Zusammenhang zwischen Aktivierungswärme und Aktivität. Z Physikal Chem 144:243

Cremer PS, Su XC, Shen YR, Somorjai GA, (1996) *Ethylene hydrogenation on Pt(111) monitored in situ at high pressures using sum frequency generation.* J. Am. Chem. Soc., 118, 2942–2949

Davisson C, Germer LH (1927) Reflection of electrons by a crystal of nickel. Nature 119:558–560

Davy H (1817) Some new experiments and observations on the combustion of gaseous mixtures, with an account of a method of preserving a continued light in mixtures of inflammable gases and air without flame. Philos Trans 107:77–85

Davy E (1820) On some combinations of platinum. Philos Trans 110:108–125

Deisenhofer J, Epp O, Miki K, Huber R, Michel H (1984) X-ray structure analysis of a membrane protein complex. Electron density map at 3 A resolution and a model of the chromophores of the photosynthetic reaction center from *Rhodopseudomonas viridis*. J Mol Biol 180:385

Döbereiner JW (1822) Zur Gährungs-Chemie und Anleitung zur Darstellung verschiedener Arten künstlicher Weine, Biere u.s.w. Aug. Schmid, Jena

Döbereiner JW (1823) Propriétés nouvelles et remarquables reconnues au sus-oxide de platine, au sulphure oxidé et â la poussiere du même mêtal. Ann Chim Phys 24:91

Dowen DA (1950) Heterogeneous catalysis. Part I. Theoretical basis. J Chem Soc 242–265

Dulong PL, Thenard L (1823a) Note sur la Propriété que Possèdent Quelques Métaux de Faciliter la Combinaison des Fluides Élastiques. Ann Chim 23:440–444

Dulong PL, Thenard L (1823b) Nouvelles Observations Sur la Propriété Dont Jouissent Certains Corps de Favoriser la Combinaison des Fluides Élastiques. Ann Chim 24:380–387

Eley DD, Polanyi M (1935) Catalytic interchange of hydrogen with water and alcohol. Trans Faraday Soc 32:1388–1397

Eley DD, Rideal EK (1940) Parahydrogen conversion on tungsten. Nature 146:40

Eley DD, Rideal EK (1941) The catalysis of parahydrogen conversion by tungsten. Proc R Soc Lond A178:429–451

Ertl G, Huber M (1980) Mechanism and kinetics of ammonia decomposition on iron. J Catal 61:537–539

Evans MG, Polanyi M (1935) Some applications of the transition state method to the calculation of reaction velocities, especially in solution. Trans Faraday Soc 31:875

Evans MG, Polanyi M (1937) On the introduction of thermodynamic variables into reaction kinetics. Trans Faraday Soc 33:448

Evans MG, Polanyi M (1938) Inertia and driving force of chemical reactions. Trans Faraday Soc 34:11–28
Eyring H (1935) The activated complex in chemical reactions. J Chem Phys 3:107
Eyring H, Polanyi M (1931) Über einfache Gasreaktionen. Z Phys Chem Abt B 12:279–311
Faraday M (1834) Experimental researches in electricity. VI. Philos Trans R Soc 124:55
Farkas A (1931) Aktivierungsenergie der p-H_2-Umwandlung an Nickel. Z Phys Chem B 14:371
Farkas A, Farkas L (1934) Experiments on heavy hydrogen. Part I. Proc R Soc Lond A 144:467–480
Farkas L, Sachsse H (1933) Über die homogene Katalyse der Para-Orthowasserstoff umwandlung. Z Phys Chem B, 23, 1, 19
Farkas A, Farkas L, Harteck P (1934) Experiments on heavy hydrogen. II. The ortho-para conversion. Proc R Soc Lond A 144:481–493
Fusinieri A (1824) Sulla Causa delle Combustione di Sostanze Gasose per Mezzo delle Superfice di Alcuni Metalli. Giorn Fis **7**:371–376, 443–449
Gadsby J, Hinshelwood CN, Sykes KW (1946) The kinetics of the reactions of the steam-carbon system. Proc R Soc Lond A 187:151–187
Gibson CH, Hinshelwood CN (1928) The influence of nitrogen peroxide on the union of hydrogen and oxygen. A problem of trace catalysis. Trans Faraday Soc 24:559
Guldberg CM, Waage P (1867) Études sur les Affinités Chimiques. Christiania University Press, Oslo
Guldberg CM, Waage P (1879) Concerning chemical affinity. Erdmann's Journal für Praktische Chemie 19: 69–114. Translated in German by Abegg in Ostwald's Klassiker der Exacten Wissenschaften, vol 104. Wilhelm Engleman, Leipzig, pp 126–171, 1899
Habas SE, Lee H, Radmilovic V, Somorjai GA, Yang P (2007) Shaping metal nanocrystals through epitaxial seeded growth. Nat Mater 6:692–697
Haber F, Le Rossignol R (1910) Über die Darstellung des Ammoniaks aus Stickstoff und Wasserstoff. Z Elektrochem 16:244
Harcourt AV, Esson W (1867) On the laws of connexion between the conditions of a chemical change and its amount. Philos Trans 157:117
Hauffe K (1950) Reaktionen in und an festen Stoffen B, vol Band 2. Springer, Berlin, p 696
Heitler W, London F (1927) Wechselwirkung neutraler Atome und homöpolare Bindung nach der Quantenmechanik. Z Phys 44:455–472
Henry W (1824) On the action of finely divided platinum on gaseous mixtures and its application to their analysis. Philos Trans 114:266–289
Herschbach DR (1966) Reactive scattering in molecular beams. Adv Chem Phys 10:319–393
Herschbach DR (1973) Reactive scattering. Faraday Discuss Chem Soc 55:233–251
Herschbach DR, Johnston HS, Pitzer KS, Powell RE (1956) Theoretical pre-exponential factors for twelve bimolecular reactions. J Chem Phys 25:736–741
Hinshelwood CN (1940) The kinetics of chemical change. Clarendon, Oxford
Hinshelwood CJ (1952) Reaction patterns in bacterial metabolism. Bull World Health Organ 6:3–17.
Hinshelwood CJ (1953) Chem. Soc, *Autosynthesis*, 1947–1956
Kassel LS (1928) Studies in homogeneous gas reactions. II. Introduction of quantum theory. J Phys Chem 32:1065–1079
Kilkenny B. C. and Hinshelwood C. (1952) An Investigation of some Adaptive Changes in Yeast Cells, Proc. R. Soc. Lond. B vol. 139, 575–583
King DL, Herschbach DR (1973) Facile four centre exchange reactions. Faraday Discuss Chem Soc 55:331–343
Kirchhoff SK (1811) Die Entdeckung der leichten Umwandlung der Stärke in Zucker. Bull Neusten Wiss Naturwiss 10:88–92
Kolbe H. (1887) Zeichen der Zeit, Journal für Praktische Chemie 15, 473
Kühne W (1877) Über das Trypsin (Enzym des Pankreas). In: Verhandlungen des naturhistorisch-medicinischen Vereins zu Heidelberg, vol 1. Neue Folge, Heidelberg, pp 194–198
Landolt H. (1902), The optical rotating power of organic substances and its practical applications: Easton, Pa., Chemical Pub. Co.

Langmuir I (1916) The constitution and fundamental properties of solids and liquids. I. Solids. J Am Chem Soc 38:2221

Langmuir I (1917) The constitution and fundamental properties of solids and liquids. II. Liquids. J Am Chem Soc 39:1848

Langmuir I (1918) The adsorption of gases on plane surfaces of glass, mica and platinum. J Am Chem Soc 40:1361–1402

Le Bel J. A. (1874) Sur les rélations qui existent entre les formules atomiques des corps organiques et le pouvoir rotatoire de leurs dissolution, Bulletin de la Société Chimique de Paris 22, 337

Le Chatelier HL (1884) Sur un énouncé général des lois des équilibres chimiques. Comptes Rendus 99:786

Lee YT, McDonald JD, LeBreton PR, Herschbach DR (1969) Molecular beam reactive scattering apparatus with electron bombardment detector. Rev Sci Instrum 40:1402

Levine RD, Bernstein RB (1987) Molecular reaction dynamics and chemical reactivity. Oxford University Press, New York

Liebig J (1829) Über Edmund Davy's schwarzen Platinniederschlag und über die Eigenschaft des Platinschwamms, das Wasserstoffgas zu entzünden. Ann Phys 93:101–114

Liebig J (1839) Über die Erscheinungen der Gährung, Fäulniss und Verwesung, und ihre Ursachen. Ann Pharm 30:250–287

London F (1928a) Probleme der moderne Physik (Sommerfeld Festschrift), vol 104. Hirzel, Leipzig

London F (1928b) Zur Quantenmechanik der homöopolaren Valenzchemie. Z Phys 46:455

London F (1929) Quantenmechanische Deutung der Vorgänge der Aktivierung. Z Elektrochem 35:552–555

Malaguti F (1853) Exposition de Quelques Faits relatifs a l'action réciproque, des sels solubles. Ann Chim 37:198–206

Marcus RA (1952) Unimolecular dissociations and free radical recombination reactions. J Chem Phys 20:359

Marcus RA (1956) On the theory of oxidation-reduction reactions involving electron transfer. I. J Chem Phys 24:966

Marcus RA (1965) On the theory of electron-transfer reactions. VI. Unified treatment of homogeneous and electrode reactions. J Chem Phys 43:679

Marcus RA (1968) Theoretical relations among rate constants, barriers, and Bronsted slopes of chemical reactions. J Phys Chem 72:891

Mayer JR (1876) Die Torricellische Leere und über Auslösung. Verlag der J. G. Cotta'schen Buchhandlung, Stuttgart

McDonald JD, LeBreton PR, Lee YT, Herschbach DR (1972) Molecular beam kinetics: reactions of deuterium atoms with halogen molecules. J Chem Phys 56:769–788

Miller WB, Safron SA, Herschbach DR (1967) Exchange reactions of alkali atoms with alkali halides: a collision complex mechanism. Discuss Faraday Soc 44:108–122

Miller JR, Calcaterra LT, Gloss GL (1984) Intramolecular long-distance electron transfer in radical anions. The effects of free energy and solvent on the reaction rates. J Am Chem Soc 106:3047

Mitscherlich E (1844) Lehrbuch der Chemie, 4th edn. Mittler, Berlin, p 617

Moissan H, Moureu Ch (1896) Action de l'acétyléne sur le fer, le nickel et le cobalt réduits par l'hydrogène. Comptes Rendus 122:1241

Nernst W (1918a) Zur Anwendung des Einsteinschen photochemischen Äquivalentgesetzes. I. Z Elektrochem 24:335–336

Nernst W (1918b) Des neuen Wärmesatzes. Verlag von Wilhelm Knapp, Halle

Nollet L'Abbé (1748) Recherches sur les Causes du Bouillonnement des Liquides. Mem Acad R Sci Paris

Ostwald W (1894) Definition der Katalyse. Z Phys Chem 15:705–706

Ostwald W (1902) Über Katalyse. Nature 65:522 (Summary of a lecture hold by Ostwald at Amburg in 1901 at the Deutsche Naturforscher Versammlung)

Ostwald W (1910) Über Katalyse. Ann Naturphilos 9:1–25

Payen A, Persoz J-F (1833) Mémoire sur la diastase, les principaux produits de ses réactions et leurs applications aux arts industriels. Ann Chim Phys 53:73–92

Pfaundler von Hadermur L (1867) Beiträge zur chemischen Statik. Pogg Ann, 131
Pfeffer WFP (1887) Osmotische Untersuchungen. W. Engelmann, Leipzig
Pisarzhevsky LV (1955) Selected works on catalysis. Izd Acad Nauk USSR Kiev, in Russian
Polanyi M (1932) Developments of the theory of chemical reactions. Naturwissenschaften 20:289–296
Polanyi JC (1963) Infrared chemiluminescence. J Quant Spect Rad Trans 3:471
Polanyi JC (1967) Dynamics of chemical reactions. Discuss Faraday Soc 44:293
Polanyi JC, Zewail AH (1995) Direct observation of the transition state. Accounts Chem Res (Holy Grail Issue) 28:119
Raja R, Thomas JM, Jones MD, Johnson BFG, Vaughan DEW (2003) A constraining asymmetric organometallic catalysts within mesoporous supports boosts their enantioselectivity. J Am Chem Soc 125:14982–14983
Rice OK, Ramsperger HC (1927) Theories of unimolecular gas reactions at low pressures. J Am Chem Soc 49:1617–1629
Rice OK, Ramsperger HC (1928) Theories of unimolecular gas reactions at low pressures. J Am Chem Soc 50:617–620
Rideal EK (1939) Parahydrogen conversion on tungsten. Proc Camb Philos Soc 35:130
Sabatier P (1897) Action du Nickel sur l'Éthylène. Compt Rend 124:616–618
Sabatier P, Senderens J-B (1897) Action du nickel sur l'éthylène. Synthèse de l'éthane. CR Acad Sci Paris 124:1358–1360
Sabatier P, Senderens JB (1902a) New synthesis of methane. Compt Rend 134:514–516
Sabatier P, Senderens JB (1902b) New synthesis of methane. J Chem Soc 82:333
Sato S (1955) Potential energy surface of the system of three atoms. Chem Phys 23:592, 2465
Schönbein CF (1848) On some chemical effects produced by platinum. Mem Proc Chem Soc 3:17
Schwab GM (1950) On compact-dispersed silver. J Phys Chem 54:576–580
Schwab GM, Pietsch E (1926) Thermische Spaltung von Methan am Glühdraht. Z Elektrochem 32:430–434
Schwab GM, Pietsch E (1929) Zur Topoehemie der Kontakt-Katalyse. II. Über einen experimentellen Fall der Adlineation. Z Phys Chem B 2:262–264
Semyonov NN (1928) Zur Theorie des Verbrennungsprozesses. Z Phys 48:571
Semyonov NN (1929) Kinetics of chain reactions. Chem Rev 6:347–379
Semyonov NN (1934) Chain reactions (in Russian). Goskhimizdat, Leningrad. English edition: Semyonov NN (1935) Chain reactions. Oxford Press
Semyonov NN (1940) Study of properties of solutions of kinetic equations. Usp Fiz Nauk 23:251
Shorter J (1980) A. G. Vernon Harcourt. J Chem Educ 57:411–416
Skottke M, Behm RJ, Ertl G, Penka V, Moritz W (1987) LEED structure analysis of the clean and (2 × 1)H covered Pd(110) surface. J Chem Phys 87:6191
Somorjai G (1960) A small angle x-ray study of metallized catalysts. Ph.D. thesis, University of California, Berkeley
Somorjai GA, McCrea KR (2000) Sum frequency generation: surface vibrational spectroscopy studies of catalytic reactions on metal single-crystal surfaces. Adv Catal 45:385–438
Somorjai GA, Park JY (2008) Evolution of the surface science of catalysis from single crystals to metal nanoparticles under pressure. J Chem Phys 128:182504
Spencer ND, Schoonmaker RC, Somorjai GA (1982) Iron single crystals as ammonia synthesis catalysts: effect of surface structure on catalyst activity. J Catal 74:129–135
Stohmann F (1894) Über dem Wärmewerth der Bestandtheile der Nahrungsmittel. Z Biol 31:364–391
Stubbs FJ, Hinshelwood CN (1951) The thermal decomposition of hydrocarbons. Discuss Faraday Soc 10:129
Thomas JM, Raja R (2005) Design of a "green" one-step catalytic production of ε-caprolactam (precursor of nylon-6). Proc Natl Acad Sci 102:13732–13736
Thomas JM, Raja R, Sankar G, Johnson BFG, Lewis DW (2001) Solvent-free routes to clean technology. Chem-Eur J 7:2972–2978
Thompson HW, Hinshelwood CN (1929) The mechanism of the homogeneous combination of hydrogen and oxygen. Proc R Soc Lond A122:170

Thomson GP, Reid AA (1927) Diffraction of cathode rays by a thin film. Nature 119:890

Van't Hoff (1874), Voorstel tot Uitbreiding der Tegenwoordige in de Scheikunde gebruikte Structuurformules in de Ruimte: benevenseen daarmeê samenhagende Opmerking omtrent het Verband tusschen Optisch Actief Vermogen en chemische Constitutie van Organische Verbindingen, Archives neerlandaises des sciences exactes et naturelles, 9, 445–454, Greven, Utrecht, 1874

van't Hoff JH (1877) Die Grenzebene ein Beitrag zur Kenntniss Esterbildung. Chem Ber 10:669

van't Hoff JH (1878) De verbeeldingskracht in de wetenschap. P.M. Bazenkijk, Rotterdam

van't Hoff JH (1884) Etudes de Dynamique Chimique. Frederik Muller, Amsterdam, pp 114–118

van't Hoff JH (1887) Die Rolle des osmotischen Druckes in der Analogie zwischen Lösungen und gases. Z Phys Chem 1:481–508

van't Hoff JH (1894) Wie die Theorie der Lösungen entstand. Ber Deut Chem Ges 27:6–19

van't Hoff JH (1898) Vorlesungen über Theoretische und Physikalische Chemie. Friedrich Viewig und Sohn, Braunschweig

Vol'kenshteĭn FF (1960) Élektronnaya teoriya kataliza na poluprovodnikakh. English edition: Fizmatgiz M (1963) The electronic theory of catalysis on semiconductors (trans: Anderson NG). Pergamon Press, Oxford

Waage P, Guldberg CM (1864) Forhandlinger: Videnskabs-Selskabet i Christiana 35

Wenzel KF (1777) Lehr von der Verwandtschaft der Korper. (V. D. H. Gründel) Dresden

Wigner EP (1933) Über die paramagnetische Umwandlung von Para-Orthowasserstoff. Z Phys Chem B 23:28

Wilhelmy L (1850) Über das Gesetz, nach welchem die Einwirkung der Säuren auf den Rohrzucker stattfindet. Ann Phys 81:413–428, 499–526

Williamson AW (1850) Theory of aetherification. Philos Mag 37:350–356

Wright PA, Thomas JM, Cheetham AK, Nowak AK (1986) Localizing active sites in zeolitic catalysts: neutron-powder-profile analysis and computer simulation of deutero-pyridine bound to gallozeolite-L. Nature 318:611–614

Zhang J, Lee YT (1997) Crossed molecular beam study of the reaction Cl + O_3. J Phys Chem A 101:6485–6495

Chapter 3
Matter and Electricity

In a great number of the cosmogonic myths the world is said to have developed from a great water, which was the prime matter. In many cases, as for instance in an Indian myth, this prime matter is indicated as a solution, out of which the solid earth crystallized out.

(Svante Arrhenius)

3.1 The Association of Matter and Electricity

The beginning of the nineteenth century witnessed the appearance of Volta's pile that gave rise to an explosion of research on the decomposition of chemical substances by means of electricity, although the mechanisms of production of the electric current and of electrolysis still remained a mystery. However, the research of Humphry Davy and Jöns Jacob Berzelius and, above all, the quantitative laws of Michael Faraday, establishing the direct proportionality between the quantity of decomposed compound and the amount of electricity used, made clear that the electric charges were directly attached to the atoms.

Faraday, a great specialist in neologisms, in collaboration with William Whewell (1794–1866), scientist, historian, and science philosopher, invented a number of technical terms used in electrochemistry such as *electrolysis, electrolyte, electrode, anode, cathode, ion, anion,* and *cation*. Nevertheless, the passage of current in electrolytic cells was far from understood. The most difficult problem concerned the electrolysis of aqueous solutions – the electrolysis of molten salts did not present great theoretical complications. Davy in fact succeeded in 1807 in preparing sodium and potassium by means of electrolysis using melted soda and potash at high temperature and platinum electrodes, and in 1808 he isolated barium and strontium by electrolysis using mercury electrodes. Strangely enough, by electrolysis of aqueous solutions, for instance, of sodium sulfate, instead of the elements, one always obtained oxygen and hydrogen! Furthermore, it was difficult to accept that two substances like chlorine and sodium, fiercely reactive if put into

S. Califano, *Pathways to Modern Chemical Physics*,
DOI 10.1007/978-3-642-28180-8_3, © Springer-Verlag Berlin Heidelberg 2012

contact, could peacefully coexist in solution in the form of ions, even if one was negatively and the other positively charged.

A somehow acceptable explanation had been put forward by the London chemist and physicist John Frederic Daniell (1790–1845), a great friend of Faraday, who in 1831 became the first professor of chemistry at King's College of London. Daniell was well known because in 1836 had invented an electrical battery more powerful than that of Volta. Convinced, as Davy, that acidity did not depend on the presence of hydrogen and oxygen in the molecule but only on the reciprocal positions of the parts, he suggested that the appearance of hydrogen and oxygen at the electrodes was not due to the current but instead was caused by a secondary reaction at the electrodes.

Adopting the nomenclature of the time, he wrote that sodium sulfate was formed from Na and (S + 4O) ions. The latter reacted with water HO at the anode producing oxygen according to the chemical equation

$$(\mathrm{S}+4\mathrm{O})+\mathrm{HO}=[(\mathrm{S}+4\mathrm{O})+(\mathrm{H})]+\mathrm{O}\uparrow$$

whereas, at the cathode, the reaction taking place was

$$\mathrm{Na}+\mathrm{HO}=(\mathrm{Na}+\mathrm{O})+\mathrm{H}\uparrow .$$

The first formulation of a theory of electrolysis that took into account atomic motion in solution was presented in 1806 by Theodor von Grotthuss (1785–1822), descended from an old family of the Kurland nobility. Grotthuss, who spent a good part of his short life in his mother's estate in Latvia, grew up in the cultural tradition of the Kurland aristocracy but, thanks to the education received later in Paris during the period 1803–1805, he discovered the existence of Volta's pile and became a confirmed admirer of French science.

However, Grotthuss developed the greatest part of his research in Italy at Naples and Rome in the period 1805–1806. He wrote his theory of electrolytes in French during his stay in Rome in 1805 and published it in 1806 (Grotthuss 1806). In this paper, immediately translated into German and English, Grotthuss proposed a theory of current conduction in water that seems astonishing for the historical period in which it was formulated if one takes into account that the concept of ions was not yet clear and that the formula for water was written HO rather than H_2O. Grotthuss considered a solution as made by chains of polarized water molecules

$$\mathrm{O^-H^+}\cdots\mathrm{O^-H^+}\cdots\mathrm{O^-H^+}\cdots\mathrm{O^-H^+}\cdots\mathrm{O^-H^+}\cdots\mathrm{O^-H^+}\cdots\mathrm{O^-H^+}\cdots\mathrm{O^-H^+}\cdots\mathrm{O^-H^+}\cdots\mathrm{O^-H^+}$$

connecting the anode to the cathode. Current flow and the transfer of matter at the electrodes took place due to the instantaneous and cooperative displacement of a hydrogen atom from one molecule to the next. At the anode, the first oxygen ion of the chain was discharged, leaving one oxygen atom free and the hydrogen atom following it in the chain was then bound to the oxygen ion on its right in the chain

$$H^+O^- \cdots H^+O^- \cdots H^+O^- \cdots H^+O^- \cdots H^+O^- \cdots H^+O^- \cdots H^+O^- \cdots H^+O^- \cdots H^+O^- \cdots H^+$$

which, in turn, produced at the right extremity of the chain a free hydrogen ion that was discharged at the cathode, eventually creating a hydrogen atom.

Grotthuss's conduction mechanism offered important hints to the theory of electric conduction in solution. First of all the model of the force field that Faraday modified and transformed in the concept of electric and magnetic fields and of lines of force arises from it. In addition it contributed to the modern formulation of electrical conductivity that correctly explains the great mobility of hydrogen ions in aqueous solutions. The mechanism consists in a series of cooperative jumps of H^+ ions from one molecule of water to the next by means of the formation of H_3O^+ molecules according to the scheme

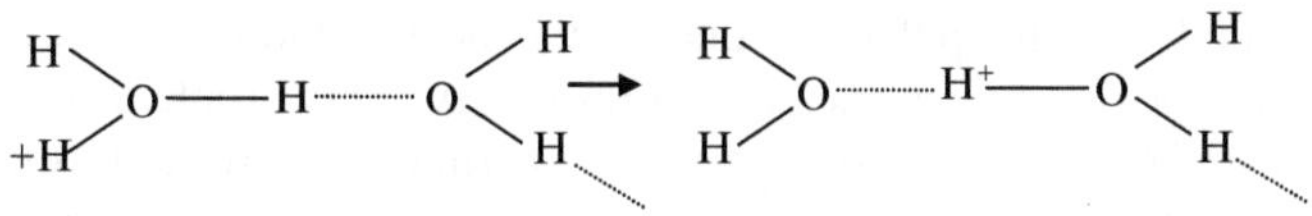

which is completely different from the classical Brownian motion of the H^+ ion.

A direct consequence of the cooperative jumps of H^+ ions from one water molecule to another in the Grotthuss model was that the ions moved with the same speed in both directions. In 1844, John Daniell, in collaboration with William Allen Miller (1817–1870), new professor at King's College, using a cell divided into three sections – anodic, cathodic, and central – separated by porous walls, found that the solute concentration around the electrodes was different and that the number of cations at the cathode was larger than the number of anions at the anode. The conclusions reached by Daniell and Miller were experimentally confirmed by Johann Wilhelm Hittorf (1824–1914), professor of physics at the University of Münster, who in the years 1853–1859 studied the ionic motion due to an electric current and proved that in electrolysis the current was not transported with the same speed by anions and cations. On the basis of these results, he defined the transport number as the fraction of current transported by each ionic species present in solution (Hittorf 1853), computed the transport number for different types of ions from measurements of the variation of concentration at the electrodes during the electrolysis, and in 1869 formulated his law concerning the migration of ions in solution (Hittorf 1859) These measurements, made at very low current intensity, convinced electrochemists that it was unnecessary to spend energy to ionize the molecules in solution.

The concept of electrolytic dissociation was, however, in the air even if not yet directly formulated. It was, however, anticipated in 1851 in a famous paper of Williamson on the formation of ethers (Williamson 1850). Williamson had suggested that, in an aggregate of molecules of any compound, there was always a continuous exchange of the relative positions of the constituent elements.

For instance, in hydrochloric acid, no hydrogen atom was ever motionless, bound to a given chlorine atom, but was in continuous motion, exchanging position with other hydrogen atoms. In 1857, Clausius, used to thinking in terms of the kinetic theory of gases of molecules in motion, continuously bumping into each other, had no difficulty in accepting this idea and proposing that in solution some ions were already there before the passage of current. The merging of the ideas of Williamson with those of Clausius gained the enthusiastic agreement of Hittorf who saw perfect parallelism between the kinetic theory of gases and solutions, parallelism that also influenced the work of van't Hoff on osmotic pressure (van't Hoff 1887, 1888).

The measurement of conductivity of a solution at first presented two significant difficulties. The first was the high resistivity, a difficulty overcome in the 1830s with the invention of the Wheatstone bridge in 1833 by Samuel Hunter Christie (1784–1865) and improved in 1843 by Sir Charles Wheatstone (1802–1875). With the help of the Wheatstone bridge, it became much simpler to measure the resistivity (inverse of conductivity), even of dilute solution with very low conductivity.

The second difficulty was connected with the fact that, using direct current, the electrolytic cell became polarized due to the accumulation of material at the electrodes with a consequent increase in resistivity. This problem was brilliantly resolved in 1868 by one of the greatest experimental physicists of all times, Friedrich Wilhelm Georg Kohlrausch (1840–1910) by the use of alternating current. Kohlrausch was a central figure in the history of electrochemistry at the end of the nineteenth century. The experimental techniques that he accurately devised and described spread from his laboratories at Göttingen, Zurich, and Darmstadt to the whole of Europe (Kohlrausch 1880). Arrhenius, Ostwald, and van't Hoff, the most important creators of ion theory, developed their research along Kohlrausch's line of experimental measurements.

In 1874, Kohlrausch showed that electrolytes possess a definite and constant value of molecular conductivity λ and by studying conductivity variation with concentration, determined the speed of ionic transfer in solution.

From 1875 to 1879, he examined several solutions of salts and acids and established that each type of ion has its own specific electric resistance, independent of the type of molecule from which it derives (Kohlrausch and Grotrian 1875). He also showed that in the case of slightly dissociated weak electrolytes, molar conductivity increases with dilution due to the increase of dissociation up to a limiting value Λ_0 (Kohlrausch 1885). By studying the conductivity of two electrolytes with a cation or an anion in common, he proved that the limiting conductivity Λ_0 at infinite dilution is the sum of two constant values, λ_0^+ relative to the cation and λ_0^- to the anion, since in these conditions the behavior of an ion is independent from that of its counterions. In addition, he proved that ions move with their own definite speed, independent of other ions present, defining the law of the independent migration of ions that bears his name (Kohlrausch 1880).

The development of the theory of electrolytic dissociation is, however, tied to the names of two great personalities that played a central role in the development of chemical physics at the end of the nineteenth century: Wilhelm Ostwald and his pupil Svante Arrhenius.

Wilhelm Ostwald (1853–1932), born at Riga, at that time part of the Russian Empire, was educated at Dorpat University (today Tartu) at the extreme limit of the Baltic zone of Europe. During that period, Latvia was a Russian territory, but the dominant culture was German, and teaching was strongly oriented toward the study of physics even in chemistry courses. He therefore avoided the teaching of organic analysis and synthesis that led to the success of the great French and German schools of chemistry of the time. This type of unconventional university training at the periphery of German academic culture was fundamental to his scientific career and responsible for his successes and failures.

As he said later, if he had been born in Germany he would have unavoidably become a boring organic chemist. Ostwald was in fact convinced that his isolation from the principal stream of European research, organic chemistry, had pushed him into getting interested in an unfashionable topic of his time, chemical physics.

After graduation, he obtained the position of assistant at the Physics Institute of Riga directed by Arthur von Oettingen (1836–1920), later moving to the chemistry laboratory of Carl Schmidt (1822–1894), a student of Liebig and Wöhler. Oswald was always very attached to his teachers for the kind of teaching received. Schmidt's assistant, Johann Lemberg (1842–1902), taught him chemistry and introduced him to the concepts of affinity, chemical equilibrium, and reaction rate.

Ostwald selected just the affinity problem for his *Kandidatenschrift* in 1875, based on the study of bismuth chloride solutions. In 1877, he was accepted as a teacher without salary in his own university and in 1881 was hired as ordinary professor of chemistry at the Polytechnic of Riga. In 1887, he became professor of chemical physics at Leipzig where he settled for life, apart for a short interruption in the period 1904–1905 when he was invited to the United States as first "Exchange Professor" at the University of Harvard.

Ostwald was the true founder of modern chemical physics, not only as a researcher but also as a great organizer and research talent scout. His teaching and research activity was impressive. At 50 years, he had already supervised the theses of 147 students, 34 of who later became university professors. His school attracted researchers from all over Europe as well as from the United States. Among his best known pupils there were three Nobel laureates, Van't Hoff, Arrhenius, and Nernst; Wislicenus, one of the fathers of stereochemistry; Gustav Tammann (1861–1922) a Russian also educated at Dorpat who gave origin to modern metallurgy; and the American Arthur Amos Noyes (1866–1936), founder of Caltech and of the Chemical Abstracts; and Willis Rodney Whitney (1868–1958), creator of the electrochemical theory of corrosion and later director of General Electric.

Ostwald published during his lifetime as many as 45 books, among them the *Lehrbuch der Allgemeinen Chemie* (1885–1887) and the *Grundriss der Allgemeinen Chemie* (1889); more than 500 scientific papers; and about 5,000 reviews and reports. In addition, he was editor of six scientific journals, and in 1889 he organized the reprinting of a large series of papers of historical interest, the famous *Klassiker der exakten Wissenschaften*, of which as many as 250 volumes were published.

The foundation in 1887 – in collaboration with van't Hoff – of the *Zeitschrift für physikalische Chemie*, first true journal of chemical physics that he edited until 1922, rapidly made him one of the key masters of the new discipline.

In 1902, he founded the *Annalen der Naturphilosophie* and in 1894 the *Deutsche Elektrochemische Gesellschaft*, the German society of electrochemistry that in 1902 transformed into the *Deutsche Bunsen-Gesellschaft für angewandte physikalische Chemie,* one of the most important European scientific societies.

In 1875, he started his experimental activity by measuring the variation of the physical parameters in chemical reactions that for several years he combined with the study of affinity, a vague expression representing the tendency of atoms to join and to react, at that time a problem central to all chemical theories.

Ostwald succeeded in coordinating in a single general framework several different ideas and measurements used to rationalize affinity (Ostwald 1880), taking up the research line of Julius Thomsen, but exploiting a different technique – that of measuring volume variations in chemical reactions that he considered simpler and more exact than calorimetry. The work object of his Ph.D. thesis in 1880 was based on the comparative study of the volume variation of couples of acid–base reactions in dilute solutions (Ostwald 1878), with a common component. If A_1 and A_2 are two acids and B is a base, the mixture of 1 L of acid A_1 with 1 L of solution of base B does not give as a result 2 L of solution of the compound A_1B, but a total volume slightly different by an amount v_1. In the same way, the mixture of 1 L of solution of acid A_2 with 1 L of solution of base B will give rise to a total volume differing from 2 L by the amount v_2. Finally, the mixture of 1 L of the A_1B compound with 1 L of solution of A_2 will not give as a result 2 L of mixture, since in the solution equilibrium takes place between A_1B, A_2B and the fractions of A_1 and A_2 free acids. The volume difference in this case furnishes an exact measure of how much of the A_1B compound has been transformed into A_2B.

At the end of the 1870s, he had already obtained significant results measuring the volume variation and other physical quantities in reactions and showing that these were in agreement with the mass action law formulated earlier by Peter Guldberg and Cato Waage. It was actually only, thanks to the citing of the papers of the two Norwegians by Ostwald, in 1875 that their law, practically ignored for more than 15 years, became known all around the world.

A decisive turn in the scientific activity of Ostwald began with the publication in 1884 of the Ph.D. thesis of Arrhenius discussing the development of the theory of electrolytic dissociation. Oswald realized right away the importance of the theory, realizing that it presented the possibility of using conductivity measurements to quantify chemical affinity in acid–base reactions. In the summer of 1884, he went to Sweden to meet Arrhenius and the other Scandinavian chemists Guldberg, Waage, and Thomsen, owing to his interest in their works.

In the meantime, van't Hoff had deduced the law of mass action from basic thermodynamics principles and had shown that molecules in solution behaved as if they were in the gas phase. At first, the contribution of Ostwald to the dissociation theory was only to clarify and improve the ideas of van't Hoff and Arrhenius. In 1887, however, he made a fundamental contribution with his law of dilution that connects the equilibrium constant to the degree of dissociation of electrolytes (Ostwald 1887). Starting from the Arrhenius finding that at infinite dilution the conductibilities of all acids are identical, Ostwald realized that dilute solutions of electrolytes behave as do the ideal gases, as he clearly stated in his famous paper of 1888 (Ostwald 1888):

> The researches of van't Hoff, Planck, and Arrhenius on dilute solutions have in recent times led to the recognition of a complete analogy of these with gases. One of the most valuable advances of these studies is that the compounds usually spoken of as held together by the strongest affinities, such as, for example, potassium chloride, hydrogen chloride, or potassium hydroxide, must actually be regarded in dilute solutions as very largely dissociated.

Ostwald also proved the importance of applying the mass action law to these solutions (Partington 1964). For a generic dissociation reaction of a molecule HA:

$$\underset{(1-\alpha)/v}{\mathrm{HA}} \rightleftarrows \underset{\alpha/v}{\mathrm{H}^+} + \underset{\alpha/v}{\mathrm{A}^-}$$

if one denotes with v the volume in liters of the solution containing one mole of solute, i.e., the inverse of the concentration c and with α the degree of dissociation, then the concentration of nondissociated molecules HA will be equal to $(1 - \alpha)/v$, whereas the concentration of dissociated particles H^+ and A^- will be equal to $2\alpha/v$ and, according to the mass action law, the equilibrium constant will be equal to

$$K = \frac{[\mathrm{H}^+] \cdot [\mathrm{A}^-]}{[\mathrm{HA}]} = \frac{\alpha^2}{v^2} \frac{v}{(1-\alpha)} = \frac{\alpha^2}{v(1-\alpha)}$$

Expressing the degree of dissociation as a function of the limiting Kohlrausch conductibility $\alpha = \Lambda/\Lambda_0$, this equation yields

$$K = \frac{\Lambda^2}{v\Lambda_0(\Lambda_0 - \Lambda)}$$

which is known as Ostwald's dilution law that allows one to compute the equilibrium constant from conductivity measurements. Ostwald and coworkers showed that hundreds of solutions of organic acids and bases obey the dilution law, confirming the validity of the ionic theory. Ostwald dilution law, the weak point of which was that it only holds for weak acids and bases, had a great number of applications (Ostwald 1889). Ostwald himself realized that the equilibrium constant K could be utilized to measure the strength of acids and bases. In the case of pure water, weakly dissociated,

$$\mathrm{H_2O} \rightleftarrows \mathrm{H}^+ + \mathrm{OH}^-,$$

the action of mass law reduces to $K = [\mathrm{H}^+]\,[\mathrm{OH}^-]$ since the concentration of water molecules is constant. From conductivity measurements, the value of K, known as the ionic product of water, was found to be equal to 10^{-14} $(mol/L)^2$, and from this one obtains a value of 10^{-7} mol/L for the concentration of hydrogen ions in pure water (Ostwald and Nernst 1889).

Several chemists of the time proposed that the concentration of hydrogen ions in solution could be used as an indicator of the strength of acids and bases.

In 1909, the Danish biochemist Soren Sørensen (1868–1939), who for 8 years had been a director at the chemical laboratories of the firm Carlsberg, one of the most important companies for the production of beer, pointed out in a paper in a biochemistry journal (Sørensen 1909) that the quantity of an acid used expressed in moles/L did not correctly represent the acidity of the solution owing to chemical interactions with other chemical species present. Sørensen suggested instead that the degree of acidity should be measured by the exponent of the ionic concentration and proposed a scale that he called pH, from the French *pouvoir hydrogéne*. The pH scale soon became very useful for biochemists and analytical chemists, $pH = 7$ meaning that a solution was neutral, whereas lower or higher values indicated acidity or basicity, respectively.

Ostwald was also interested in thermodynamics and thermochemistry. In 1892, he realized the importance of Willard Gibbs' papers and the difficulties that the European chemists had in following the complex mathematical treatment and decided to make them accessible, translating them into German, simplifying the mathematical formalism, and reformulating the concepts in simpler forms (Ostwald 1892).

In the 1890s, Ostwald concentrated his efforts completely on the theoretical and philosophical problems of energy, convinced that matter was only a mirage created in our minds to embrace the energy activity.

For him, atoms and molecules were only symbols of statistical regularities, being convinced that the deep truths of science should be expressed only in terms of energy; for this reason he always objected to the atomic theory. Only in 1909, after having evaluated Perrin's experiments on Brownian motion, did he reluctantly accept the existence of atoms.

Ostwald received honors, awards, and honorary doctorates from more than 60 universities throughout the world and in 1909 the Nobel Prize for chemistry for his works on catalysis and reaction rates.

After his retirement from academic life, he dedicated himself to intense philosophical, political, and social activity aimed at research for peace in the world and to a better understanding between countries. He was also interested in educational reforms and in monism and tried without success to promote a new universal language "Ido," a new version of Esperanto. Outbreak of the First World War in 1914 dampened his dreams of great international collaboration. A confirmed patriot, but not a militarist, he vainly hoped for a rapid and rightful peace. Criticized by his compatriots for lack of enthusiasm for the actions of his country and by his international colleagues for not having condemned the war, he left active politics and concentrated on the study of colors and speculating on esthetic and philosophical problems.

The other father of electrochemistry was the Swede Svante Arrhenius (1859–1927) who developed the theory of electrolytic dissociation in solution. Enrolled at the Uppsala University in 1876, he started to study chemistry under the supervision of Per Cleve, the chemist who isolated holmium and thulium in 1879. Experimental teaching at Uppsala was not among the best and in 1881 Svante decided to move for a time to Stockholm to work with Eric Edlung (1819–1888) of the Swedish Academy of Sciences to measure electromotive forces. At Stockholm, he started his Ph.D. work which he continued after returning to Uppsala in 1884, without much success. The thesis project of Arrhenius, *Recherches sur la conductibilité galvanique des électrolytes* (Arrhenius 1884), was to determine the molecular weight of nonelectrolytes such as cane sugar, measuring the decrease of electrical conductivity of mixed solutions with electrolytes of known conductivity. Arrhenius did not succeed in solving the problem experimentally and was only able to prove that the molar conductivity increased as the concentration decreased, a result that only confirmed those obtained previously by Kohlrausch. Instead, it was the theoretical content of the second part of his thesis entitled *Théorie chimique des electrolytes* that assured his success.

Arrhenius concluded that the only way to explain his results and those of Kohlrausch was to assume that the electrolytes in solutions were a mixture of "complex" inactive molecules in dynamic equilibrium that did not conduct the current and of "active" parts that were responsible for conduction. As the dilution increased so did the number of dissociated molecules. Very probably, Arrhenius was convinced of the theory developed by Mendeleev (Brock 1993) that hydrates involving particles of solute and of water were formed and for this reason did not use the term "ion" at this stage.

It was only in 1887 that Arrhenius completed the development of his theory in terms of dissociation into positive and negative ions. The theory was first explained in a letter to his friend van't Hoff and then published (Arrhenius 1887) with Ostwald's support, receiving its definitive acceptance in 1903 with the speech of Arrhenius at Stockholm during the award of the Nobel Prize.

Arrhenius' thesis was not received with favor by the committee of the science faculty of the University of Uppsala which approved it with only a low vote. According to the tradition of Uppsala University, Arrhenius sent a copy of the thesis to the most renowned electrochemists of the time, among them Ostwald and van't Hoff. Happily for him, Otto Pettersson, professor of chemistry at the Högskola of Stockholm, defended the originality of the thesis and Ostwald, who had a preeminent position in European chemistry, was so enthusiastic that he even went to Uppsala to meet the young author and convince him to follow him to Riga, offering him the position of *Privatdozent* for the years 1884–1886. Ostwald's offer caused sufficient embarrassment to the scientific community of Uppsala which eventually saved face by offering to Arrhenius a chair of chemical physics, the first created in Sweden, at the end of 1884. Thanks to Edlund's influence, the Swedish Academy of Sciences awarded him a 5-year fellowship, enabling him to work with Ostwald in Riga in 1886, with Kohlrausch in Wurzburg, with Boltzmann in Graz in 1887, and with van't Hoff at Amsterdam in 1888.

In 1891, Arrhenius became teacher of physics at the Högskola of Stockholm, after having refused a position of professor at the University of Giessen. During this period, he became interested in the production of ions in flames (Arrhenius 1891). In 1895, he was made professor and in 1905, after receiving the Nobel Prize for chemistry, he became director of the Nobel Institute of Chemical Physics of Stockholm, inaugurated in 1909. In 1889, Arrhenius studied the inversion of cane sugar induced by weak acids, a process investigated before by Ostwald in 1884 (Ostwald 1884, 1885) and confirmed Ostwald's equation of reaction rates that since then has improperly been called the Arrhenius equation.

Arrhenius published two treatises on electrochemistry – the *Lärobok i teoretisk elektrokemi* (treatise of theoretical electrochemistry) in 1900 and *Theorien der Chemie* (theory of chemistry) in 1906 – and became interested in several other fields of science such as geophysics, meteorology, astronomy, and cosmology (Arrhenius 1903), developing theories on the influence of carbonic anhydride (Arrhenius 1896) in the atmosphere and on the possibility that the radiation pressure could be responsible for the dispersion of living spores in the universe (panspermia).

The electrolytic dissociation theory did not have an easy life, disputed between supporters and adversaries, both with valid arguments. The theory explained well in an intuitive and simple manner the existence of a dynamic equilibrium, controlled by the mass action law, between dissociated molecules and ions, electrical conduction in the solutions, and anomalies of the colligative properties of electrolytes. Strangely, however, it worked well for weak electrolytes and high dilutions, and worst for strong electrolytes and higher concentrations. In addition, it presented a conceptual difficulty that left many chemists doubtful of the solidity of its theoretical grounds. It was in fact difficult to accept the idea that, in a solution, positive and negative ions could freely coexist, unaffected by each other, when it was well known that in crystalline solids these ions formed stable and long-lasting structures with strong ionic bonds.

Additional political and academic difficulties added to those of a purely technical nature. Both in Germany and France, organic chemistry had seen a century of great splendor and it was difficult to accept that chemical physics could assert its authority as the only science able to explain what was happening in the test tubes and flasks of organic chemists. Furthermore, ionic theory was a typical outcome of German culture and after the Franco-Prussian war it received little support in the French-speaking world. On the other hand, different ideas on the interaction of solute with solvent molecules had arisen and these did not encourage the idea of ionization. In Russia, for instance, Mendeleev, the country's most influential and respected chemist, had already developed in 1865 his theory of hydrates that explained the anomalies of the colligative properties as due to the formation of hydrate structures as a function of temperature (Mendeleev 1877).

In England, Henry Edward Armstrong, supporter of the theory of residual affinities, thought that these residual affinities would associate the solute molecules to those of the solvent and explained the anomalies in the colligative properties of electrolytes as due to the formation of a connective network among water molecules caused by residual affinities.

The antipathy that Armstrong showed for the new chemical physicists is testified by his writing (Armstrong 1936):

> . . .people without knowledge of the laboratory arts and with sufficient mathematics at their command to be led astray by curvilinear agreements; without the ability to criticise, still less of giving any chemical interpretation. The fact is, the physical chemists never use their eyes and are most lamentably lacking in chemical culture. It is essential to cast out from our midst, root and branch, this physical element and return to our laboratories.

The eccentric Spencer Umfreville Pickering (1858–1920), for some time professor of chemistry at the Bedford College and since 1887 owner of a private laboratory in London, was also convinced that in electrolytic solutions the dissociation in ions did not really happen and that the observed anomalies for osmotic pressure and colligative properties of solutions of electrolytes were due to residual affinities that connected the solute and solvent molecules. Pickering is known for the discovery of the stabilization of emulsions by small particles.

The reception of chemical physics was different in American universities, which were undergoing great expansion and were ready to accept promising young scientists from Ostwald's laboratory at Leipzig.

Josiah Willard Gibbs and the school that he created at Yale were in favor of the development of chemical physics. Among the young people from Ostwald's laboratory, Louis Kahlenberg (1870–1941), professor of chemistry at the University of Wisconsin, was one of the few who, having departed from Leipzig as a convinced supporter of the ionic theory, rapidly became an active opponent. In 1901, Kahlenberg measured the dielectric constant of liquid cyanidric acid and found that, although higher than that of water, it was less effective in favoring electrolytic dissociation. This result was in contrast to those previously obtained by Nernst and Thomson showing that the increase in the dielectric constant favored dissociation. Kahlenberg, having realized that nonaqueous solvents also conducted electricity, started a long-term research project aimed at measuring the conductivity and the colligative properties of nonaqueous solutions (Kahlenberg 1902). He collected a series of "anomalies" published in articles in the *Journal of Physical Chemistry* that he considered countered the ionic theory, despite Ostwald clearly stating in 1899 that the ionic theory applied only to solutions (Ostwald 1899). Kahlenberg's anomalies showed that conductivity variations with concentration did not follow the mass action law and thus the role of the solvent needed to be taken into account in the ionic theory. Kahlenberg's ideas did not, however, find favor in the American electrochemical community, owing essentially to the strong support that the ionic theory had received from Noyes and Lewis, although they stimulated research on solutions of strong electrolytes and of inter-ionic interactions. In 1909, the Dane Niels Bjerrum (1879–1958) presented at a meeting held in London a report (Bjerrum 1909) on the spectroscopic investigation of chromium chloride, nitrate, and sulfate at different concentrations in aqueous solutions. Bjerrum proved that, since light absorption did not change with concentration, these salts had to be completely dissociated in solution, even at high concentration, and therefore the mass action law could not be applied in this case since there was no dynamic equilibrium. The conductivity variations and the decrease of the freezing point were therefore due to inter-ionic forces. Bjerrum's paper infuriated Arrhenius who considered any small revision of his theory as a personal insult. After a few years, however, Bjerrum's conclusions received unquestionable confirmation from x-ray analysis which proved that ions already existed in ionic crystals as independent entities.

3.2 Solutions

Solutions were always important to handle chemical compounds, especially when the conduction of current was important such as in electrochemical processes.

In the Middle Ages, the only solvent used by the alchemists and the "new" chemists was water. The dissolution of a chemical compound in water was for the medieval alchemists a way of respecting the will of Nature that all matter should be transformed in water, as maintained at the end of the sixteenth century by the Dutchman Johannes Isaac Hollandus in his treatise *De Lapide Philosophorum*. Hollandus's ideas agreed with those of van Helmont that all substances were obtainable from water and of Boyle who in the "Sceptical Chymist" wrote that it was possible to transform water into all other elements. The laboratory experience of medieval alchemists was condensed in the sentence *Corpora non agunt nisi soluta* (substances do not act if not dissolved) or in the more specific one *Salia non agunt nisi dissoluta, nec agunt si dissoluta nimis* (salts do not act if not dissolved and if not very diluted). Alchemists had also elaborated a series of practical rules derived from laboratory practice. A well-known rule, going back to medieval times, was expounded in the sentence *Similia similibus solvuntur* (similar substances will dissolve similar substances) that, in terms of the affinity theory, corresponds to the fact that the more similar the particles of the solvent to those of the solute, the greater the possibility that these could be dissolved.

In the search for solvents suitable to dissolve chemical compounds, the alchemist's main problem became that of discovering an ideal solvent, capable of dissolving everything. The existence of this ideal solvent, *summus atque felicissimus salium* (Porto 2002), was assumed in the sixteenth century by Paracelsus, who even invented the name *alkahest*, a name perhaps derived from *alkali est* (it is an alkali) or perhaps even from the German *al-geist* (all-spirit). However, Paracelsus never described the chemical composition of the alkahest. van Helmont, instead, said that he had truly obtained it and that it was a liquid resembling water so he suggested the name "ignisaqua." The liquid that van Helmont mentioned was probably a concentrated solution of caustic potash in alcohol, know to alchemists as *sal alkali*. Alkahest was considered the universal solvent par excellence and the alchemists, followers of Paracelsus, tried to outdo each other with their preparations, despite Johann Kunckel having ridiculed its existence by stressing that a universal solvent would first of all dissolve the receiver in which it was contained.

The alchemic dream of the alkahest even outlived the scientific revolution, entering the world's literature. In the Comédie humaine, Honoré de Balzac tells in *La recherche de l'absolu ou de l'alkahest* the tragic history of Balthazar Claës living at Douai in Flanders. A pupil of Lavoisier, after a meeting with a Polish soldier who told him of the alchemic dream, he decided to hide himself in his laboratory, quitting his family and squandering his fortune in research of the absolute, basic principle of the Creation.

For some time even the dissolution of a metal in an acid was considered as a type of solution, even if it was evident that it was fundamentally different from solution of a salt in water. The capability of acids to act as solvents of metals was explained assuming that the acid particles had sharp points that allowed them to trickle between those of the metal, separating them through violent pushes. These sharp points were even considered as responsible for the acid taste.

The first clear classification of the nature of solutions, separating the solution process from the dissolution process, appeared in the 1789 Lavoisier *Traité élémentaire de Chimie*. According to Lavoisier, in solutions of salts the salt particles were simply separated one from another but neither the salt nor the water was decomposed and both could be recuperated in the same amounts as before. The same was true for solutions of resin in alcohol or in other solvents.

In contrast, in the dissolution of a metal in an acid, a true decomposition of both acid and metal took place and neither of the two substances preserved its initial state; the metal was transformed into an oxide while a gaseous substance was produced. Dissolution was therefore a chemical transformation, whereas solution was a purely physical process.

A different distinction, strictly bound to the nature of electrochemical processes, was proposed by Theodor von Grotthuss at the beginning of the nineteenth century. For him, solutions and dissolutions differed because a solution was the conversion of a solid or a liquid into a new liquid, not separable with electricity into the original components. Dissolution was instead the conversion of a solid or a liquid into a new liquid, electrically decomposable into its elements (examples: acids and bases). Still different was the position of Berzelius, for whom a solution differed from a chemical process in that it preserves the original electrical charge distribution. The dissolved substance was, according to him, more active in a solution thanks not to a specific variation of affinity but preserved electrical differences between the two partners.

The alchemist speculations on solutions concerned the modus operandi (behavior) of the solvents rather than their nature. The most diffused opinion, developed within the framework of the corpuscular theory, was that the solute particles simply entered the pores or interstices existing in the solvent, an idea derived from Plato's Timaeus and taken up later by Pierre Gassendi. Since the salt particles were considered as small cubes reproducing the external form of the substances, it was necessary to assume that in water there were pores with a cubic shape to host the small salt mini-cubes. When all water pores were filled, the solution was saturated and no more salt could be dissolved in it.

Since, however, alum possessed an octahedral crystalline structure it was evident that water should also contain pores of this shape that only the alum crystallites and not those of the common salt could enter. The extension of this theory to all possible shapes of solute particles obviously presented too many difficulties and required too many ad hoc hypotheses to be easily accepted.

The development of a new theory of solution on a completely different basis started with Newton who extended to solutions the idea of attractive and repulsive forces. Newton supposed that the universal attraction force acting among celestial bodies was also responsible for the interaction between minute particles of different substances; a salt dissolved in water because the salt particles exerted on water particles an attractive force stronger than on other salt particles. When a solid was dissolved in a liquid it distributed throughout the whole volume of the liquid owing to the repulsive forces between the solute particles that tried to move away from each other while being attracted by those of the solvent.

This concept was taken up again by Herman Boerhaave who, in *De Artis Theoria*, went so far as to state that the action was reciprocal in the sense that the solute also dissolved the solvent and that the particles of solvent and solute were bound together to form a new homogeneous substance (Boerhaave 1732):

> post solutionem peractam, particulae solventes et solutae denuo se affinitate suae naturae colligant in corpora homogenea[1]

In other words, the nature of both substances changed since the particles' extension was smaller than their *minima naturalia*.

Georges Louis Leclerc, Count of Buffon, connected Newton attraction and repulsion forces between particles to their shape in solution. The form of the solvent and solute particles was, according to him, particularly relevant, since they were in close contact, in contrast with the attraction between celestial bodies, so far apart that their shapes had no importance. Johan Gottschalk Wallerius (1776) and Martin Heinrich Klaproth (1792–1793) were convinced that a solution was the result of a chemical affinity between solvent and solute and that the attraction between their particles was greater than the "cohesion" between particles of the same substance.

The ideas of Claude-Louis Berthollet on the composition of solutions, dictated by his polemics with Proust relative to the law of the definite proportions, were different again. In his 1803 book *Essai de Statique Chimique*, Berthollet called all stable compounds with a fixed composition "combinations" and those less stable with a variable composition "dissolutions" (Berthollet 1803). Even after having accepted the validity of the law of definite proportions, he insisted in distinguishing between chemical combinations always occurring in definite proportions and solutions occurring instead in any possible proportion, at least up to a given limit. According to him, a solute could dissolve in a solvent only if the attraction forces between them could overcome the cohesion force of the solute, and therefore this

[1] Once the solution had taken place, the solvent and solute particles, due to the nature of their affinity, were connected again in a homogeneous body.

had to be considered to be a true chemical process. The true difference between a compound and a solution was to be found only in the strength of the union of the parts.

The idea that solutions were true chemical compounds was shared by several members of the nineteenth-century chemical community such as Thomas Thomson and Herman Kopp, who proposed interesting variations to Berthollet's position. Thomson, for instance, distinguished between solids which dissolved in liquids and becoming liquids themselves and solids which instead combined with part of the solvent while remaining in hydrated solid forms. Kopp distinguished between compounds obeying the law of definite proportions and compounds with variable amounts of constituents.

The beginning of the nineteenth century witnessed intense research activity on the chemical physics of solutions, ranging from the study of the properties of electrolytes, discussed in detail in the next section, to the examination of the colligative properties of solutions depending only on temperature, pressure, and concentration and not on chemical composition.

The dependence on temperature of the colligative properties of solutions, in particular their freezing points and vapor tensions, proved to be very important to furnish precise methods of determining the molecular weight of chemical compounds. It had been known for a long time that, by dissolving chemical substances in water, their freezing point could be decreased and this gave rise to the production of frigorific mixture for the preparation of ice creams.

The Greeks and Romans used minced ice granules with honey or fruit juice. Sherbets (*sharbat*, fresh drink) were probably invented by the Arabs and later taken up by Sicilians and Neapolitans who exported them to Europe. In the sixteenth century, a courtesan called Ruggeri at the Court of Catherina of the Medicis prepared an ice cream made of water and perfumed ice, and the architect Bernardo Buontalenti (1531–1608) invented an ice cream made of ice and zabaglione (a *dessert made of egg yolks, sugar, and marsala*). It was, however, the Sicilian Francesco Procopio dei Coltelli who first developed the industry and commerce of ice creams, opening the *Café Procope* in Paris in 1686, where ice creams were already being sold around the end of the sixteenth century.

The first description of a freezing mixture made with ice, salt, and water goes back to the writings of the Islamic physician Ibn Abu Usaybi'a (1203–1270) and the technique probably reached Europe through Spain during the Arab domination. Freezing mixtures were already known in the sixteenth century and the Italian humanist and philosopher Gian Battista della Porta (1535–1615) mentioned them in his treatise *Magiae naturalis sive de miraculis rerum naturalium* of 1558. In 1724, Gabriel Daniel Fahrenheit (1686–1736), a German instrument-maker living in the Netherlands, found that a mixture in equal parts of water, ice, and Ammonium Chloride salt melts at −32°C, a temperature that he selected as the zero of a new temperature scale that bears his name (Fahrenheit 1724). In addition, in 1788 Charles Blagden (1748–1820), assistant and coworker of Cavendish, found that the melting point of water decreases with dissolved chemical substances and that the decrease is proportional to the quantity of dissolved substance (Blagden 1788).

Blagden gave a mathematical form to his numerous measurements, showing that the ratio between the decrease C of the freezing point and the quantity P of dissolved substances was expressed by the equation

$$\frac{C}{P} = K$$

where K is a constant specific to any kind of solute (Blagden law).

Research on the decrease of the freezing point of aqueous solutions merged with that on the variation of thc density of water with temperature and the extension to solutions. In 1805, the Scot Thomas Charles Hope (1766–1844) showed that water density had a maximum at about 4°C (Hope 1805), and in a short while the Belgian César Mansuète Despretz (1789–1863), professor of physics in Paris, realized that the same phenomenon occurred with solutions, which the Paduan Francesco Rossetti (1833–1885) was able to connect with that of the freezing point decrease (Rossetti 1869).

The question was reconsidered by the French-Swiss Louis de Coppet (1841–1911) who in 1871 proved that solutions with the same freezing temperature had equal concentrations of dissolved substances (de Coppet 1872).

Frenchman François-Marie Raoult (1830–1901) conducted more than 20 years of research on the physical properties of solutions, and in particular on cryoscopy (from the Greek κρυοσ, ice, and σκοπεω, to examine). He started by studying electrical processes and then considered strictly chemical problems. From 1878 onward he studied the colligative properties of solutions, properties that depend only on the number and not on the chemical composition of dissolved particles. In the same year he published his first paper on the depression of the freezing point of solvents as a function of the concentration of the solutes (Raoult 1887). In this field he published about 60 papers with freezing temperature tables of a large number of solutions that made him known all over Europe. At first he attributed the phenomenon to the effect of the affinity of the solutes for water, but soon, working with organic solvents such as benzene and acetic acid, he realized that a simple relationship existed between the concentration of the solute and the decrease of the freezing point of the solvent. In 1882, he proposed (Raoult 1882) the *loi générale de la congélation* which establishes that if a molecule of a compound is dissolved in 100 molecules of a solvent, the freezing point decreases by 0.63°C:

> Il est donc permis de dire, dès à présent: Dans une multitude de cas, l'abaissement du point de congélation d'un dissolvant ne dépend que du rapport entre le nombres de molécules du corps dissous et du dissolvant; il est indépendant de la nature, du nombre, de l'arrangement des atomes qui composent les molécules dissoutes

Raoult's work had important implications for the extension of the theory of solutions to chemical kinetics and electrochemistry. A surprising result was that the depression of the freezing point of water with inorganic salts was almost double that of the same concentration with organic substances. This result was interpreted by Raoult as distinguishing between physical and chemical molecules, namely, between the molecules and the items into which these were dissociated in solution.

De Coppet had also realized that the freezing point depression of a solvent was related to the molecular weight of the solute, but he did not reach the conclusions that Raoult had, thereby transforming cryoscopy measurements into a convenient method for the determination of molecular weights. At that time the definition of atomic and molecular weights was still a source of confusion. The chemical formula H_2O for water implied an atomic weight of 16 for oxygen, but leading chemists like Berthelot continued to maintain the HO formula, leading to an atomic weight of 8 for oxygen. Raoult's measurements definitely showed that the atomic weight of oxygen was 16.

In parallel with cryoscopy, Raoult worked on another colligative phenomenon, the variation of the vapor tension of solvents. He found that for diluted solutions, this depression is proportional to the solute concentration and in 1887 calculated (Raoult 1887) the reduction of the vapor tension K due to a molecule of substance dissolved in 100 g of liquid from the relationship

$$K = \frac{f - f'}{fP} M$$

where f is the vapor tension of the pure solvent and f' that of the solution, M the molecular weight of the solute, and P the weight of solute dissolved in 100 g of solvent. Similar results had previously been obtained by others such as Adolph Wüllner (1835–1908) and Gustav Tammann (1861–1938).

Vapor pressure measurement became a very precise method thanks to the invention made by Ernst Otto Beckmann (1853–1923), a pupil of Kolbe and Wislicenus, of a very high precision thermometer (Beckmann 1888) that made it possible to measure temperature differences of about 0.001°C. Beckman was an organic chemist who, after having discovered the acid catalyzed rearrangement of an oxime to an amide (Beckmann 1886), investigated the determination of molecular weights to evaluate the dimension of the oximes he had synthesized. First, he used Raoult's law of vapor tension depression, but soon realized that it was simpler and more precise to measure boiling point increase, giving rise to the birth of ebullioscopy, a technique that had developed parallel to cryoscopy for molecular weight determinations. The mathematician William Esson (1838–1916), a coworker of Harcourt, interpreted Harcourt's kinetic data in terms of differential equations not too different from those used today (Harcourt and Esson 1867). In 1878, John J. Hood found empirically that the reaction kinetics of oxidation of ferrous sulfate with potassium chlorate catalyzed by sulfuric acid is regulated by an exponential law (Hood 1878) and only a few years later Max Bodenstein integrated these results with those concerning the reaction between hydrogen and iodine (Bodenstein 1894, 1897, 1899).

Of equal interest was the research at the end of the eighteenth century concerning another colligative property of solutions, the osmotic pressure, namely, the hydrostatic pressure produced by a solution in contact with the pure solvent through a semipermeable membrane (which allows only the solvent molecules to pass through). Osmotic pressure was discovered in 1748 as the result of a curious experiment of the French clergyman known as the Abbé Jean Antoine Nollet (1700–1770), tutor of natural philosophy to King Louis XV and a strong adversary of Franklin's electric fluid theory. Nollet immersed in water in a container a bottle filled with alcohol sealed with a pig's bladder; after some hours he saw that the bladder had expanded inside the water container, showing that some water had entered the bottle (Nollet 1748). By switching the water with alcohol, the opposite reaction occurred and the bladder contracted into the water bottle, showing this time that some water had left the bottle and entered the alcohol in the container. It was easy for Nollet to deduce that the bladder was more permeable to water than to alcohol. Nollet's experiment remained a scientific curiosity for a while until it was taken up by the French botanist René Joachim Henri Dutrochet (1776–1847), a descendant of a noble family which had fallen out of favor during the French revolution, who, by investigating the motion of liquids in plants and diffusion through semipermeable membranes, gave full credence to the study of osmotic pressure. Dutrochet built an osmometer, a device to measure osmotic pressure quantitatively, and studied the transport of water and of solutions through animal membranes (bladders and pigs' intestine walls) or in vegetable tissues. He was also one of the first to understand that cellules were minute bags of liquid (Dutrochet 1837) and to appreciate their importance for the activity of living organisms. For the passage of liquids across cellular membranes he coined the words "endosmosis" and "exosmosis" (Dutrochet 1828), and since then the word osmosis has entered the scientific language.

Research on osmotic pressure became an important topic for botanists interested in the circulation of fluids in plants such as the Dutch Hugo de Vries (1848–1935) who studied plasmolysis in sugar beet, the Polish-German Moritz Traube (1826–1892), inventor in 1866 of a fragile chemical membrane permeable only to water, or the German Wilhelm Pfeffer (1845–1920) who, in order to measure osmotic pressure, built in 1887 a cell in which a porous septum held a semipermeable membrane made of copper ferrocyanide separating the solution from the pure solvent (Pfeffer 1887).

For some time, very few chemists were interested in osmotic pressure, among them Justus von Liebig, who showed interest in the subject in relation to his research of biological interest, and Thomas Graham, who had studied the diffusion of solutions as a function of concentration and molecular mass. Osmotic pressure thus remained of little interest until it attracted the attention of van't Hoff, the rising star of chemical physics. As all chemists of his time, van't Hoff was interested in measuring the affinity that appeared to be responsible for reactions among molecules, but was still not easily quantifiable and translatable in measurable quantities. An attempt to find a numerical value for a quantity somehow connected to affinity had been made in 1844 by Eilhard Mitscherlich (1794–1863), the discoverer of isomorphism, who, while working with Berzelius, became interested in the affinity problem. Mitscherlich attempted to measure the attractive force of hydration water in salts controlling the decrease of the vapor tension in these crystals. By introducing some crystals of Glauber salt ($Na_2SO_4 \cdot 10H_2O$) into the empty space on top of a barometer, he observed that the mercury height decreased by 5.45 mm with respect to pure water that produced a decrease of 8.72 mm. He thus deduced that the difference of 3.27 mm represented the affinity of sodium sulfate for hydration water, corresponding to a force of 12 g/cm^2 (Mitscherlich 1844). van't Hoff read Mitscherlich's paper and was impressed by the low value of the hydration force, believing that even the weakest chemical forces should have been much larger.

As he stated, one day in the streets of Amsterdam he met his friend Hugo de Vries who told him that his colleague Wilhelm Pfeffer, had observed a dependence of osmotic pressure on temperature, finding that each degree of temperature decrease reduced the osmotic pressure by about 1/270.

van't Hoff immediately realized the importance of osmotic pressure for the evaluation of chemical affinity and decided to use the kind of osmotic cell invented by Pfeffer to confirm his ideas. Using a diluted solution of cane sugar in water, van't Hoff proved that the attraction between water and sugar was more than 100 times stronger than that measured by Mitscherlich between sulfate and hydration water. From these measurements he deduced a result far more important and general, namely, that in diluted solutions the solute particles behave like particles of an ideal gas and therefore that osmotic pressure is linearly proportional to absolute temperature as is the pressure of an ideal gas. In 1886, van't Hoff published a new text in French entitled *l'Équilibre chimique dans* I'*État dilué gazeux ou dissous* that presented his ideas on the chemical physics of diluted solutions, establishing a perfect parallelism between ideal gases and diluted solutions in the form of the equation $\pi = \text{í}cRT$ where π is the osmotic pressure, R the universal gas constant, T the absolute temperature, c the concentration, and í an empirical corrective coefficient. This equation, practically identical to the state equation of gases, $PV = nRT$, was perfectly valid (Van't Hoff 1887) for dilute solutions of organic compounds such as sugar with a numerical value of í = 1. Determining í by different physical methods, such as the increase of boiling temperature or using Raoult's data on the decrease of freezing point, van't Hoff realized that for solutions of acids, bases, and inorganic salts, í assumed instead values close to 2. Van't Hoff did not present a valid explanation for the meaning of his empirical constant í in these cases. Only in 1887 did Arrhenius, with whom van't Hoff was in contact, suggest (Arrhenius 1887) that a value of í of the order of 2 was nothing other than proof of the fact that electrolytes were dissociated in solution and therefore that the concentration of solute particles was almost the double. van't Hoff accepted Arrhenius explanation with enthusiasm, reinterpreting his data in terms of dissociation in ions in a paper that he published in the same year with the favorable agreement of Ostwald (van't Hoff 1887).

The analogy between dilute solutions and ideal gases turned out to be very seminal for the extension of the second principle of thermodynamics to solutions (van't Hoff 1894). The parallelism allowed the full thermodynamical machinery developed for gases to be rapidly extended to solutions as well as the general relationships controlling the thermodynamic properties of diluted solutions to be found.

After the fundamental work of Gibbs on the thermodynamics of condensed systems and on chemical potentials, the application to solutions was effectively developed by the Americans John Kirkwood and Frank P. Buff in 1940–1960.

John Gamble Kirkwood (1907–1959) was a theoretical physicist who has made fundamental contributions to statistical mechanics and thermodynamics and to the theory of liquids. Kirkwood developed his treatment of the liquid state and of solutions in terms of very refined and elegant mathematics that makes use of the formalism of molecular distribution functions, today a normal component of physical-chemical research but very pioneering and of difficult application then. His most important contribution was to identify an intuitive technique of approximation that allowed the basic equations describing the structure and properties of liquid systems to be solved. In collaboration with Frank P. Buff (1924–2009), he published in 1951 a fundamental paper that represented the basis of all future developments (Kirkwood and Buff 1951).

3.3 The Debye–Hückel Theory

The beginning of the twentieth century saw a number of attempts to extend the validity of the ionic theory to strong electrolytes by developing either the treatment of the inter-ionic forces or the theory of the hydrates. The idea that nonideal behavior depended on long-range Coulombic interactions between ions was universally accepted but, owing to the complexity of the problem, an adequate theoretical treatment of these effects was still missing. Different attempts with scant results were made with more or less complex models such as that of William Sutherland (1859–1911), a Scot who had immigrated to Australia, author of a diffusion equation known as the Einstein–Sutherland equation, who in 1886 became interested in the interactions between molecules. In 1905, Sutherland wrote an obscure paper (Sutherland 1905) on inter-ionic forces starting from the assumption of complete dissociation of strong electrolytes. In addition, Englishman Samuel Roslington Milner (1875–1958) developed in a series of papers from 1912 to 1919 mathematical models of inter-ionic interactions between solute ions and solvent molecules (Milner 1919–1920). Another attempt to introduce inter-ionic forces in theory was made in 1918 by Jnanendra Chandra Ghosh (1893–1959), an attempt that after a brief initial success proved to be wrong and was rapidly forgotten (Chandra Ghosh 1918). In 1923, the treatment of inter-ionic interactions was, however, correctly resolved by Peter Debye (1884–1966) and his assistant Erich Hückel at the Eidgenössische Technische Hochschule of Zurich, who found a final solution to clarify the properties of strong electrolyte solutions, starting a new era for the study of electrochemistry.

The Debye–Hückel theory (Debye and Hückel 1923) also started from the idea that strong electrolytes were completely dissociated in solution. The tendency of ions to migrate and to give rise to an electric current was, however, counterbalanced by the electrostatic interactions between ions of opposite charge and between ions and solvent molecules, an effect that increased with solute concentration.

The basic idea of the theory was that, as a consequence of attraction between opposite charges, there were on average more ions of different than of the same sign surrounding an ion. Each ion was therefore surrounded by an atmosphere of charges of opposite sign that in the absence of an external field had central symmetry.

When, however, a current started to circulate in the solution and all positive ions moved in the direction of the negative pole and vice versa, each ion rebuilt an atmosphere of opposite sign in its motion, while that left behind slowly dissolved. Since the two processes were not instantaneous, namely, did not take place at the same time, the result was that an attractive force was exerted in the direction opposite to that of the ion motion that slowed it down.

The influence of the cloud of opposite charges on the velocity of an ion is called the relaxation effect. There is in addition a further factor opposed to ion movement, due to the fact that its ionic cloud tends to move in the direction opposite to that of the ion by consequence of the external field; this is called the electrophoretic effect since it is equivalent to that which opposes the motion of colloidal particles in an electric field. Finally, ion motion is also hindered by the viscosity of the medium, an effect controlled by Stokes' law and depending on viscosity and particle radius. The idea of the theory was that every ion creates a potential that exerts a Coulomb force on all ions in its neighborhood. This force decreases exponentially as the distance increases owing to the shielding effect of the surrounding ions. In order to obtain a reasonable expression for the potential, Debye and Hückel utilized a known theorem of electrostatics, analytically expressed by the Poisson equation that allows the variation in time of the electrostatic potential at any point of a system to be calculated as a function of the charge density at that point.

By solving Poisson's equation with the approximation that the disordered motion of the ions can be described as a diffuse density of charge around the central ion, they obtained for the average potential ψ the expression $\psi = \frac{e^{-\kappa r}}{r}$ where the shielding factor $\kappa = \sqrt{\frac{8\pi e^2}{DkT}\mu}$ depends on the dielectric constant D and on the ionic force μ that measures the total ionic concentration in solution.

A great success of the Debye–Hückel theory was to explain the behavior of strong electrolytes in diluted solutions and to prove the importance of ionic force in determining activity coefficients of electrolytes, namely, the effective concentrations of the species involved in liquid-phase reactions.

The theoretical previsions were, however, verified only for dilute solutions of salts and acids of univalent elements (HCl, NaCl, etc.) and became less and less valid as the electrolyte concentration increased.

In 1923, when the Debye–Hückel theory was published, Lars Onsager was still a student at the *Norges Tekniski Hoslashgskole* in Trondheim that he had joined in 1920. In 1925, he realized that molar conductivity values predicted by the Debye–Hückel theory were very different from those obtained experimentally and understood the reason for these differences. In the same year, after his graduation in technical engineering, Onsager accompanied his professor Johan Peter

Holtzmark in a trip to Denmark and Germany and eventually to Zurich, where he stayed for about 2 months. During this period, he had the opportunity to present his ideas on the theory of strong electrolytes directly to Debye. When he met Debye for the first time, he simply turned to him saying: "Professor Debye, your theory is wrong!" Debye, instead of being offended and without batting an eyelid, offered him a position of researcher at Zurich that Onsager accepted for 2 years.

Debye and Hückel had based their treatment on the idea that the electrostatic field due to an ion in motion in a solution was "shielded" by the atmosphere of opposite charges and that the effective shielding distance was inversely proportional to the square root of the ionic force. Their equation for the calculation of the molar conductibility of the form

$$\Lambda = \Lambda_0 + \Lambda_1 \sqrt{c}$$

was formally correct but the computed values of Λ_1 differed appreciably from the observed ones. This result seemed strange to Onsager who believed that Debye and Hückel had correctly considered the presence of the electrophoretic and relaxation shielding effects.

In 1923, Onsager verified the consequences of the Debye–Hückel treatment, and in 1925, before visiting Debye, he had already found the reasons for the disagreement between observed and computed molar conductivity Λ. As he later wrote in 1968 (Onsager 1968):

> The relaxation effect ought to reduce the mobilities of anion and cation in equal proportion. Much to my surprise, the results of Debye and Hückel did not satisfy that relation, nor the requirement that wherever an ion of type A is 10 Å west of a B, there is a B 10 Å east of that A. Clearly something essential had been left out in the derivation of such unsymmetrical results.

Debye and Hückel had computed the conductivity Λ in the assumption that an ion moved along a straight line toward the pole of opposite sign among all the other ions in random Brownian motion in the field created by their charge distribution. Onsager realized that in order to obtain good agreement with experimental data, it was sufficient to eliminate the restriction that the central ion moved in a uniform motion, leaving it instead free to move randomly through the neighboring ions, under the effect of their presence and of the external field. Onsager developed his modification of the Debye–Hückel law in two famous papers known as the Onsager limiting law (Onsager 1926, 1927) and later in another significant paper in collaboration with his pupil Raymond Matthew Fuoss (Onsager and Fuoss 1932).

During 1955–1965, Onsager considered again the problem of the behavior of electrolytes in collaboration with his old student Raymond Fuoss (1905–1987) who, in 1995 had also obtained a position of professor at Yale. Together they reexamined the approximations of the Debye–Hückel–Onsager theory of 1927 and reformulated it in a series of highly sophisticated papers (Fuoss and Onsager 1955, 1958) without, however, reaching any truly innovative result.

3.4 Acids and Bases

The year 1923 was a year of grace for electrochemistry. Within a few months two independent theories relative to the electrochemical nature of acids and bases, albeit almost coincidental, were published to complete the framework of the study of electrochemical solutions supplied by the Debye–Hückel theory and by Onsager's modifications. The first among them was by the Dane Johannes Nicolaus Brønsted (1879–1947) who graduated in chemistry in 1903 at the University of Copenhagen and obtained his Ph.D. in 1908 with a thesis on the affinity of binary mixtures of sulfuric acid and water. In the same year, he became professor at the same university, beating in the competition another famous candidate, Niels Bjerrum, his inseparable colleague and friend.

Brønsted's theory of acids and bases (Brønsted 1923) defined an acid as a compound tending to give a proton and a base as a compound tending to acquire one.

This new idea explained all known experimental facts and allowed the concept of acid and base to be extended to compounds that could not previously be classified as such, for instance, water. New concepts such as those of acid and conjugated base or those of base and conjugated acid thereafter became integral parts of the electrochemical language.

A conjugated base was that part of a molecule of an acid remaining after having taken out a hydrogen atom (proton) whereas a conjugated acid was the molecule of a base that had accepted a proton:

$$HA + OH^- \rightleftarrows \underset{\text{basis}}{A^-} + \underset{\text{acid}}{H_2O}$$

Thanks to Brønsted's approach a new ion appeared in acid–base reactions, the hydronium ion H_3O^+, formed by the union of a hydrogen atom and a water molecule:

$$HA + H_2O \rightleftarrows H_3O^+ + A^-$$

According to Brønsted, when a hydrogen atom is created, it is immediately attracted by a polar molecule of water, giving rise to a symmetrical structure with three hydrogen atoms around a central oxygen atom.

The second theory was developed in the same year (Lowry 1923) by the Englishman Thomas Martin Lowry (1874–1936), professor of chemical physics at the University of Cambridge. Starting from the study of the variations of the rotatory power of camphor derivatives produced by acid and base catalyzed reactions, he reached independently the same result, formulating a theory practically identical to that proposed by Brønsted.

These theoretical results were in a short time associated with important experimental developments that transformed electrochemical techniques into powerful analytical methods – polarography and electrophoresis.

Polarography was invented by Jaroslav Heyrovsky (1890–1967), born in Prague into a family of the university milieu. After having studied chemistry in his native town, Heyrovsky went to study at the University College of London from 1910 to 1914 where, in 1913, he obtained a bachelor degree under the supervision of Sir William Ramsay and, above all, of Frederick George Donnan (1870–1956), who encouraged the young Czech to take up the study of electrochemistry.

As a matter of fact, while Heyrovsky attended the University College laboratories, Donnan studied the equilibrium between two solutions of electrolytes separated by a semipermeable membrane that allowed only solute molecules to go through, thereby giving rise to a concentration difference that could be detected by measuring a potential difference between the two sides of the membrane (Donnan effect).

During the First World War, the young Heyrovsky worked as dispensing chemist and radiologist in a medical hospital, a situation which enabled him to continue to pursue his studies and to obtain his Ph.D. first in Prague in 1918 and then in London in 1921. He then started his academic carrier as assistant to Bohuslav Brauner in analytical chemistry and became associate professor in 1922 until in 1926 he held the first chair of chemical physics at the Carl I University in Prague. In 1922, he invented the polarograph and then concentrated his research activity on the development of polarography, a technique that became an important branch of electrochemistry when, in collaboration with his student Masuzo Shikata (1895–1964), he developed the automatic polarograph (Heyrovský and Shikata 1925). For this invention, Heyrovsky received the 1959 Nobel Prize for chemistry.

Polarography is an analytical technique based on the measurement of the electrical current flowing in a controlled voltage circuit made from a capillary dripping mercury droplets and a reference electrode immersed in an electrochemical solution.

The polarographic technique originated from a previous analytical method in which, for a given applied voltage at the electrodes, the mercury droplets falling from the capillary were counted in a definite time interval. The collected mercury droplets were accurately dried and weighted and the operation was repeated for different values of applied voltage. The mercury weights obtained were eventually reported in a graph as a function of the voltage. The innovative idea of Heyrovsky was that, instead of counting the mercury droplets, it was more convenient to measure the diffusion current flowing in the cell, proportional to the concentration of the chemical species present in solution. In this way, he obtained current–voltage curves for each concentration of the electrolyte.

Another important chemical physics technique connected with the motion of electric charges in liquids is electrophoresis which uses an electric field to move the charged molecules in the fluid. Particles with a positive charge are displaced toward the cathode (cataphoresis) and those with a negative charge toward the anode (anaphoresis).

The idea that it might be possible to displace electrically charged molecules, even of great dimensions, by means of an electric field was quite widespread in the first part of the twentieth century. Electrophoresis became an important analytical technique for the study and separation of proteins, thanks to the researches of the Swede Arne Wilhelm Kaurin Tiselius (1902–1971) born in Stockholm and educated at the University of Uppsala. An assistant in 1925 of Theodor Svedberg (1884–1971), he obtained his Ph.D. in chemistry in 1930 with a thesis on the electrophoresis of proteins (Tiselius 1930). An associate professor at the University of Uppsala, Tiselius became full professor and director of the institute in 1938. Under Tiselius' direction, the Uppsala Biochemistry Institute contributed to the development of a large number of important biochemical techniques such as electrophoresis, chromatography, gel filtration, etc. The Tiselius group also applied these techniques with great success to the study of molecules of very high molecular weight (Tiselius 1937), essentially enzymes and proteins (Tiselius 1939), but also polysaccharides and nucleic acids.

References

Armstrong HE (1936) Ionomonia in extremis. Chem Ind 14:917

Arrhenius S (1884) Recherches sur la conductibilité galvanique des électrolytes. Thesis Uppsala

Arrhenius S (1891) Über die Leitung von Elektrizität durch heisse Salzdämpfe. Wied Ann 43:18

Arrhenius S (1887) *Über die Dissociation der im Wasser gelösten Stoffe*, Zeitschrift fur physikalische Chemie, 1, 631

Arrhenius S (1896) On the influence of carbonic acid in the air upon the temperature of the ground. Philos Mag 41:237–276

Arrhenius S (1903) Lehrbuch der kosmischen physik, vol I und II. Hirschel, Leipzig

Beckmann E (1886) Zur Kenntniss der Isonitroso verbindungen. Ber Deut Chem Gesell 19:988–993

Beckmann E (1888) Über die methode der Molekulargewichtsbestimmung durch Gefrierpunkterniedrigung. Z Physik Chem 2:638–715

Berthollet CL (1803) Essai de statique chimique. Firmin Didot, Paris

Bjerrum N (1909) A new form for the electrolytic dissociation theory. In: Proceedings of the seventh international congress of applied chemistry, London, Section X, pp 55–60

Blagden Ch (1788) Experiments on the effect of various substances in lowering the point of congelation in water. Philos Trans 78:125–277

Bodenstein M (1894) Zersetzung und Bildung von Jodwasserstoff. Z Physik Chem 13:22

Bodenstein M (1897) Zersetzung und Bildung von Jodwasserstoff. Z Physik Chem 22:1

Bodenstein M (1899) Zersetzung und Bildung von Jodwasserstoff. Z Physik Chem 29:295

Boerhaave Hermannus (1732) Elementa Chemiae Historia et artis Theoria. Tome I. Pars II: De Artis Theoria. Lugduni Batavorum, Imhof

Brock WH (1993) The Norton history of chemistry. W.W. Norton & Company, New York/London

Chandra Ghosh J. (1918) *The Abnormality of Strong Electrolytes. Part 1, Electrical Conductivity of Aqueous Salt Solutions*, J. Chem. Soc. 113, 449

Brønsted JN (1923) Some remarks on the concept of acids and bases. Recl Trav Chim Pay-B 42:718–728

de Coppet M (1872) Recherches sur la température de congélation des dissolutions salines. Ann Chim Phys 25(4):502–553

Debye P, Hückel E (1923) Zur Theorie der Elektrolyte. I. Gefrierpunktserniedrigung und verwandte Erscheinungen. Phys Z 24:185–206

Dutrochet H (1828) Nouvelles recherches sur l'endosmose et l'exosmose. JB Ballière, Paris

Dutrochet H (1837) Mémoir pour servir a l'hitoire anatomiques et physiologiques des végetaux et des animaux. JB Baillière, Paris
Fahrenheit DG (1724) Experimenta et observationes de congelatione aquae in vacuo factae. Philos Trans (London) 33:78
Fuoss RM, Onsager L (1955) Conductance of strong electrolytes at finite dilutions. Proc Natl Acad Sci USA 41:274–283
Fuoss RM, Onsager L (1958) The kinetic term in electrolytic conductance. Phys Chem 62:1339–1340
Grotthuss T (1806) Sur la décomposition de l'eau et des corps qu'elle tient en dissolution à l'aide de l'électricité galvanique. Ann Chim 58:54–73
Harcourt, A. V. and Esson, W. (1867) *On the Laws of Connexion between the Conditions of a Chemical Change and Its Amount*, Phil. Trans. 157, 117
Heyrovský J, Shikata M (1925) Researches with the dropping mercury cathode. Rec Trav Chim Pay-B 44:496
Hittorf W (1853) Über die Wanderungen der Ionen während der Elektrolyse. Pogg Ann 89:177
Hittorf W (1859) Über die Wanderungen der Ionen während der Elektrolyse. Pogg Ann 106:513
Hood JJ (1878) Rate of chemical change as a function of temperature. Philos Mag 6:371
Hope TC (1805) Experiments and observations upon the contraction of water by heat at low temperatures. Trans Roy Soc Edinb 5:379
Kahlenberg L (1902) Instantaneous chemical reactions and the theory of electrolytic dissociation. J Phys Chem 6:1–14
Kirkwood JG, Buff FP (1951) Statistical mechanical theory of solutions. J Chem Phys 19:774–777
Klaproth MH (1792–1793) Rapport touchant l'ouvrage intuitulé Traité élementaire de Chimie par Lavoisier, T. 1 et 2. Paris 1789 et la tradution Allemande de cette Ouvrage par Hermbstädt intitulée: System der Antiphlogistischen Chemie 2 Bände. Berlin 1798, Mémoires de l'Académie Royale des Sciences et Belle-Lettres
Kohlrausch F (1880) Einfache Methoden und Instrumente zur Widerstandbestimmung inbesondere in Electrolyten. Ann Phys 11:653
Kohlrausch F (1885) Über das Leitungsvermögen einiger Electrolyte in äußerst verdünnter wässeriger Lösung. Ann Phys Chem 26:161–226
Kohlrausch F, Grotrian F (1875) Das elektrische Leitungsvermogen derChlor-Alkalien und alkalinischen Erden, sowie der Salpetersaure in wasserigen Losungen. Ann Phys 154 (1–14):215–239
Lowry TM (1923) The uniqueness of hydrogen. Chem Ind (London) 42:43
Mendeleev D (1877) L'origine du petrole. Rev Sci VIII:409–416
Milner SR (1919–1920) Note on the effect of interionic forces in electrolytes. Trans Farad Soc 15:148
Mitscherlich, E. (1847) "Lehrbuch der Chemie," 1829 (4th ed.,); p 565
Nollet J.A., (1748) Lecons de physique experimentale, Hippolyte-Louis Guerin and Louis-Francios Delatour, Paris
Onsager L (1926) Zur Theorie der Elektrolyte. I Phys Z 27:388
Onsager L (1927) Zur Theorie der Elektrolyte II. Phys Z 28:277
Onsager L, Fuoss RM (1932) Irreversible processes in electrolytes. Diffusion, conductance, and viscous flow in arbitrary mixtures of strong electrolytes. J Phys Chem 36:2689
Onsager L. (1968) *The motion of ions: principles and concepts*, Nobel lecture
Ostwald W (1878) Volumchemische und optisch-chemische Studien. Tesi di dottorato, Università di Dorpat
Ostwald W (1880) Chemische Affinitätsbestimmungen. J Prakt Chem 22:251–260
Ostwald W (1884) Studien zur chemischen dynamik III. Die inversion des rohrzuckers. J Prakt Chem 29:385–408
Ostwald W (1885) Studien zur chemischen dynamik III. Die inversion des rohrzuckers. J Prakt Chem 31:307–317
Ostwald W (1887) Elektrochemische Studien. Z Phys Chem 1:75–86

Ostwald W (1888) Über die Dissociationstheorie der Elektrolyte. Z Phys Chem 2:36
Ostwald W (1889) Über die Affinitätsgössen organischer Sauren und ihre Beziehungen zur Zusammensetzung und Konstitution derselben. Z Phys Chem 3:170
Ostwald W (1892) Thermodynamische Studien, von Willard Gibbs. W. Engelmann, Leipzig
Ostwald W (1899) Periodisch veränderliche Reaktionsgeschwindigkeiten. Phys Z 1:87–88
Ostwald W, Nernst W (1889) Über freie Ionen. Z Phys Chem 3:120–130
Partington JR (1964) A history of chemistry, vol 4. Macmillan, London
Porto PA (2002) Summus atque felicissimus salium. The medical relevance of the liquor alkahest. Bull Hist Med 76:1–29
Pfeffer, W. F. P, (1887) *Osmotische Untersuchungen*, Leipzig, W. Engelmann
Raoult M (1882) Loi générale de la congélation. Comptes Rendus 95:1030–1033
Raoult M (1887) Loi générale des tensions de vapeur des dissolvants. Comptes Rendus 104:1430
Rossetti MF (1869) Sur le maximum de densite et la dilatation de l'eau distillee, de l'eau de l'Adriatique et de quelques solutions salines. Ann Chim Phys 17:370–384
Sørensen S (1909) Enzyme studies II. The measurement and meaning of hydrogen ion concentration in enzymatic processes. Biochem Z 21:131–200
Sutherland W (1905) A dynamical theory of diffusion for non-electrolytes and the molecular mass of albumin. Philos Mag 9(6):781–785
Tiselius A (1937) A new apparatus for electrophoretic analysis of collodal mixtures. Trans Faraday Soc 33:524
Tiselius A (1939) The chemistry of proteins and amino acids. Ann Rev Biochem 8:155–184
Tiselius A (1930) The moving-boundary method of studying the electrophoresis of proteins. Nova Acta Reg Soc Sci Upsaliensis Ser IV **7**
Van't Hoff JH (1887) Die Rolle des osmotischen Druckes in der Analogie zwischen Lösungen und gases. Z Phys Chem 1:481–508
Van't Hoff JH (1888) Law of osmosis. Philos Mag 26:81
Van't Hoff (1894) Wie die Theorie der Lösungen entstand. Ber Deutsch Chem Ges 27:6–19
Wallerius JG (1776) Der Physischen Chemie, bey Christian Mevius. Reprint in English, Kessinger Publishing, Paperback, 2009
Williamson AW (1850) Theory of aetherification. Philos Mag 37:350–356

Chapter 4
Chemical Physics Structural Techniques

There are two possible outcomes: if the result confirms the hypothesis, then you've made a measurement. If the result is contrary to the hypothesis, you've made a discovery. (Enrico Fermi)

4.1 Development of Physical Techniques in Chemistry

The prodigious growth of chemistry in the twentieth century would not have been possible without the development of chemical physics techniques allowing one to investigate molecular structures and dimensions and to define the relative positions of atoms in space, connecting molecular dynamics to reactivity and to energy transformations.

The construction of chemical physics instruments was strictly associated in the first half of the century with the ability of researchers and technicians to plan and build sophisticated instruments with relatively limited means at their disposal. The availability of expert built instruments capable of high-level performance was limited to few avant-garde laboratories in which the collaboration between chemists, physicists, and engineers enabled groups of young researchers to develop original techniques, borrowing the competencies accumulated in the physics laboratory and integrating them with those used in investigating molecular behavior.

The use of homemade instrumentation allowed them to acquire sufficient experimental data which the rising quantum mechanics theoretical developments interpreted and organized.

At the end of the Second World War, industry in the most advanced countries realized the enormous possibilities offered by the exploitation of the technical competences developed during the war for the construction of military devices and electronic and optical instruments, radars, night vision and tracking systems, etc., to develop a promising market in sophisticated laboratory instrumentation, more and more in demand for chemical and clinical analysis as well as medical care applications.

S. Califano, *Pathways to Modern Chemical Physics*,
DOI 10.1007/978-3-642-28180-8_4,

The number of chemical physics techniques developed in the twentieth century were considerable. Several of them have been discussed earlier; here we shall consider those of a spectroscopic nature of great importance in the determination of atomic and molecular structures. Their interpretation in terms of the complex theoretical apparatus from the developments in quantum mechanics has played a fundamental role in the understanding of the interaction of matter with electromagnetic fields.

4.2 X-Ray Diffraction

At the end of the nineteenth century, the attention of several physicists had been attracted by the radiation emitted by the cathode tubes developed by Plucker, Hittorf, Crookes, Hertz, and Lénárd. In the period 1881–1883, the Ukrainian physicist Ivan Puluj (1845–1918), professor at Prague University, published a series of papers, later collected in a book, in which he noted the presence of an obscure radiation emitted by cathode tubes. Puluj's papers, written in an obscure language which often made use of obsolete expressions, were almost completely ignored until translated into English and published by the Royal Physical Society in 1899 (Puluj 1899). In 1895, Wilhelm Conrad Röntgen (1845–1923), professor of physics at the University of Würzburg, who had been Puluj's master, while studying electrical discharges in cathode tubes, realized that a barium platinocyanide screen took on a glowing fluorescence if hit by the obscure radiation emitted by the tubes, even if covered with black cardboard. Röntgen realized the importance of this discovery and, after having used photographic plates to take the very first picture of his wife's hand, Anna Bertha, bearing a ring on her finger, gave preliminary notice to the secretary of the physical-medical society of Würzburg, which soon published the paper *Über eine neue Art von Strahlen* (Röntgen 1895, 1896) that made him famous, and in 1901 he was honored with the first Nobel Prize for physics.

X-rays soon found important applications in medicine for the radiography of the human skeleton and became a fundamental instrument for the study of the structure of molecules and crystals when, after few years, it was realized that they were in fact an electromagnetic radiation of wavelengths much smaller than those of visible radiation.

At the beginning of his research, Röntgen suggested that x-rays were longitudinal waves, but this hypothesis was soon rejected when Charles Glover Barkla (1877–1944), professor at the University of Liverpool, showed in 1904 that they could be polarized (Barkla 1904), earning for his research on x-rays the Nobel Prize for physics in 1917. At first, Einstein disagreed with the idea that they were electromagnetic waves of wavelengths thousands of times smaller than those of visible light and, on the basis of his theory of photoelectric effects, favored a particle nature for them. Of the same opinion, i.e., a supporter of the particle nature of x-rays, was W.H. Bragg, one of the physicists who would later contribute to the experimental investigation of x-ray diffraction.

The theoretical research of Arnold Sommerfeld during the period 1912–1915 (Sommerfeld 1912, 1913, 1915) shifted interest toward the wave nature of x-rays which opened the route to unexpected developments. These were essentially due to the German physicist Max von Laue (1879–1960) who, after having obtained his Ph.D. in physics at the University of Berlin with a thesis on interference phenomena (Laue 1904) worked as assistant to Max Planck in Berlin until 1909, when he moved to Munich. There he went to work in Sommerfeld's institute where he became fully involved in the discussion on the nature of x-rays. During late 1911 and early 1912 the young physicist Peter Paul Ewald (1888–1985), working on his Ph.D. thesis under the supervision of Sommerfeld, asked for his help to rationalize the interaction mechanism of an electromagnetic wave with a crystalline structure. Von Laue did not find a plausible answer for Ewald right away, but, turning it over in his mind, reached the right solution after a few months.

During his thesis, von Laue had accumulated considerable competence on wave interference processes. He started his approach to this new problem assuming that if the x-ray wavelength was of the order of Ångstroms as suggested by Sommerfeld, it had to be of the same order of magnitude as the distance between the grating planes of the crystals. These then would behave as diffraction gratings for the x-rays, giving rise to diffraction patterns. von Laue explained his idea to Sommerfeld who, although skeptical about the success of the experiment and having in mind other research, did not hesitate to assign to von Laue his assistant Walter Friedrich (1883–1969) and the Ph.D. student Paul Knipping (1883–1935) to work on the project.

After some failed attempts, Friedrich and Knipping succeeded on April 23, 1912, in obtaining the first diffraction spectrum of the x-rays from a crystal of copper sulfate (Friedrich et al. 1912). Max von Laue presented the results to the Bavarian Academy of Sciences (Laue 1912) and in 1914 was honored with the Nobel Prize for physics. In his Nobel lecture, as well as in several interviews and articles, von Laue emphasized the fundamental contribution of Friedrich and Knipping to the success of the experiment and even decided to share the conspicuous sum of the Nobel Prize with his two coworkers.

The importance of von Laues discovery of x-ray diffraction was immediately realized by the English physicist William Henry Bragg (1862–1942) who immigrated to South Australia as professor of mathematics and physics in the University of Adelaide, where he built an excellent reputation as an instrument maker. In 1896, having learned of Röntgen's discovery, he started to be interested in the x-ray physics and even succeeded in obtaining a picture of the broken arm of his 6-year-old son with a self-made x-ray tube.

In 1904, Bragg started a series of research on the penetration of α and β particles in matter, succeeding in distinguishing the penetrating power of α particles produced by different radioactive elements and in proving that the ability of different chemical elements to stop their progress was roughly proportional to the square root of their atomic weights.

These results drove him to suggest that the x-rays and γ-rays were both particles, a position that put him at the center of a long-lasting polemic until in 1912 the von Laue experiment cut the matter short. Bragg returned to England in 1909 as professor of physics at the University of Leeds, again became interested in x-rays, and built the first true spectrograph for the study of their diffraction from crystals. Very useful for this work was the collaboration with his son William Lawrence Bragg (1890–1971) working at Cambridge, who realized an important simplification of the von Laue theory by formulating his famous Bragg's law (Bragg 1914) that relates the maxima of the diffraction patterns to the wavelength λ of the radiation, to the distance d between two crystallographic planes, and to the angle θ between the incident radiation and these planes:

$$n\lambda = 2d \sin\theta.$$

With their new X spectrograph, the Braggs determined in 1913 the structure of diamond (Bragg 1913a, b, 1914) and of a series of minerals and organic molecules (Bragg 1913b) obtaining in 1915 the Nobel Prize for physics for having opened up the new field of crystallographic structure determination with their research.

Since then the x-ray diffraction of single crystals has been used by chemists essentially for the determination of the stereochemistry of simple molecules using photographic detection. In 1916, Peter Debye and Paul Scherrer (1916, 1917) in Switzerland and Albert Hull (1917) in the United States developed the powder method that made it possible to study materials not available as single crystals. Molecular structures were determined starting from experimental diffraction patterns, assuming a possible model for the structure and using the method of successive approximations (trial and error technique). In 1934, Arthur Lindo Patterson (1902–1966) introduced the Fourier analysis technique that allows one to obtain maps of electronic density with maxima where the atoms are localized (Patterson 1934). Patterson's method is based on the idea that diffraction patterns are nothing else that the tridimensional Fourier transform of the crystal structure and therefore that the problem could be solved by computing the inverse Fourier transform. The structure of complex molecules thus became easily accessible thanks also to technical improvements both in detection and in x-ray production.

An important contribution to molecular structure determination also came from the papers of the Dutchman Johannes Martin Bijvoet (1892–1980), professor of chemistry and crystallography at the University of Utrecht, who in 1949 developed the method to establish the space structure (absolute configuration) of chiral molecules (Bijvoet 1949, 1951). Bijvoet's method relies on the analysis of the anomalous dispersion of the x-rays due to the interaction between the elastic x-rays scattered by the atoms (with the same energy but a phase shift with respect to the incident radiation) and the radiation generated by the inner electron shells of the atoms near the atomic absorption edge by a partial ionization process. This second radiation interferes with the elastic scattered rays, altering both the amplitude and the phase. The total scattered radiation is then a complex quantity made of a real part, $\Delta f'$ and an imaginary part, $\Delta f''$. Whereas the real part is either positive or negative, the imaginary part is always positive, resulting in an addition to the phase angle. In 1951, using an x-ray tube with a zirconium target, Bijvoet and his coworkers Antonius Franz Peerdeman and Adriaan van Bommel achieved the first experimental determination of the absolute configuration of sodium rubidium tartrate (Peerdeman et al. 1951). In this compound, rubidium atoms were the ones close to the absorption edge.

In 1953, the Viennese Max Ferdinand Perutz (1914–2002) developed the technique of heavy atoms to solve the phase problem. The heavy atom method consists in incorporating in a molecule of large dimension an atom with several electrons like mercury or gold without altering the space distribution of the atoms (Perutz 1956). This technique allows one to create an important reference spot in the electronic density map to localize better the atomic positions. With this technique, the x-ray structural analysis was easily extended to very complex molecules. Max Perutz studied the x-ray diffraction of proteins extensively in his life and in particular resolved in 1960 the structure of hemoglobin (Perutz et al. 1960). For his research he received the 1962 Nobel Prize for chemistry together with his colleague at the University of Cambridge, the Englishman John Kendrew (1917–1997) who resolved in 1953 the structure of myoglobin (Bluhm et al. 1958).

In 1950, the Jewish American mathematician Herbert Aaron Hauptmann (1917–2011) and the chemical physicist Jerome Karle (1918–...) started in collaboration a complex research project in crystallography to elaborate statistical methods, called direct methods, for the determination of the tridimensional structure of molecules using the intensity of the diffraction spots which supply an important body of information on the phases of the structure factors (Hauptman and Karle 1956). For their research they obtained the 1985 Nobel Prize for chemistry.

4.3 Neutron Diffraction

A new source of radiation for the study of the structure of molecules and crystals became available at the end of the Second World War with the construction of the first neutron sources for scientific applications. These radiation sources became important for structure determination when in 1938 Enrico Fermi proved that thermal neutrons, i.e., neutrons properly slowed down, are more efficient than the fast ones. While x-rays are scattered by the electron cloud, neutrons are scattered by directly interacting with the nuclei and therefore represent a convenient method to localize the nuclear positions in the crystal gratings. Furthermore, their diffusive power, technically called cross section, does not depend on the number of electrons but is a specific property of the nuclei. This, for instance, makes it easier to find the positions of hydrogen atoms that X-rays have difficulty in localizing. After a famous paper by the Italian physicist Franco Rasetti (1901–2001), who immigrated to Canada in 1939 and who in 1940 proved the phenomenon of thermal neutron diffusion from crystal gratings (Rasetti 1940), the American physicist Ernest Omar Wollan (1902–1984) began a research project in 1944 on the diffusion of neutrons by crystals using the neutron beam produced by the Oak Ridge National Laboratory reactor in Tennessee. In 1946, the physicist Clifford Shull (1915–2001) joined his group and contributed in collaboration with him to develop the neutron diffraction technique (Davidson et al. 1947). Shull later became director of the group and developed the technique of neutron diffraction from organic crystals containing hydrogen and deuterium atoms. In the 1950–1960 period, Bertram Neville Brockhouse, (1918–2003) ran similar projects, using the Chalk River Nuclear Laboratory reactor in Canada, and also contributed to the study of phonon modes in crystals which are delocalized grating vibrations, using the inelastic neutron diffusion technique (Brockhouse and Stewart 1958). If fast neutrons with energy of the order of 10 MeV (1 MeV = a million electronvolts), are used instead of slow thermal neutrons, new types of physical processes take place in which the neutrons give part of their energy to the crystal grating, giving rise to collective vibrations of the atoms (phonons). However, these experiments require monochromatic neutron beams (namely, neutrons all with the same energy) and sophisticated electronics for measuring the energy of the anelastically diffused neutrons. The Shull and Brockhouse research, including in particular that relating to inelastic diffusion processes, was honored in 1994 with the award of the Nobel Prize for physics. Neutron scattering experiments are widely used in chemical physics experiments to study the dynamics of molecular crystals and in particular the frequencies of the crystal normal modes in which the molecules oscillate in phase or in contrast of phase to give rise to acoustic (in phase) or optical (in contrast of phase) phonons representing lattice vibrations propagating in the crystal (Califano et al. 1981).

4.4 Vibrational Spectroscopy

Vibrational spectroscopy, which investigates the molecular vibrations due to oscillation of atoms about their equilibrium positions, was developed, on the one hand, for the study of molecular dynamics and the forces binding atoms together and, on the other, as an analytical tool for identification of molecular species in solids, liquids, and gases.

Molecular vibrations give rise to the occurrence of absorption bands in the infrared region of the electromagnetic spectrum with wavelengths ranging from about 1 to roughly 300 μm, due to the presence of oscillating dipoles associated with the vibrations which interact with the electromagnetic field. The assignment of the infrared absorption bands to characteristic vibrations of chemical groups has given chemistry a fundamental tool for the determination of the structure and chemical composition of large molecules, and, from this point of view, the infrared spectrum can be considered as a true fingerprint of molecular structure.

The discovery of infrared radiation can be traced to the observation made by William Herschel that solar radiation has two parts, one visible and the other invisible. The concept of radiant heat, the "warm" radiation coming from the sun, was, however, known long before since a first recorded account of a far infrared experiment appears to be that reported by the Italian philosopher and scientist

Giovanni Battista della Porta (1535–1615) in his book *Magiae Naturalis* published in 1589. Della Porta described how he was able to focus with the help of a concave mirror both the "cold and the warm" using his "natural senses" as detectors. As soon as the thermometer was invented it was used in 1612 by Balthasar Sartorius, a physician and mathematician in Breslau during Galileo's time, to measure better the heat coming from the sun. Other significant measurements of the heat from solar radiation were made almost at the same time by members of the short-lived Accademia del Cimento in Florence using homemade sealed glass thermometers. However, it was Sir William Herschel who first carried out in 1800 a true quantitative measurement of the heating power of the sun in both the visible and near-infrared portion of its spectrum. In 1829 in Florence, the Italian Leopoldo Nobili (1784–1835) reported the construction in collaboration with Macedonio Melloni of the first infrared detector, the *thermomultiplier*, a combination of a thermopile and a galvanometer, before being appointed professor of physics at the Regal Museum of Physics and Natural History in Florence, where he worked with Antinori on electromagnetic induction. Macedonio Melloni (1798–1854) extended the range of infrared detectors by discovering that rock salt crystals were transparent to infrared radiation and then building a prism of this material. The progress on infrared detectors took a further step forward when the astronomer Samuel Pierpont Langley (1834–1906), future secretary of the Smithsonian Institution, invented in 1878 the bolometer, a radiant-heat detector sensitive to differences in temperature of one hundred thousandth of a degree Celsius (0.00001°C).

The interpretation of the infrared spectra of complex molecules began in 1872 when Walter Noel Hartley (1845–1913), member of the Royal College of Science in Dublin, and his coworker A.K. Huntingdon examined, using a spectrograph equipped with ruled gratings and photographic detection, the infrared spectra of a series of organic molecules (Hartley and Huntingdon 1879). In 1881, Sir William Abney (1843–1920) and his friend Colonel Edward Robert Festing (1839–1912), the first director of the Science Museum in London, studied the infrared spectra of a large number of chemical compounds and found a full series of spectral correlations among them (Abney et al. 1881). The correlation between molecular structure and absorption bands in infrared was further improved and transformed in a true diagnostic criterion by William Coblentz (1873–1962) who from 1905 to 1945 directed, first at the Cornell University and after at the National Bureau of Standards, a systematic research project involving the analysis of the spectra of hundreds of organic molecules, proving that the presence in molecules of characteristic groups such as OH, NH_2, NH, CN, CH_3, CHO, COOH, C_6H_5, etc., was associated with the occurrence of characteristic infrared bands in well-defined spectral regions. Charles F. Mabery of the Case School of Applied Science in Cleveland, a well-known petroleum chemistry specialist in the United States, helped Coblentz in obtaining samples of pure distilled petroleum. A large group of researchers contributed all over the world to the assignment of infrared band to the vibrations of specific molecular groups. Harrison McAllister Randall (1870–1969), David Mathias Dennison (1900–1976), Nelson Fuson (1913–2006), and George Claude Pimentel (1922–1989) in the United States, Norman Sheppard and John Lionel Bellamy in England, Jean Lecomte (1898–1979), Josette Vincent-Geisse, and Marie Louise Josien in France, Reinhard Mecke (1895–1969) in Germany, Giobatta Bonino in Italy, and Sanichiro Mizushima (1899–1983) in Japan were the pioneers of the technique of vibrational assignments to the normal modes of vibration of molecules. The use of infrared spectroscopy as a diagnostic tool in analytical chemistry became widespread thanks to the publication of a series of books such as *Le Spectre Infrarouge* (Leomte 1928) and especially the Bellamy text *Infrared Spectra of Complex Molecules* (Bellamy 1958). The application of infrared spectroscopy was strongly supported by the oil companies in the United States and by the availability of the first commercial infrared spectrographs produced by the Perkin Elmer and the Beckman industrial companies. Perkin Elmer, founded in 1937 by the banker Richard S. Perkin, grabbed an important share of the market with the production of the first double-beam spectrograph, the famous model 21, which became one of the most common instruments in research laboratories. The Beckman company, specializing more in the production of commercial spectrographs for the visible and near-ultraviolet regions, launched in 1945 its first true commercial infrared spectrograph, the model IR-2. At the end of the 20th century, Fourier transform spectrographs became available, ensuring higher resolving power and larger frequency intervals.

In 1928, the Indian physicist Chandrasekhara Venkata Raman (1888–1970) and his student Kariamanikam Srinivasa Krishnan (1898–1961) discovered an effect of the inelastic scattering of light that took the name Raman effect, a kind of spectroscopy complementary to the infrared type since it allows one to observe molecular vibrations often absent (inactive) in the infrared spectrum (Raman 1928).

The Raman effect, theoretically predicted (Smekal 1923) by the Austrian Adolf Smekal (1895–1959) in 1923, and independently developed in Russia by Grigorii Samuilovich Landsberg (1890–1957) and Leonid Isaakovich Mandel'shtam (1879–1944) (Landsberg et al. 1928), is a scattering process of electromagnetic radiation due to the presence of a dipole moment induced in the molecule by the impinging radiation. The theory of Raman scattering, developed by the Czech George Placzek (1905–1955), shows that if radiation of frequency ν_0 is applied to a sample without being absorbed, it gives rise, in addition to the emission of diffused radiation at the same frequency (Rayleigh diffusion effect), to radiations at frequencies $\nu_0 \pm \nu_k$, where ν_k is a vibrational frequency of the molecule (Placzek 1929). The spectral bands appearing in the scattered spectrum at frequencies $\nu_0 - \nu_k$ are called Stokes bands and those at frequencies $\nu_0 + \nu_k$ anti-Stokes bands. For this discovery, Raman obtained the 1930 Nobel Prize for physics. For several years Raman spectroscopy was confined to a few specialized laboratories until in 1952 Harry Lambert Welsch (1910–1984) developed in Canada a mercury vapor lamp called the *Toronto arc* (Welsh et al. 1952) which started the commercial production of Raman spectrographs (the model 81 of the Cary company was famous) and their spread throughout research laboratories. With the availability of laser sources, after the first experiments in 1962 of Sergio Pereira da Silva Porto (1926–1979) with a ruby laser (Porto 1962) and of Boris Peter Stoicheff (1924–2010) with an He-Ne laser (Stoicheff 1963), Raman spectroscopy became one of the most used techniques for the study of molecular structure thanks to the directionality and power of laser radiation. Starting from the 1970s, the availability of ultrafast pulsed lasers of picosecond (1 ps = 10^{-12} s) and femtosecond (1 fs = 10^{-15} s) durations has led to powerful techniques of nonlinear optics, such as CARS (Coherent Anti-Stokes Raman Spectroscopy), that have opened up new techniques for molecular structure determination, ultrafast kinetics, and determination of unstable chemical species with very short lifetimes.

The theoretical treatment of coupled molecular vibrations and rotations started with a 1914 paper of Niels Bjerrum (1914) which showed how one could explain in the framework of the old quantum theory the fine structure of the vibrational bands of carbon dioxide, assuming that the molecule could vibrate and rotate contemporaneously. The work in the Bjerrum paper was soon taken up by the Harvard Ph.D. student Edwin Crawford Kemble (1889–1984). Using Bjerrum's model, Kemble developed a semiclassical treatment (Kemble 1916), obtaining the expression $\nu = \nu_0 + \nu_r$ for the frequencies where $\nu_r = nh/2I$ (I = inertia moment). In a further paper he introduced vibrational anharmonicity, adding higher terms to the potential and also taking into account the interaction effects between rotations and vibrations (Coriolis interaction).

The quantum theory of the vibro-rotational structure of molecular spectra was further developed by Robert Oppenheimer in May 1926 while working in England at Cambridge (Oppenheimer 1925–1927). In this paper, Oppenheimer discussed the calculation of the frequencies and of the vibro-rotational structure of infrared bands of simple small molecules using the formalism of Dirac operators. A similar result was obtained almost at the same time by Lucy Mensing (1926, 1927) using instead the matrix formalism of Werner Heisenberg. In a short period of time, the papers of John Hasbrouck van Vleck (van Vleck et al. 1928; van Vleck 1935) and Paul Ehrenfest, and after those of Gerhard Herzberg and Edgar Bright Wilson Jr. (Wilson and Howard 1936), completed the initial picture. The quantum theory of molecular vibrations reached its full maturity with the applications of group theory developed by Eugene Wigner (1902–1995) (Wigner 1931) and Hermann Weyl (1885–1955) (Weyl 1927, 1928) and with the definition of the vibrational selection rules (Sommerfeld and Schönflies 1928–1929) of Arthur Moritz Schönflies (1853–1928).

In 1941, Edgar Bright Wilson Jr. (1908–1992) developed (Wilson 1941) the theoretical calculation of the normal modes of vibration of complex molecules in terms of internal coordinates (variations of bond lengths, bond angles, etc.) to compute the amplitudes of the harmonic vibrations of the atoms in the molecules. His method of calculation, known as Wilson's GF matrix method, was largely used for the calculation of force constants (Wilson et al. 1955) and of intramolecular potentials (Califano 1976). A significant contribution to the theory of molecular vibrations and to the calculation of force constants from intermolecular potentials arose from the work of Bryce L. Crawford Jr. (1914–2011) and of the research group that he directed at the university of Minnesota (Crawford 1940), which included many outstanding spectroscopists from all parts of the world.

After the Second World War, vibrational spectroscopy had a significant development also in Japan thanks to the work of San-ichiro Mizushima (1899–1983), former pupil of P.J. Debye at the University of Leipzig. Professor of chemical physics at the University of Tokio and director of the Yawata Research Institute (1959–1969), he was Member of the Pontifical Academy from 1961. He and his coworkers supplied original contributions to different fields of chemical physics, including infrared and Raman spectroscopy, rotational isomerism, and conformational analysis of polypeptides and proteins. Mizushima summarized his results in his famous book entitled *Structure of Molecules and Internal Rotation* (Academic Press, New York, 1954). His discovery that the internal rotation of dichloroethane is not free but involves zigzag motions due to the presence of restrictions marked the beginning of a concept that led to the understanding of the α-helix and other molecular conformations. Moreover, this result introduced the idea of flexibility of molecular structures along with the discovery of the "wobbliness" of cyclohexane, involving the inverted isomers of chair and boat molecular structures (Nakagawa and Mizushima 1953).

The pupil of Mizushima, Takehiko Shimanouchi (1916–1980) made in Japan a very important group of vibrational spectroscopy and force constant calculations (Shimanouchi and Suzuki 1961; Shimanouchi 1972) that in a relatively short time became as important as those of the leading groups existing in the United States.

4.5 Rotational Spectroscopy

Conventionally, the part of the electromagnetic field extending from about 1 m to about 1 cm below the infrared region is called the microwave region of the spectrum. In this region fall all absorption bands due to the interaction of rotations of polar molecules with an electromagnetic field. Rotational bands carry important information on molecular inertia moment and molecular bond lengths and angles, and thus rotational spectroscopy represents a convenient technique for the study of molecular structure.

Use of the microwave region of the electromagnetic spectrum developed as a high precision spectroscopic technique for the determination of molecular structure only after the Second World War. Until the first part of the twentieth century, practically no significant progress had been achieved using the long wavelength region of the electromagnetic spectrum due to the absence of proper long wavelength sources and detectors. Experiments on far-infrared radiation were oriented toward what we now call the far infrared (FIR) or THz region by the work of Heinrich Rubens and his colleagues in Berlin using the reststrahlen technique to produce far-infrared radiation. A paper on the refraction of rays of long wavelength in rock salt, sylvine, and fluorite is probably the first significant paper on long wavelength radiation (Rubens 1894). In 1894, Rubens was joined by the American Herbert Hollnagel with whom he built in 1910 a Fourier transform spectrometer for conducting measurements in the far-infrared region of the spectrum. Rubens was also in contact by letter with the Italian physicist Augusto Righi (1850–1920) who performed similar measurements using a homemade oscillator able to produce waves of 10 cm wavelength. However, in 1934, the barrier was overcome by Claud. E. Cleeton and Norman H. Williams (1933) at the University of Michigan who developed a new source of microwave radiation consisting of a magnetostatic generator of waves in the region from 1 to 3 cm. Using this source of microwave radiation, they made the direct observation, for the first time, of a radio frequency change of state in the ammonia molecule (Cleeton and Williams 1934). This achievement opened up the whole new field of microwave spectroscopy and the possibility of using it for the determination of the structure of molecules took a giant step forward.

In 1934, the sources and detectors of microwave radiation that were available were extremely limited and difficult to work with. For this reason, additional progress in microwave spectroscopy was delayed until the end of the Second World War in 1945. By that time, advances in radar technology had provided excellent sources and detectors and a period of rapid progress followed in the field of microwave spectroscopy, including extensive studies of the microwave spectrum of ammonia (Bleaney et al. 1946).

An important step toward the production of microwave sources and detectors was realized in 1920 by Albert Hull at General Electric's Research Laboratories with the invention of the magnetron, the first simple source of microwave radiation. In 1924, the Czech August Žáček (1886–1961), professor at Prague's Charles University, discovered that the magnetron could generate waves of 100 MHz to 1 GHz (Žáček 1924) and at the same time the German Erich Habann student at the university of Jena, while investigating the magnetron for his doctoral dissertation, reached the same conclusion. A more efficient multi-cavity version of the magnetron was reported in 1934 by Theodor V. Ionescu, professor at the Bucharest University, followed in 1937–1940 by a similar multi-cavity magnetron built by the British physicist, Sir John Turton Randall, in collaboration with a team of coworkers for the British and American military radar installations in the Second World War. After the war the situation changed drastically due to the intensive development of microwave electronics and more sophisticated microwave spectrophotometers became available. In 1954, Charles Andrew Burrus and Walter Gordy extended microwave spectroscopy to 770 μm and Ludwig Genzel and Wilfried Eckhardt at the University of Frankfurt (Genzel and Eckhardt 1954) made measurements up to 990 μm.

With the development of the electronically variable klystron oscillators as sources of monochromatic microwave radiation, and with the realization of the Stark modulation technique, a variety of microwave spectrographs became available in several advanced laboratories all over the world, especially in the United States. A large number of important papers were published especially on gases, thanks to a large number of high-level scientists including Walter Gordy (1948) at the Duke University, E. Bright Wilson at Harvard (Wilson 1957), David R. Lide Jr. (1959) at the National Bureau of Standards, Richard J. Myers and William Dulaney Gwinn at the University of California Berkeley (Myers and Gwinn 1954), and

Dudley R. Herschbach (1956; Swalen et al. 1957) and William Klemperer, also at Harvard (Fraser et al. 1986).

Since then, many millions of cavity magnetrons have been manufactured; while some have been for radar, the vast majority have been for microwave ovens. The use in radar itself has dwindled to some extent, as more accurate signals have generally been needed and developers have moved to klystron and traveling-wave tube systems for these needs.

As the first half of the twentieth century drew to a close, the rate of investment in further refinements of the techniques of atomic and molecular structure determination decreased and research oriented in a more technical direction with the construction of the maser and then of its noble son, the laser, the most spectacular invention of the century that radically changed all research fields in physics, chemistry, and biology. The word maser is the acronym of *Microwave Amplification by Stimulated Emission of Radiation.* The stimulated emission of electromagnetic radiation was postulated in a famous paper by Albert Einstein in 1917 (Einstein 1917), proposing that when a photon interacts with an excited atom, it stimulates the emission of a second coherent photon of the same frequency, phase, polarization, and direction of propagation. The occurrence of stimulated emission, at that time called *negative absorption*, was experimentally verified in 1928 by Rudolf Lademburg and his student Hans Kopfermann at the university of Berlin by studying the spectroscopic emission of neon in the visible spectral region, excited by an electric discharge (Kopfermann et al. 1928). In 1939, Valentin Aleksandrovich Fabrikant (1907–1991) at the Lebedev Institute of Moscow supplied new evidence for the existence of stimulated emission and even suggested the possibility of utilizing the process as a method of amplification of electromagnetic radiation. In 1953, the American Joseph Weber (1919–2000) discussed in a public lecture the importance of light amplification and suggested that it could be realized through an inversion of population between two quantum levels. The first inversion of population was observed in 1946 in a nuclear magnetic resonance experiment by Felix Bloch and his coworkers William Webster Hansen and Martin Packard (Bloch et al. 1946a, b) while applying radiofrequency to a sample of water in the presence of a variable magnetic field. By sweeping the magnetic field around the proton resonance frequency, they observed emission of radiofrequencies. An even more convincing NMR experiment was realized by Edward Mills Purcell and Robert Vivian Pound. They proved that if the magnetic field applied to a crystal of lithium fluoride is very rapidly inverted, the magnetization is unable to follow the fast field variation in a sufficiently rapid time, and for a short period a temporary inversion of population is realized. The project was realized after a short while at the Columbia University by Charles Hard Townes (1915–) who developed a broad research program of rotational spectroscopy looking for new microwave sources at wavelengths much lower than those currently utilized in radar technology. After several attempts, Townes realized that, as a source of sub-centimetric waves, it was necessary to use a specific molecule and he selected ammonia (NH_3) the spectroscopic properties of which were well known:

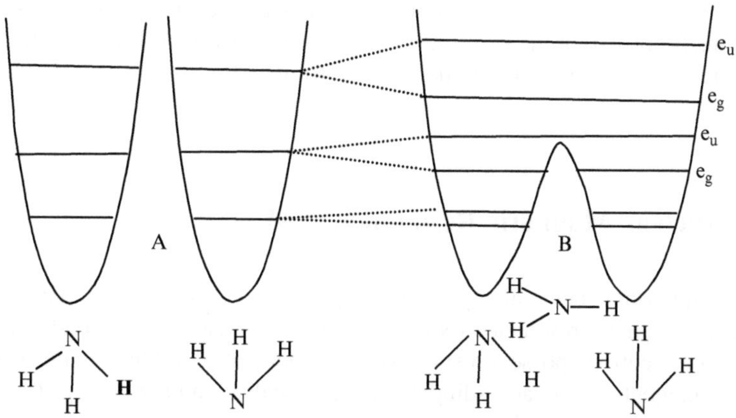

A.Two separated potential surfaces B. Two interacting potential surfaces

With his enthusiasm, he convinced his coworker Herbert Zeiger and James Gordon, a young student working on his Ph.D. thesis, to get involved in this adventure.

Townes selected ammonia since it possesses two equivalent conformations, one with the nitrogen atom above and one below the plane of the three hydrogen atoms. Since the potential barrier to the ammonia inversion is relatively weak, the two conformations interact, giving rise to a splitting of the levels into two components e_g and e_u that increases with increasing quantum number. For the two lowest levels, the energy difference (ε_u–ε_g) is equal to 0.79 cm^{-1}, corresponding to a frequency of 23,830 MHz.

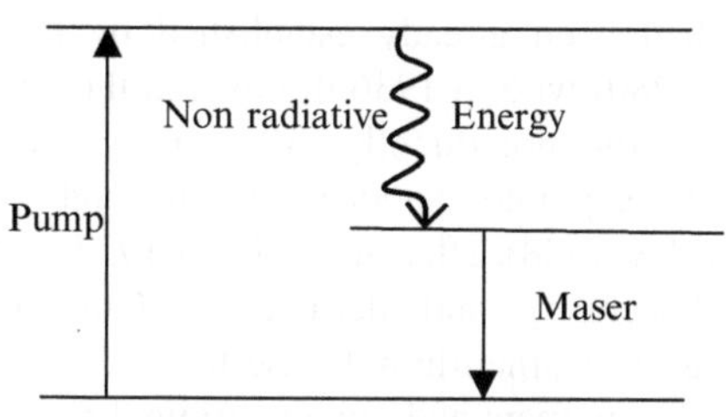

Three levels Energy scheme

Townes and Gordon realized the population inversion in ammonia in April 1953 using a molecular beam instrument in which the molecules crossed an asymmetric quadrupolar electric field which separated the ε_u from the ε_g molecules in a resonant cavity with the 23,830 MHz frequency.

The Townes experiment marked the birth of the first maser, which became officially known through the publication of two consecutive papers in 1954 (Gordon et al. 1954a, b). Almost at the same time the maser was also constructed in Russia by Aleksandr Prokhorov and Nikolay Gennadiyevich Basov at the Lebedev Institute of Moscow (Basov et al. 1954). In the following years, new methods of realizing population inversion were formulated, the most important being the one called the three-level method, which and schematically represented in the figure, proposed for the first time by Basov and Prokhorov (Basov et al. 1955) in 1955 and successively by Bloembergen (1956) in 1956.

4.6 Nuclear Magnetic Resonance (NMR)

Another important chemical physics technique for the study of molecular structures is nuclear magnetic resonance (NMR). All atomic nuclei with an odd number of protons and neutrons possess a spin magnetic moment μ and thus interact with an external magnetic field according to the classical formula $E = -\mu.B$ (Zeeman interaction) thereby experiencing a torque, i.e., a couple of forces trying to align the moment with the field.

NMR measurements are normally made by aligning the nuclear spins with a constant magnetic field and perturbing this alignment with an oscillating field in a perpendicular direction. Nuclear magnetic resonance is based on the fact that the nuclear quantum spin states are separated by relatively weak energies, of the order of the energies of photons in the radio frequency region (megahertz). The separation between the spin levels is a function of the intensity of the applied external magnetic field. By properly changing the radio frequency and the magnetic field intensity, the nuclear separation becomes resonant with the field and the radiofrequency is absorbed.

The nuclear magnetic resonance theory was developed thanks to the work of two Jewish physicists, Isidor Rabi and Felix Bloch at the end of the 1930s, even though the theoretical basis had been already established by the Dutchman Cornelis Jacobus Görter (1907–1980) who in 1936 discovered the paramagnetic relaxation process. Görter actually did not directly discover nuclear magnetic resonance himself, but his suggestions pointed Isidor Rabi in the right direction to do so.

Isidor Isaac Rabi (1898–1988), after his Ph.D. in 1927, spent 2 years in Europe working with Bohr, Heisenberg, and Stern. In the Otto Stern laboratory at the University of Hamburg, he learned how to use the molecular beam technique to measure nuclear magnetic moment and saw the original Stern apparatus in which a magnetic field was used to deflect the atoms. He imported the Stern technique to Columbia University in New York where he worked on his return to America in 1929 and where in 1933 he measured the nuclear magnetic moment of sodium in a molecular beams experiment (Rabi et al. 1933). These measurements were, however, very imprecise and Rabi and his coworker Cohen spent most of their time attempting to improve the measurements with rather disappointing results. In 1937, Görter, passing through New York, went to visit Rabi and suggested that he use radiofrequencies to send the nuclear spin into a higher energy state. This suggestion opened the route to the discovery of nuclear magnetic resonance.

Following Görter's suggestion, Rabi had the brilliant idea to use not a single intense magnetic field as in the original paper of Stern and Gerlach, but three deflecting fields, one intense and two weaker, in order to realize three different deflections on the atoms.

This simple expedient allowed him to solve the hyperfine structure of the resonant signal into $2(2I + 1)$ components where I is the quantum number of the nuclear spin of the atom. Rabi's experiments, which marked the birth of nuclear magnetic resonance, were published in 1937 (Rabi 1937) and 1938 (Rabi et al. 1938) and were followed by a series of important papers on the nuclear spin of different atoms. For these experiments, Rabi was honored in 1944 with the Nobel Prize for physics. The Swiss Felix Bloch (1905–1983) understood the importance of using transitions between nuclear spin states to perform precise measurements. After having studied at the ETH of Zurich he moved to Germany in 1927 to the University of Leipzig, where he obtain the Ph.D. in physics in 1928 with a thesis on the quantum mechanics of electrons in crystals and on the theory of metallic conduction. After the Ph.D., he remained in Germany where he had the opportunity of collaborating with Heisenberg, Pauli, Bohr, and Fermi. In 1933, in order to escape the Nazi race laws, he left Germany and immigrated to the United States where in 1934 he obtained a position at Stanford University. At Stanford, even though trained as a theoretical physicist, he started experimenting in neutron physics and realized that proof of the existence of magnetic moment of spin could be obtained by studying their diffusion from iron samples. In 1939, he succeeded in performing the experiment at the Berkeley cyclotron in collaboration with Luis Walter Alvarez (1911–1988), measuring (Alvarez et al. 1940) the magnetic moment of the neutron with a precision of about 1%. At the end of the Second World War, during which he was involved in the Manhattan project at Los Alamos and then with the development of radar at Harvard, he returned to Stanford where in 1946 he developed, in collaboration with William Webster Hansen (1909–1949) and Martin E. Packard, the technique of electromagnetic measurement of the magnetic moments of atomic nuclei by resonance with a radiofrequency technique (Bloch et al. 1946) contemporaneously discovered by Edward M. Purcell (Purcell et al. 1946) at Harvard.

Edward Mills Purcell (1912–1997) graduated in electronic engineering at the Purdue University in Indiana, after a year in Germany at the Technische Hochschule of Karlsruhe, and joined Harvard University in 1934 where he obtained a Ph.D. in 1938. After 2 years he was hired at the MIT Radiation Laboratory, where he became director of a research group studying new microwave technologies. At the end of the war he went back to Harvard where, in collaboration with Henry C. Torrey and Robert Vivian Pound, he improved the techniques applied to resonance of atomic nuclear spins with radiofrequencies in a magnetic field. The precession frequency ν of the spin magnetic moment μ with respect to the direction of an applied constant magnetic field H_0 is given by the expression $\nu = \mu H_0/Ih$ where h is Planck's constant and I the nuclear spin quantum number that can be an integer or a half integer, depending on the kind of nucleus. From the measurement of the resonance with radiofrequency in the megahertz spectral region, both Bloch's and Purcell's groups obtained the nuclear spin moment of several atomic nuclei with high precision. For these measurements, representing the first true experiments in nuclear magnetic resonance, Bloch and Purcell together obtained the Nobel Prize in physics in 1952. Nuclear magnetic resonance has also made it possible to measure the relaxation times of magnetic polarization, i.e., of the time necessary to reach thermal equilibrium between nuclear spins and their environments. The Dutch American naturalized Nicolaas Bloembergen (1920–) added further contributions to the knowledge of relaxation, and was awarded the 1981 Nobel Prize for physics for his research on laser spectroscopy. Bloembergen worked for his Ph.D. thesis at Harvard on nuclear magnetic resonance, building an NMR instrument in Purcell's laboratory and collaborating with Purcell on the spin relaxation problem (Bloembergen et al. 1948).

The nuclear magnetic field is very sensitive to the field created by the electrons which depends on the type of bonding in the molecule and thus on the relative positions of the atoms. The interest of nuclear magnetic resonance in chemistry is therefore bound to the fact that not only the nuclei but also the electrons interact with an external magnetic field. The electrons, rotating around the direction of the field, give rise to a small field oriented either in the field or in the opposite direction that alters its effect on the nuclei. As a consequence of the shielding effect, the value of the magnetic field B on a given nucleus will be greater or smaller than the value of the magnetic field B_0 in vacuum by a small amount σ, such that $B = B_0(1 - \sigma)$. Different nuclei in a polyatomic molecule will therefore experience slightly different fields and will be in resonance at different radiofrequencies. Thus by continuously sweeping the radiofrequency applied to the sample, one obtains a spectrum with peaks characteristic of the different nuclei with a nonzero spin present in the molecule. Since the resonance frequency varies with the intensity of the applied magnetic field, it would be difficult to compare different results collected in different laboratories without using a common reference standard. This has given rise to the practice of using relative quantities called chemical shifts δ, defined as the difference between the resonance frequency ν of a given nucleus and the normalized resonance frequency ν_r of a reference nucleus I, given by $\delta = (\nu - \nu_r)/\nu_r$.

The chemical shift theory was developed by Norman Foster Ramsey (1915–) (Ramsey 1950), Warren Proctor (Proctor and Yu 1950), and W.C. Dickinson (1950) in 1950. Chemical shift became an important and widespread analytical tool in organic chemistry when, in 1952, the first commercial instrument for nuclear magnetic resonance, working at a frequency of 30 MHz, was produced by the Varian Company. At the end of the 1950s, Dr. James Schoolery of Varian Associates started an information campaign in all laboratories involved with organic chemistry to convince researchers of the great possibilities offered by the new technique. In 1961, Varian developed a new NMR spectrograph, the A-60 model, which became the first routine instrument in organic analysis. With the development of superconducting magnets and the complete automation of the Fourier transform technique, very high resolution has been achieved in NMR spectroscopy. New powerful 300–750 MHz spectrographs are now possible. NMR represents today the most important technique available to organic chemists for the identification of molecular species.

Nuclear magnetic resonance has more recently become of great interest in physics, chemistry, and biology, with the realization of two- and three-dimensional imaging of NMR outputs.

4.7 Electron Spin Resonance Spectroscopy EPR

The study of nuclear spin interaction with a magnetic field found an important extension with the development of the technique of electron spin resonance spectroscopy (EPR) dealing with electron spin interaction with an external magnetic field. The basic principle of pulsed EPR is similar in many respects to NMR spectroscopy, the only important difference being associated with the relative sizes of the magnetic interactions and of the relaxation rates which are orders of magnitudes larger in EPR than NMR.

In the presence of an applied external magnetic field B_0, electron spins are characterized by two quantum mechanical states, corresponding to a spin orientation parallel or antiparallel to the field direction, the parallel state occurring at lower energy than the antiparallel one. An unpaired electron can jump from the lower to the higher state by absorbing electromagnetic radiation of energy $\Delta E = g_e \mu_B B_0$ where g_e is the electron's Zeeman factor and μ_B is the Bohr magneton.

In a magnetic field, an electron spin also experiences torque which causes its magnetic moment to precess around the magnetic field. The precession frequency is known as the Larmor frequency ω_L given by the equation $\omega_L = -\gamma B_0$, where γ is the gyromagnetic ratio and B_0 the magnetic field. This process results in a net magnetization, which is the vector sum of all magnetic moments in the sample, parallel to the direction (conventionally the z-axis in a reference system) of the magnetic field. EPR experiments usually use a microwave resonator designed to create a linearly polarized microwave field B_1, perpendicular to the applied magnetic field B_0.

Electron spin resonance spectroscopy, often called electron paramagnetic resonance spectroscopy (EPR), was discovered in 1944 by the Soviet physicist Yevgeny Konstantinovich Zavoisky (1907–1976) observing a radiofrequency absorption line from a solid sample of $CuCl_2 \cdot 2H_2O$ (Zavoisky 1945a, b, 1946). In 1941, Zavoisky had also realized, well before Felix Bloch and Edward Mills Purcell, the first experiment of nuclear magnetic resonance, but did not publish his results that he considered too unstable and not reproducible, owing to the lack of spatial homogeneity of the magnetic field. Zavoisky instead focused his attention in 1943 on the study of electron spins in a magnetic field, a technique that is less demanding than NMR for the homogeneity of magnetic field, although requiring a much more sensitive detection system. Zavoisky used in his experiments a very sensitive electronic technique of grid current and realized a further improvement by adding a small a.c. magnetic field to the main static magnetic field that dramatically increased the detection sensitivity and allowed easy amplification of the resonance signal. In 1944, he detected EPR signals in several salts, including hydrous copper chloride ($CuCl_2{\cdot}2H_2O$), copper sulfate, and manganese sulfate. His revolutionary results were first not accepted even by Soviet scientists, until the doubts were dispersed when Zavoisky, visiting Moscow, assembled an EPR spectrometer from scratch and reproduced his results there. In May 1944, after the presentation of his Ph.D. thesis at the University of Moscow, Zavoisky discussed his results with Frenkel who became very interested in this new phenomenon and in the space of few days worked out the theory of paramagnetic resonance in condensed matter, deriving for the imaginary part of the paramagnetic susceptibility the expression

$$\chi = \frac{2{\nu_0}^2 \nu \tau}{\left({\nu_0}^2 - \nu^2\right)^2 \tau^2 + 4\nu^2} \chi_0$$

where ν_0 is the Larmor precession frequency, τ the relaxation time, and χ_0 the static susceptibility. This equation created serious problems for Zavoisky who was unable to fit his data for some time until he succeeded in obtaining better data on the basis of new measurements at higher frequencies (approximately 0.5 and 1.0 GHz).

At almost the same time, electron spin resonance was discovered by the English physicist Brebis Bleaney (1915–2006), professor at the University of Oxford who during the Second World War had gained considerable experience in radar technology and in particular in microwave generation with Klystrons tubes.

After some early experiments on magnetic salts showing clear but too broad and structureless resonances, Bleaney, together with Roger Penrose, and a student, Betty Plumpton, who later became his wife in 1949, succeeded in measuring spin–lattice relaxation times at low temperature in crystals with isolated magnetic sites diluted in a host lattice, obtaining beautiful spectra showing beautiful hyperfine structures associated with the interaction between the electronic and nuclear spins (Bleaney and Penrose 1946). From this time, Bleaney was able to explore a wide range of materials containing transition metals and rare earths. The theoretical interpretation of these results was enhanced by close collaboration with the group directed by Professor M.H.L. Pryce, who became one of the leading scientists in the field of magnetic resonances and Anatole Abragam professor of "magnétisme nucléaire" at the Collège de France in Paris. Bleaney and Abragam (1914–2011) together wrote a classic book *Electron Paramagnetic Resonance of Transition Ions* (Abragam and Bleaney 1970).

After 1946, rapid exploitation of the EPR technique was achieved thanks to the availability of efficient microwave systems realized as a consequence of the research on electronics and microwave generation for radar equipment for military use. Of central importance in the rapid growth of research in this field were the discovery of electron spin echo and the realization of pulsed electron resonance instruments in which the alignment of the net magnetization vector of the electron spins is perturbed by applying a short oscillating field, usually a microwave pulse. This creates a time-resolved microwave signal whose Fourier transform yields the EPR spectrum in the frequency domain. In 1958, Richard Blume at the Watson Laboratory, Columbia University, reported the first observation of an electron spin echo (Blume 1958) following the discovery of the spin echo realized by Erwin Hahn in 1950 (Hahn 1950) in the case of nuclear magnetic resonance. In the same year, James P. Gordon and Klaus Bowers reported the detection of microwave spin echoes from dopants in silicon crystals (Gordon and Bowers 1958).

During the 1960s, a considerable amount of pioneering early pulsed EPR research was conducted by the group of William B. Mims at the Bell Laboratories (Bowers and Mims 1959) after the first observation of electron spin echo envelope modulation (ESEEM) made in 1961 by Mims, Nassau, and McGee (1961). Pulsed electron nuclear double resonance (ENDOR) was also invented in 1965 by Mims. ESEEM and pulsed ENDOR are important for the study of the coupling between nuclear and electron spins.

In the 1980s, the arrival of the first commercial pulsed EPR and ENDOR spectrometers in the X-band frequency range led to fast growth in the field. In the 1990s, in parallel with the upcoming high-field EPR, pulsed EPR and ENDOR became new, fast advancing magnetic resonance spectroscopy tools and the first commercial pulsed EPR and ENDOR spectrometers appeared on the market.

References

Abney W, Festing ER (1881) The influence of the atomic grouping in the molecules of organic bodies on their absorption in the infra-red region of the spectrum. Philos Trans Roy Soc 172:887–918

Abragam A, Bleaney B (1970) Electron paramagnetic resonance of transition ions. Oxford University Press, Oxford/England

Alvarez LW, Bloch F (1940) A quantitative determination of the neutron moment in absolute nuclear magnetons. Phys Rev 57:122

Barkla CG (1904) Energy of secondary Röntgen radiation. Philos Mag 7:543–560

Basov NG and Prokhorov AM (1954) First Russian ammonia maser; in Russian Zh. Eksperim i Teor Fiz 27:431

Basov NG, Prochorov AM (1954) Possible methods of obtaining active molecules for a molecular oscillator. Sov JETP 27:431–438, 28, 249, 1955

Bellamy L (1958) Infrared spectra of complex molecules. Wiley, New York

Bijvoet JM (1949) Phase determination in direct Fourier-synthesis of crystal structures. Proc K Ned Akad Wet B52:313–314

Bijvoet JM (1951) X-ray analysis of crystals. Butterworths, London

Bjerrum N (1914) The ultra-red spectra of gases. The configuration of the carbon dioxide molecule and the laws of intramolecular forces. Ber Deutch Phys Ges 116:737–753

Bleaney B, Penrose RP (1946) Ammonia spectrum in the 1 cm. Wave-length region. Nature 157:339–340

Bloch F, Hansen WW, Packard ME (1946a) Nuclear induction. Phys Rev 70:460–473

Bloch F, Hansen WW, Packard M (1946b) The nuclear induction experiment. Phys Rev 70:474

Bloembergen N, Purcell EM, Pound RV (1948) Relaxation effects in nuclear magnetic resonance absorption. Phys Rev 73:679

Bloembergen N (1956) Proposal for a new type solid state maser. Phys Rev 104:324

Bluhm MM, Bodo C, Dintzis HM, Kendrew JC (1958) The crystal structure of myoglobin. IV. A Fourier projection of sperm-whale myoglobin by the method of isomorphous replacement. Proc Roy Soc Lond A246:369

Blume RJ (1958) Electron spin relaxation times in sodium-ammonia solutions. Phys Rev 109:1867
Bowers KD, Mims WB (1959) Paramagnetic relaxation in nickel fluosilicate. Phys Rev 115:285
Bragg WH, Bragg WL (1913a) The reflection of X-rays by crystals. Proc Roy Soc A88:428
Bragg WH, Bragg WL (1913b) The structure of the diamond. Proc Roy Soc A89:277–291
Bragg WL (1913a) The diffraction of short electromagnetic waves by a crystal. Proc Camb Philos Soc 17:43–57
Bragg WL (1913b) The structure of some crystals as indicated by their diffraction of X-rays. Proc Roy Soc Lond A 89:248–277
Bragg WL (1914) The analysis of crystals by the X-ray spectrometer. Proc Roy Soc A89:468–489
Brockhouse BN, Stewart AT (1958) Normal modes of aluminum by neutron spectrometry. Rev Mod Phys 30:236–249
Califano S (1976) Vibrational states. Wiley, London
Califano S, Schettino V, Neto N (1981) Lattice dynamics of molecular crystals, Lecture notes in chemistry. Springer, New York
Cleeton CE, Williams NH (1933) A magnetostatic oscillator for the generation of 1 to 3 cm waves. Phys Rev 44:421
Cleeton CE, Williams NH (1934) Electromagnetic waves of 1.1 cm wave-length and the absorption spectrum of ammonia. Phys Rev 45:234
Crawford BL (1940) Infra-red and Raman spectra of polyatomic molecules XII. Methyl acetylene. J Chem Phys 8:526
Davidson WL, Morton GA, Shull CG, Wollan EO (1947) Neutron diffraction analysis of NaH and NaD, vol 842, Report number MDDC. Oak Ridge, US Atomic Energy Commission (AEC), April 28
Debye P, Scherrer P (1916) Interferenzen an regellos orientierten Teilchen im Röntgenlicht I. Phys Z 17:277–83
Debye P, Scherrer P (1917) Interferenzen an regellos orientierten Teilchen im Röntgenlicht III (Über die Konstitution von Graphit und amorpher Kohle). Phys Z 18:291
Dickinson WC (1950) Hartree computation of the internal diamagnetic field for atoms. Phys Rev 80:563
Einstein A (1917) Zur Quantentheorie der Strahlung. Phys Z 18:121–128
Fraser GT, Lovas FJ, Suenram RD, Nelson DD, Klemperer W (1986) Rotational spectrum and structure of CF_3H–NH_3. Chem Phys 84:5983
Friedrich W, Knipping P, Laue M (1912) Interferenz-Erscheinungen bei Röntgenstrahlen, Eine quantitative Prüfung der Theorie für den Interferenz-Erscheinungen bei Röntgenstrahlen. Sitzungsberichte Bayer Akad Wiss 303–322; reprint in: Ann Phys (1913) 41:971–988
Genzel L, Eckhardt W (1954) Spektraluntersuchungen im Gebiet um 1 mm Wellenlänge. Z Phys 139:578–591
Gordon JP, Zeiger HJ, Townes CH (1954a) Molecular microwave oscillator and new hyperfine structure in the microwave spectrum of NH_3. Phys Rev 95:282
Gordon JP, Zeiger HJ, Townes CH (1954b) The maser–new type of microwave amplifier, frequency standard, and spectrometer. Phys Rev 99:1264
Gordon JP, Bowers KD (1958) Microwave spin echoes from donor electrons in silicon. Phys Rev Lett 1:368–369
Gordy W (1948) Microwave spectroscopy. Rev Mod Phys 20:668–717
Hahn E (1950) Spin echoes. Phys Rev 80:580
Hartley WN, Huntingdon AK (1879) Spectra of organic compounds. Proc Roy Soc 28:233
Hauptman H, Karle J (1956) Structure invariant and semivariants for non-centrosymmetric space groups. Acta Cyst 9:45–55
Herschbach DR (1956) Internal barrier of propylene oxide from the microwave. Spectrum, I. J Chem Phys 25:358
Hull AW (1917) A new method of X-ray crystal analysis. Phys Rev 10:661–696
Kemble E (1916) The distribution of angular velocities among diatomic gas. Phys Rev 8:689

Kopfermann H, Ladenburg R (1928) Bildung und Vernichtung angeregter Atome. Z Phys 48:26–50

Landsberg GS, Mandelstam LI (1928) Über die Lichtzerstreuung in Kristallen. Z Phys 50:769–780

Laue M (von) Ph.D., Universität Berlin (1903) Über die Interferenzerscheinungen an planparallelen Platten. Ann Phys 318:163–181, 1904

Laue M (1912) Eine quantitative Prüfung der Theorie für die Interferenzerscheinungen bei Röntgenstrahlen. Berichte Bav Acad Sci 363–373 reprinted in Ann. Phys. (1913), 41, 989-1002

Leomte LJ (1928) Spectre Infrarouge. Les Presses Universitaires, Paris

Lide DR Jr (1959) Microwave spectrum of trimethylarsine. Spectrochim Acta 15:473–476

Mensing L (1926) The rotational-oscillation bands according to quantum mechanics. Z Phys 36:814

Mensing L (1927) Zur Theorie des Zusammenstoßes von Atomen mit langsamen Elektronen. Z Phys 45:603–609

Mims WB, Nassau K, McGee JD (1961) Spectral diffusion in electron resonance lines. Phys Rev 123:2059–2069

Myers RJ, Gwinn WD (1954) The microwave spectra of gases. Annu Rev Phys Chem 5:385–394

Nakagawa I, Mizushima S (1953) The assignments of the Raman and infrared frequencies of 1,2-Dichloroethane observed in the gaseous, liquid and solid states. J Chem Phys 21:2195–2198

Oppenheimer R (1925–1927) On the quantum theory of vibration-rotation bands. Proc Camb Philos Soc 23:327–335

Patterson AL (1934) A Fourier series method for the determination of the components of interatomic distances in crystals. Phys Rev 46:372–376

Peerdeman AF, van Bommel AJ, Bijvoet JM (1951) Determination of the absolute configuration of optically active compounds by means of X-rays. Nature 168:271

Perutz MF (1956) Isomorphous replacement and phase determination in non-centrosymmetric space groups. Acta Crystallogr 9:867

Perutz MF, Rossmann MG, Cullis AF, Muirhead H, Will G, North ACT (1960) Structure of haemoglobin. A three-dimensional Fourier synthesis at 5.5Å resolution, obtained by X-ray analysis. Nature 185:416–422

Placzek G (1929) Zur Theorie des Ramaneffekts. Z Phys 58:585–594

Porto SPS, Wood DL (1962) Ruby optical maser as a Raman source. J Opt Soc Am 52:251–152

Proctor WG, Yu FC (1950) The dependence of a nuclear magnetic resonance frequency upon chemical compounds. Phys Rev 77:717

Puluj I (1899) Radiant elektrode matter and the so called fourth state. Lond Phys Mem 1:233–331

Purcell EM, Torrey HC, Pound CV (1946) Resonance absorption by nuclear magnetic moments in a solid. Phys Rev 69:37–38

Rabi II, Cohen VW (1933) The nuclear spin of sodium. Phys Rev 43:582

Rabi II (1937) Space quantization in a gyrating magnetic field. Phys Rev 51:652–54

Rabi II, Zacharias JR, Millman S, Kusch P (1938) A new method of measuring nuclear magnetic moment. Phys Rev 53:318

Raman CV, Krishnan KS (1928) A new type of secondary radiation. Nature 121:501

Ramsey NF (1950) Magnetic shielding of nuclei in molecules. Phys Rev 78:699–703

Rasetti F (1940) Scattering of thermal neutrons by crystals. Phys Rev 58:321–325

Röntgen WC (1895) Über eine neue Art von Strahlen. Ann Phys 300:1–11

Röntgen WC (1896) On a new kind of rays. Nature 53:274–276

Rubens H (1894) Zur Dispersion der ultraroten Strahlen im Fluorit. Ann Phys 51:381–391

Shimanouchi T, Suzuki I (1961) Force constants of chloro- and bromomethanes. J Mol Spectrosc 6:277–300

Shimanouchi T (1972) Tables of Molecular Vibrational Frequencies Consolidated Volume II J Phys Chem Ref Data 6, 3:993–1102

Smekal A (1923) Zur Quantentheorie der Dispersion. Die Naturwiss 43:873–875

Sommerfeld A (1912) Über die Beugung der Röntgenstrahlung. Ann Phys 38:473–506

Sommerfeld A (1913) Unsere gegenwärtigen Anschauungen über Röntgenstrahlung. Vortrag bei der Versammlung des Vereins zur Förderung des Unterrichtes in der Mathematik und den Naturwissenschaften, München. Gehalten Pfingsten 1913. Die Naturwissenschaften, 1:705–713

Sommerfeld A (1915) Über das Spektrum der Röntgenstrahlung. Ann Phys 46:721–748

Sommerfeld A, Schönflies A (1928–1929) Jahrbuch der bayerischen Akademie der Wissenschaften. Schoenflies, Spencer, pp 86–87

Stoicheff BP (1963) Theory of stimulated Brillouin and Raman scattering in gases. Phys Lett 7:186

Swalen JD, Herschbach DR (1957) Internal barrier of propylene oxide from the microwave spectrum, I. J Chem Phys 27:100–108

van Vleck JH, Hill EL (1928) On the quantum mechanics of the rotational distortion of molecular spectral terms. Phys Rev 32:250–272

van Vleck JH (1935) The rotational energy of polyatomic molecules. Phys Rev 47:487–494

Welsh HL, Crawford MF, Thomas TR, Love GR (1952) Raman spectroscopy of low-pressure gases and vapors. Can J Phys 30:577

Weyl H (1927) Quantenmechanik und Gruppentheorie. Z Phys 46:1–46

Weyl H (1928) Gruppentheorie und Quantenmechanik. Verlag S. Hirzel, Leipzig

Wigner EP (1931) Gruppentheorie und ihre Anwendungen auf die Quantenmechanik der Atomspektren. Verlag, Braunschweig

Wilson EB Jr, Howard JB (1936) The vibration rotation energy levels of polyatomic molecules.I. Mathematical theory of semirigid asymmetrical top molecules. J Chem Phys 4:260–268

Wilson EB Jr (1941) Some mathematical methods for the study of molecular vibrations. J Chem Phys 9:76–84

Wilson EB Jr, Decius JC, Cross PC (1955) Molecular vibrations. McGraw-Hill, New York

Wilson EB Jr (1957) On the origin of potential barriers to internal rotation in molecules. PNAS 43:816–820

Žáček A (1924) Nová metoda k vytvorení netlumenych oscilací [New method of generating undamped oscillations], Časopis pro pěstování matematiky a fysiky [Journal for the Cultivation of Mathematics and Physics] 53:378–380

Zavoisky E (1945a) Relaxation of liquid solutions for perpendicular fields. J Phys USSR 9:211–216

Zavoisky E (1945b) Spin-magnetic resonance in paramagnetics. J Phys USSR 9:245

Zavoisky E (1946) Spin magnetic resonance in the decimetre-wave region. J Phys USSR 10:197–198

Chapter 5
The Electron and Atomic Structure

Particles attract one another by some force, which in immediate contact is exceedingly strong, at small distances performs the chemical operations, and reaches not far from the particles with any sensible effect.
(Isaac Newton, Query 31, Opticks)

5.1 Birth of the Electron

In the nineteenth century, chemists and physicists had a completely different image of electricity. The chemists, in contact with a discontinuous and discrete world made of atoms and molecules that they handled and combined together at will in their laboratories, conceived electricity as made of charges indissolubly bound to matter and responsible for the affinities binding together the atoms in the molecules. Volta's pile, which at first appeared to be just an instrument to break molecules into pieces, soon led to a new theoretical paradigm. The Arrhenius theory of electrolytic dissociation had in fact clearly proved that even the electrical charges of ions occurred in a discrete and discontinuous form, and in 1873 Maxwell had shown that the cations all carried a positive electrical charge, always a multiple of the same quantity, and that the same situation occurred for anions but with negative charges. The fact that ions carried a "definite quantity" of electrical charge had been reinforced by von Helmholtz in a famous *Faraday Lecture* held on 5 April 1881 at the Chemical Society in London (Helmholtz 1881):

> Thus established, Faraday's law tells us that through each section of an electrolytic conductor we have always equivalent electrical and chemical motion. The same definite quantity of either positive or negative electricity moves always with each univalent ion, or with each unit of affinity of a multivalent ion, and accompanies it during all its motions through the interior of the electrolytic fluid. This quantity we may call the electric charge of the atom.

However, the proposition that electricity could consist of particles looked like heresy to the physicists, used to discussing the phenomena of electrical conduction

S. Califano, *Pathways to Modern Chemical Physics*,
DOI 10.1007/978-3-642-28180-8_5,

in terms of a continuous fluid and to master abstract concepts as waves, fields, and potentials. At the end of the century, the idea of the corpuscular nature of electricity had, however, already entered the physics world through the study of electrical discharges in rarefied gases at low pressure, a phenomenon known for some time and normally presented to show the properties of electricity in elegant soirées to ladies and gentlemen. This physical effect was discovered in 1838 by Michael Faraday who found that a flux of electrical current is observed by applying a potential difference of thousands of volts to the metallic electrodes of a void glass tube.

In 1855, Heinrich Geissler (1814–1879), a glass blower who possessed a private building laboratory for physical instruments in Bonn, invented the mercury displacement pump able to achieve sufficiently high vacuum in glass tubes. In 1858 Julius Plucker (1801–1868), professor of physics at Bonn, used Geissler's pump to produce the vacuum in a glass tube equipped with electrodes at each end. He noticed that when sufficient potential was applied to the electrodes, a greenish luminescence occurred at the cathode extending in the tube as the pressure decreased and observed that it could be shifted along the tube by means of a magnet (Plucker 1858). In 1869, Johann Wilhelm Hittorf (1824–1914), a pupil of Plucker who contributed from 1853 to 1859 to the study of electrolytic conduction, concluded that during the electrical discharge in rarefied gases some rays of unknown nature, that he named cathode rays, were emitted from the cathode and proceeded along a straight line toward the anode, giving rise to an intense fluorescence on the glass in the anodic region (Hittorf 1869).

For several years the nature of the cathode rays remained unknown. Several physicists continued to believe that they were of a wave nature until in 1879 the English chemist and spectroscopist William Crookes (1832–1919) performed a series of experiments with magnetic fields, concluding that they were made of negatively charged particles that he called "molecules" (Crookes 1879a, b).

If the cathode rays were charged particles, they should feel the effect of an electric field. New experiments made by the German physicist Heinrich Hertz (1857–1894) did not, however, confirm this hypothesis. By letting the cathode rays go through an electric field created by two metallic plates inserted in the tube, he noticed that these were not deflected as expected for charged particles. Furthermore, Hertz (Hertz 1892) proved in 1892 that one could not stop the fluorescence in the anodic region by inserting a thin metallic foil in the beam's path inside the tube, a fact that seemed incomprehensible if the beams were made of particles. Hertz's research was continued (Lénárd 1893) by his assistant Philipp Eduard Anton von Lénárd (1862–1947), a fervent Nazi and a violent anti-Semite, who in 1905 obtained the Nobel Prize in physics for his research on the cathode beams. Lénárd constructed a glass tube with a thin metallic foil soldered at one end and showed that not only did it hold the vacuum well but also that the cathode rays went easily through, a fact inconceivable for a physicist of the time if these were solid objects such as particles (Lénárd 1894). Even Eugen Goldstein, who in 1886 performed a series of experiments on cathode rays, was convinced that they were of a wave nature.

The particle-wave nature of cathode rays was also the object of the research of the Italian physicist Augusto Righi (1850–1920), professor of physics at the University of Bologna, who studied the trajectories followed by the cathode rays (Righi 1890) using an electrometer to measure their negative charge (Righi 1896). The fact that the cathode rays were made of particles with a negative charge was, however, definitively proved by the French physicist Jean-Baptiste Perrin (1870–1942), Nobel Prize in physics in 1926, who used a cathode tube in which the rays, crossing a metallic cylinder with a thin slit, charged an electroscope (Perrin 1897).

In 1897, Joseph John Thomson (1856–1940), professor of physics at the Cavendish Laboratory of Cambridge, took up the study of the mysterious cathode rays. At the beginning of his experiments, Thomson was convinced that the cathode rays were corpuscles made of fragments of atoms. Later, however, he considered the cathode rays as vortices of ether, the support that the electromagnetic theory envisaged at that time for wave propagation.

Thomson devoted himself for some times to the theoretical treatment of the dynamics of vortices (Thomson 1883, 1885), influenced by an old hypothesis of Helmholtz on atomic structure. The possibility that the atoms were vortices of ether was also maintained by his friend and roommate at Cambridge, Joseph Larmor (1857–1942).

If the cathode rays consisted of charged particles they should necessarily feel the presence of a magnetic field. The argument in favor of the particle hypothesis was, however, in contrast with the Hertz-Lénárd experiments that seemed to support the opposite idea, namely, that cathode rays were not deflected by electric fields.

Since a moving charged particle is not affected by the presence of an electric field only when located inside a metallic conductor, the only possible explanation for the Hertz experiment was that the residual gas molecules left in the tube, once ionized by the cathode rays, formed an electric shield. Pushing to the maximum the vacuum inside the tube, Thomson succeeded in proving that this hypothesis was correct and that at very high vacuum, when the shielding effect due to the residual molecules was reduced and eventually vanished, deviation of the cathode rays from the straight path occurred regularly.

By studying their deviation in both electric and magnetic fields, Thomson finally succeeded in computing the ratio e/m between the charge and the mass of the cathode rays, particles that he continued to call "corpuscles." He proved that the mass was about 1/1,000 the mass of the hydrogen atom. On 30 April 1897, at the Royal Institution theater in London, J.J. Thomson told a selected audience of ladies and gentlemen that he had discovered a particle 1,000 times smaller than an atom. As Thomson wrote (Thomson 1897):

> Could anything at first sight seem more impractical than a body which is so small that its mass is an insignificant fraction of the mass of an atom of hydrogen?

In 1881, George Johnstone Stoney (1826–1911) suggested the name *electron* for these negatively charged particles and that was soon accepted (Stoney 1881). From that moment, the electron entered the scientific world as the first-known elementary particle and as a basic constituent of matter.

At that point it was evident that if electrical discharges were able to extract negative particles from the atoms of a gas, an equivalent number of positive particles had to be produced. The existence of these particles was discovered in 1886 by Eugen Goldstein (1850–1931), pupil of Von Helmholtz at the Potsdam Observatory in Berlin. Goldstein presented his experiments on these strange rays, that he named *Kanalstrahlen* (canal rays), in several papers published in the Monatsberichte of the Berliner Academy of Sciences (Goldstein 1876, 1880, 1886). It was, however, Wilhelm Wien (1864–1928) who in 1898 proved that the canal rays were made of positively charged particles with a mass of the same dimension as that of the hydrogen atom (Wien 1898). The instrument devised by Wien to study these particles was the origin of the mass spectrograph, an instrument that later became one of the most powerful for the study of the composition of molecules.

In 1907, J.J. Thomson returned to the Wien experiments, improving the instrument, and measuring the ratio e/m for the particles H^+ and H_2^+ (Thomson 1907b). Finally, Lord Rutherford in 1919 continued with his pupil Aston the experiments of Wien and Thomson, proving that the mass of these particles was a thousand times larger than that of the electron.

5.2 Models of the Atom

The discovery of the electron represented a fundamental step in the development of the structure of matter. The indivisible atom of the Greek philosophers, whose existence had given rise to so many discussions and controversies during the nineteenth century, was now known to be made of particles of dimension smaller than that of the atom and, in addition, they were electrically charged. Electricity, long considered as a continuous fluid, now acquired a particle structure and the interaction between opposite charges became the basic interaction in the interpretation of atomic structure. Soon, models of the atom started to flourish, filling the scientific literature. A model of the atom had been already proposed in 1867, before the discovery of the electron, by Lord Kelvin (William Thomson) taking up an old paper of Helmholtz on the dynamics of vortices (Helmholtz 1858). Helmholtz's idea was that filaments of an ideal incompressible and nonviscous fluid, rolled up in the form of rings in vortex motion, would be stable and last to infinity. Of course in air and water, which are nonideal fluids, the vortices are rapidly dissolved. Ether, however, was considered a true ideal fluid and therefore vortices in the ether could possess an infinite life – nothing better for resuming the old Prout's theory on primordial matter.

Lord Kelvin started to become interested in vortices after having assisted to a lecture by his friend Peter Tait. Peter Guthrie Tait (1831–1901), professor of physics at the University of Edinburgh, was a mathematical physicist who, in addition to developing quaternion physics, had worked for long time on the vortex theory (Tait 1877, 1884, 1885). In order to prove experimentally the validity of Helmholtz's vortex theory, he even built a machine made of two receivers, each equipped with a rubber diaphragm that once compressed produced beautiful smoke rings in vortex rotation in air. These rings behaved as if made of rubber: colliding with one another they would bounce without breaking and if one tried to break them with a knife they would simply roll up around the blade. Lord Kelvin was enthused with the idea of the vortices and in the period 1867–1900 published a series of papers on the matter. Since he had always been an adversary of atoms as material objects, he ventured with great enthusiasm to represent them as vortices in the ether (Thomson 1869, 1875).

The vortices theory of the atom had a short life but the fact that a scientist of the stature of Lord Kelvin had adopted it and also that Maxwell, even without believing it without doubt, had considered it as "a marvelous example of creative interaction between mathematics and physics" excited the interest of several mathematicians especially in England, leading to important developments in hydrodynamic theory. In 1902, however, Lord Kelvin completely abandoned the idea of the vortices and proposed a new model that regarded the atom as made of a positive charge balanced by a negative one. In this model, Lord Kelvin took up again the ideas developed more than a century before by Franz Maria Ulrich Theodosius Aepinus (1724–1802), a German physicist and astronomer at the court of Catherine the Great in Russia, who in a treatise of 1759 had been the first to connect electricity and magnetism, developing a theory of the electrical fluid made of very small immaterial particles filling the space. According to him, matter was made of particles permeated by the electric fluid and of particles free of it (Aepinus 1759). Particles filled with electric fluid would repel each other but would be attracted by those free of fluid with which they would easily associate (Thomson 1901, 1902):

> According to the well-known doctrine of Aepinus, commonly referred to as the one-fluid theory of electricity, positive and negative electrifications consist in excess above, and deficiency below, a natural quantum of a fluid, called the electric fluid, permeating among the atoms of ponderable matter. Portions of matter void of the electric fluid repel one another; portions of the electric fluid repel one another; portions of the electric fluid and of void matter attract one another. My suggestion is that the Aepinus' fluid consists of exceedingly minute equal and similar atoms, which I call electrions, much smaller than the atoms of ponderable matter; and that they permeate freely through the spaces occupied by these greater atoms and also freely through space not occupied by them.

Lord Kelvin's idea of atoms balancing positive and negative charges was accepted by J.J. Thomson (Thomson 1904). Thomson's model was made of a sphere of uniform positive charge of the dimension of the atom in which the electrons were inserted as seeds in a watermelon or as raisins in a *plum-pudding*, the typical English Christmas cake. The electrons occupied equilibrium positions stabilized by the balance between their repulsion and by the attractive interaction with that part of positive charge internal to their position. Up to a given number, the electrons were disposed on a plane and for greater numbers on ring structures. In this pudding of positive charge, the electrons would oscillate with fixed frequencies around their equilibrium positions, emitting or absorbing the spectral lines characteristic of the atoms. Thomson concluded on the basis of complex calculations that few electrons would form triangular, tetrahedral, etc., structures, whereas after eight electrons concentric structures would be formed. In 1878, the American Alfred Marshall Mayer (1836–1897) of the University of Maryland, in the attempt to prove how atoms were organized in the molecules, had the idea of immersing in a water container a series of magnetized needles mounted upon corks, with their south poles upward. Hanging at the center of the receiver a powerful steel magnet with its north pole oriented toward them (Mayer 1878), he discovered that the needles were arranged in concentric circles forming regular structures. Three magnets would form a triangle and four would arrange themselves at the corners of a square. Five either formed a square with one magnet in the center, or arranged themselves into a pentagon. Six would form a pentagon with one in the center, or arrange themselves three on a side in the form of an equilateral triangle. Seven magnets would form a hexagon with a magnet at the centre and eight magnets would be arranged either in the form of a hexagon with two magnets at the center or alternatively of a heptagon with a center magnet. For higher combinations of nine or more needles, represented in the figure below, Mayer discovered that the configurations of the floating magnets may be divided into primary, secondary, tertiary, etc., classes, and that the stable configurations of a lower class form the basis of the succeeding ones. When there are two or more forms of arrangement, some are more stable than others, and only the most stable would survive in higher classes.

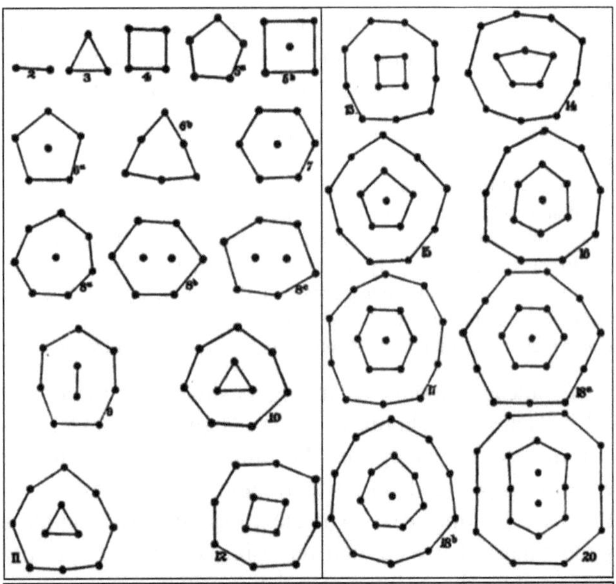

Mayer's picture of the arrangement of the magnetized needles

From 9 to 18 magnets the stable configuration had 2 central magnets and 2 concentric rings. From 19 magnets on, 3 concentric rings were formed and for a larger numbers 4, 5, and so on rings would be formed. In 1897, J.J. Thomson had considered the Mayer picture very appealing, pointing out the close resemblance with the periodic table of the elements and using Mayer's organization of magnets to build his atomic model in the framework of the Mendeleev periodic system (Thomson 1897).

In the same year, 1904, the Japanese Hantaro Nagaoka (1865–1961), professor of physics at the University of Tokyo, developed a planetary model of the atom of the type of the planet Saturn, namely, a structure made of a heavy central nucleus of positive charge surrounded by a ring of electrons orbiting around it.

The model predicted that the electron ring should be stabilized by the relatively large mass of the nucleus (Nagaoka 1904):

> The system, which I am going to discuss, consists of a large number of particles of equal mass arranged in a circle at equal angular intervals and repelling each other with forces inversely proportional to the square of distance; at the centre of the circle, place a particle of large mass attracting the other particles according to the same law of force. If these repelling particles be revolving with nearly the same velocity about the attracting centre, the system will generally remain stable for small disturbances, provided the attracting force is sufficiently great. The system differs from the Saturnian system considered by Maxwell in having repelling particles instead of attracting satellites.

This prediction, although supported by Lord Rutherford, was soon recognized to be physically incorrect since a ring of negative charges would be very unstable due to the disruptive repulsion of the electrons and was in fact abandoned by Nagaoka himself in 1908.

Thomson's atomic model also had a short life. It was not well thought of by the chemistry community of the time, which could not easily accept the idea that such a huge dissymmetry could occur between the negative charge condensed in very small particles, the electron, and the positive charge uniformly spread in a volume many orders of magnitude larger. However, even after the speech of Helmholtz at the London Chemical Society, the physicists were now convinced of this kind of corpuscular structure of electricity.

In that period of time the most important center of atomic physics was the Cavendish Laboratory of Cambridge where two great physicists, John William Strutt (Lord Rayleigh) (1842–1919) and Joseph John Thomson, had created the basis of the new atomic physics. It was a pupil of Thomson, Ernest Rutherford, who invented the crucial experiment that marked the end of the *plum-pudding* model and paved the way to the modern theory of the atom. Ernest Rutherford (1871–1937), from a Scottish family and educated in New Zealand, having obtained his degree, went in 1894, thanks to a fellowship, to work in Thomson's laboratory at Cambridge and started research on the electric discharge in rarefied gases. He soon showed great experimental ability coupled with an uncommon imagination, developing clever instruments to study the mechanism by which strong electric fields or intense electromagnetic radiation such as x-rays could ionize the gas molecules in a discharge tube.

With the discovery of Becquerel rays in 1896, his research oriented toward the study of radioactivity. Rutherford's activity in this field is extensively discussed in the following chapter, whereas here we shall concentrate on his contribution to the field of atomic structure. In 1909, he proved that the Becquerel's radiation emitted by radioactive materials was made of two types of particles, α rays that are ionized helium nuclei and β rays that are very fast electrons (Rutherford and Royds 1909):

> These experiments show that the uranium radiation is complex, and that there are present at least two distinct types of radiation – one that is very readily absorbed, which will be termed for convenience the α radiation, and the other of a more penetrative character, which will be termed the ß radiation.

At the end of the century, Rutherford's research was already recognized to be of a very high level, in particular by his master Thomson who wanted to keep him in Cambridge. The rules at Cambridge were, however, very strict and offered little hope that a stable university position could be found in the available time. An interesting opportunity was, however, offered to him when a position of full professor was made available at the McDonald Laboratory of the McGill University of Montreal in Canada. Rutherford decided to accept this offer and in 1898 left for the new position. He stayed in Canada until 1907 when he returned to England to become professor of physics at the University of Manchester. There he remained until 1919, when he took over from Sir Joseph Thomson as *Cavendish Professor* of physics at Cambridge.

When in 1907 Rutherford moved to Manchester, a young German physicist, Johannes Wilhelm (Hans) Geiger (1882–1945) went to work with him, starting a collaboration that lasted until 1912 and gave rise in 1908 to the identification of the α particles as ionized helium atoms and in 1911 to the development of the first instrument to count their number. This instrument, that later became the famous Geiger counter, used in its earlier version a screen covered by zinc sulfide that sparkled when hit by a particle.

In 1909, Rutherford asked Geiger to investigate a phenomenon that he had noticed when working at the McGill University in Canada (Rutherford and Royds 1909), namely, that a beam of α particles was broadened by traversing a thin mica foil and suggested to Geiger that he count the number of α particles diffused by metallic foils as a function of the diffusion angle.

Geiger, who was in charge of orienting young students toward research, told Rutherford that a new student, Ernest Marsden (1889–1970) was, according to him, ready to start research and Rutherford asked him to study just the problem of the diffusion of the α particles. Geiger and Marsden started to work together using thin foils of aluminum, iron, gold, and lead of different thicknesses and even very thin overlapping foils of gold. The result of their experiments was that, while the majority of the α particles easily crossed the metallic foils with slight deviation from the straight path, once in a while one of them would come back as having bumped against a solid wall. This result seemed absolutely inconceivable within the framework of Thomson's model: it was like a cannon ball shot against paper foil coming back!

Geiger and Marsden published in 1909 the result of their experiment without even attempting to supply an explanation for this strange behavior (Geiger and Marsden 1909). Rutherford, thought it through for 2 years, until he grasped the right solution that he presented at the March 7 1911 meeting of the *Literary and*

Philosophical Society of Manchester in the form of a short note in which he concluded that the only possible explanation for the Geiger and Marsden results was to assume that both the positive charge and the atomic mass were localized in a volume much smaller than the total volume of the atom, a volume that he named atomic nucleus.

On the basis of this new hypothesis, the majority of α particles was able to cross the metallic foils without encountering obstacles of dimensions such as to disturb their path. Only a very small fraction of them would, however, follow a path sufficiently close to a nucleus to feel its repulsion and to be deflected by a small angle. Finally, a very small number of α particles, of the order of about 1 every 8,000, would follow a path leading to a collision with a nucleus and thus to a deflection of 90° or more, in very few cases even close to about 180°. By measuring the fraction of α particles deflected by large angles, Rutherford was able to estimate the nuclear dimension, obtaining a value for the nuclear radius about 1,000 times smaller than that of the atom, i.e., of the order of 10^{-13} cm. Rutherford's atom was then essentially empty and the real dimensions of the atom were determined only by the orbits of the electrons distributed around the nucleus. The new model of the atomic structure which definitively eliminated Thomson's model was published the same year in a paper that is today considered as one of the classics of the scientific literature (Rutherford 1911):

> In comparing the theory outlined in this paper with the experimental results, it has been supposed that the atom consists of a central charge supposed concentrated at a point, and that the large single deflexions of the α and ß particles are mainly due to their passage through the strong central field.

In this paper, Rutherford developed in detail the theory of the diffusion of the α particles by collision with a metallic sheet, obtaining the result that the number of particles diffused at a given angle depend on the thickness of the metallic sheet, on the square of the nuclear charge, and on the inverse fourth power of the particles' speed. Geiger and Marsden verified experimentally (Geiger and Marsden 1913) Rutherford's conclusions in 1913. In the same period of time Charles Galton Darwin (1887–1962), a student of Rutherford, highly gifted as theoretician and mathematician, of noble academic origin as the son of the mathematician George Howard Darwin and grandson of the great Charles Darwin, also worked on the theory of the diffusion of the α particles (Darwin 1914a) contributing in particular to evaluate their slowing down due to the shielding effect of the external electrons of the atoms and proving that the gradual energy loss of an α particle travelling through a metallic foil depended on the number of electrons that it encountered on its path (Darwin 1914b).

5.3 The Old Quantum Theory

During the autumn of 1911, the Danish physicist Niels Bohr (1885–1962), thanks to a fellowship of the Carlsberg foundation, reached Cambridge from Copenhagen, where he had obtained his degree with a thesis on the electronic theory of metals.

He wanted to visit the laboratory directed by J.J. Thomson, intending, among other things, to discuss with him some errors that he had noticed reading Thomson's treatise *Conduction of Electricity through Gases.* Thomson, who not only was very effervescent as director of the laboratory, but had no enthusiasm for discussing his own errors, did his best to avoid the discussion and limited his intervention to assigning him an experiment on canal rays. Bohr found this of very limited interest so he spent most of his time writing a paper on the electrons in metals and playing football.

In December, Rutherford visited Cambridge to attend the ritual dinner of the end of the year and met Bohr with whom he enjoyed a pleasant evening. They met again in Manchester and after a short while Bohr decided to join Rutherford's group at Manchester. There Bohr had long discussions with Darwin concerning the role of the external electrons in slowing down and reducing the energy of α particles, a problem of considerable interest to him since he was just working on the distribution of electrons in atoms. Bohr made several suggestions to Darwin and helped him correct and improve the paper on the argument that he later published.

In July 1912, Bohr returned to Copenhagen where he started teaching, in the meantime writing down what he had seen and done at Manchester. He soon realized that the interpretation in terms of classical mechanics of Rutherford's atomic model, using a structure of electricity made of charged particles, corresponded to a planetary model of the atom in which the electrons orbited around the nucleus as the planets orbited around the sun. For a physicist, an atomic model with a positive central nucleus around which the electrons rotated was undoubtedly fascinating and presented a nice parallelism between the infinitely big and the infinitely small worlds, between electrons and planets, both forced to move in fixed orbits by the deterministic laws of classical dynamics. This model, although highly appealing, presented, however, an insurmountable difficulty, being in strong contrast with Maxwell's electromagnetic theory that dictates that an electric charge in motion in an orbit, being subjected to an acceleration, must continuously radiate energy. The atom then, losing energy, would be unstable and after an extremely short time the electron would fall in on the nucleus.

Rutherford, who was aware of the limits of the planetary model for electrically charged particles, had actually avoided in his 1911 paper to speak of orbits, specifying that in his model the atom consisted of a positively charged nucleus surrounded by a uniform distribution of negative charges.

The problem of assigning the electron to specific orbits was instead tackled by Niels Bohr in a famous series of three papers (Bohr 1913a, b, c), which soon become the basis of the whole of modern spectroscopy.

The first paper was limited to the simplest case of the hydrogen atom, one proton and one electron, whereas the two following ones extended the treatment to many electron atoms.

In these papers, Niels Bohr made a brilliant attempt to save the determinism of classical mechanics, bound to the concept of orbit, by conciliating electron dynamics with electromagnetism on the basis of a hypothesis made by Max Planck in 1900. Planck had supposed, to explain blackbody radiation, that the radiation could not be emitted or absorbed in a continuous process but only by discrete amounts that he named light quanta, thus giving rise to the theory known today as the old quantum theory. Using this quantum hypothesis he developed a new atomic model in which the electrons preserved their classical motion in circular orbits but their energy possessed only discrete values, defined by two conditions known as the quantization conditions.

The first of these conditions radically changed the mechanism of classical electromagnetism at the microscopic level, dictating that the energy difference between two orbits was equal to a multiple of the quantity hv, where h is a constant introduced by Planck and v is the frequency of the radiation emitted or absorbed in the transition between two discrete orbits. Bohr defined this first quantization condition on the basis of a discussion with his friend and former classmate, the spectroscopist Hans Marius Hansen (1886–1956), who told him of the existence of the Balmer equation that Bohr had ignored.

The Swiss Johann Jacob Balmer (1825–1898), teacher of mathematics in a girls' school in Basel and passionate about numerology, belonged to the large group of atomic spectroscopists interested in that period in finding simple relationships between the spectral lines of the elements (Balmer 1885a, b). Balmer succeeded in representing the frequency v of the hydrogen atom spectral lines (in units of cm^{-1}) with the empirical formula

$$v = R_H\left(\frac{1}{4} - \frac{1}{n^2}\right)$$

where $n = 3,\ 4,\ 5$, etc. and R_H is the so-called Rydberg constant ($R_H = 109{,}737\ cm^{-1}$).

To the same group of atomic spectroscopists belonged the Swedish physicist Johannes Robert Rydberg (1854–1919), professor of physics at the University of Lund. He was convinced that the order of the elements in the periodic table was connected to the atomic structure and should be reflected in the atomic spectra. On this basis, he generalized Balmer's empirical formula in order to classify the spectral lines of the elements (Rydberg 1886, 1890) with the expression

$$v = R_H\left(\frac{1}{n_f} - \frac{1}{n_i}\right)$$

where n_i and n_f are integers, 1, 2, 3, ... up to infinity, with $n_i > n_f$. The Balmer series for the hydrogen atom corresponds to $n_i = 2$. Other series for the hydrogen atom are a set of ultraviolet lines (the Lyman series) that fit the above relationship with $n_i = 1$. A series in the infrared region is the Paschen series that corresponds to $n_i = 3$. Other series are the Brackett and Pfund series corresponding to $n_i = 4$ and $n_i = 5$, respectively

Bohr examining Balmer's formula realized that the frequencies emitted or absorbed by the hydrogen atom were all obtained as the difference between two numbers and he concluded that only the difference between the energy of two electronic states would explain the atomic spectra.

The second condition "quantized" the angular momentum of the electron, stating that it should be a multiple of hv/c where c is the speed of light. This condition was suggested to Bohr by the papers of John William Nicholson (1881–1955), a mathematical physicist of Cambridge, who had attempted to interpret the complex emission spectrum of the solar corona with an atomic model in which rings of electrons orbited around the nucleus (Nicholson 1912). According to Nicholson, the electron oscillations in these rings gave rise to the spectrum. Even if incorrect, this theory involved an important idea that was included in Bohr's theory. Nicholson wanted to incorporate Planck's ideas in his model and, knowing that the Planck h constant had the right dimension, he decided to use it as a unit of angular momentum, stating that the atom could lose or gain angular momentum only in definite amounts, multiples of h. According to him, the angular momentum quantization was more correct and important than the energy quantization.

Bohr's genial ideas were, on the one hand to couple the energy and the angular momentum quantization, reducing in this way the number of possible circular electron orbits only to the stationary ones, and on the other to understand the importance of introducing a "foreign" quantity such as Planck's constant in the laws governing the old quantum theory:

> Whatever the alteration in the laws of motion of the electrons may be, it seems necessary to introduce in the laws in question a quantity foreign to the classical electrodynamics, i.e. Planck's constant, or as it often is called the elementary quantum of action. By the introduction of this quantity the question of the stable configuration of the electrons in the atoms is essentially changed as this constant is of such dimensions and magnitude that it, together with the mass and charge of the particles, can determine a length of the order of magnitude required.

In reality, Nicholson and Bohr did not realize that their idea to quantize the angular momentum corresponded to considering the electron not only as a particle but also as a wave, anticipating by 10 years the Louis de Broglie principle. In fact, an orbit that in order to be stable satisfies the de Broglie principle, corresponds to a stationary wave and therefore the circumference of the orbit must necessarily be an integer multiple of the wavelength. As a consequence, only special values of the circumference radius are allowed.

Bohr succeeded in this way in obtaining a stupefying agreement between his theory and the empirical relationships found by several authors, in particular by Balmer and Rydberg, for the visible frequencies of the hydrogen atom. A further significant success of his theory was the direct calculation of Rydberg's constant as a function only of the mass and charge of the electron and of Planck's constant.

The quantization of the electronic orbits had already been proposed in 1910 by the Viennese physicist Arthur Erich Haas (1884–1941), who anticipated Bohr's papers by 3 years (Haas 1910), but was never seriously considered and instead largely ignored or even ridiculed by the scientific community.

The extension of Bohr's theory to many electron systems, presented in the second and third papers of 1913, was not very satisfactory for the interpretation of their emission spectra and turned out worse as the number of electrons increased. An important improvement of the theory was developed by Arnold Sommerfeld

(1868–1951), professor at the University of Munich, and one of the founders of the great school of theoretical physics in Germany. The role played by Sommerfeld in the development of the golden period of German physics is immense and is testified by the large number of Nobel laureates that studied with him and by the fact that he reached a record of 81 nominations for the Nobel Prize in physics.

Sommerfeld introduced elliptical orbits in addition to circular ones in Bohr's theory (Sommerfeld 1916) and defined more general quantization conditions than those of Bohr. The addition of elliptical orbits turned out to be very useful for the interpretation of the periodic system of the elements in terms of electron orbits. A further correction due to Sommerfeld was to assume that both the nucleus and the electrons orbited around the atomic center of mass, not coinciding with the nucleus as in Bohr's version of the theory. This modification led to a small correction of the numerical value of the Rydberg constant and to a reasonable explanation of the difference observed between the spectrum of hydrogen and deuterium, which in the old version of Bohr were identical.

Sommerfeld also introduced a relativistic correction of the electronic motion that led to a splitting in multiplets of fine structure of the single spectral lines predicted by Bohr's version of the theory. Finally, Sommerfeld, taking into account the fact that Bohr's orbits represented electrical currents in closed loops, thus giving rise to a magnetic moment perpendicular to the plane of the orbit, introduced into the theory a further quantization condition that allowed the orbits not necessarily to be all in the same plane and to assume different orientations in a magnetic field.

Sommerfeld published a detailed presentation of the theory in his famous 1924 book *Atombau und Spektrallinien.* For him, to quantize a physical entity corresponded to isolating in his continuum of classical values only some discrete multiple of a quantum unit of measure.

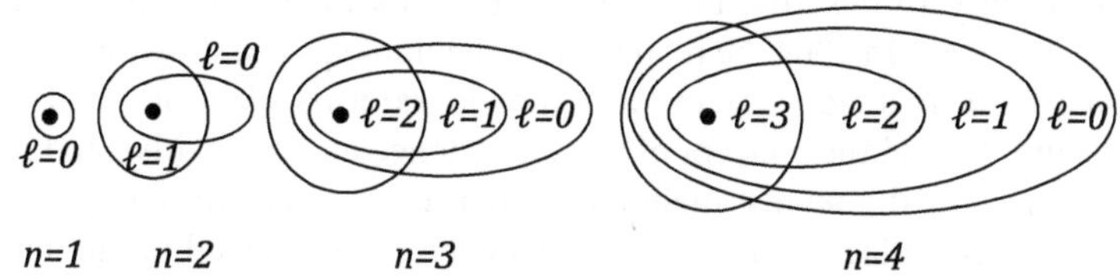

Electronic orbits according to the Bohr-Sommerfeld model

In the quantum theory that includes Sommerfeld contributions, known as the Bohr–Sommerfeld theory, the three quantization conditions of energy, angular momentum, and magnetic moment orientation were specified by three quantum numbers labeled with the letters n, ℓ, and m. The theory also specified the relations between their possible values. The principal quantum number n could assume only integer value with $n = 1, 2, 3$, etc., corresponding to energies $E = -R_H(1/n^2)$. The orbital quantum number ℓ could instead assume all values from $\ell = 0$ up to $\ell = n - 1$. Finally, the magnetic quantum m could vary from $-\ell$ up to $+\ell$.

With these quantum conditions, the Bohr–Sommerfeld theory explained relatively well the spectra of the hydrogen atom and of the hydrogenoid atoms, i.e., of atoms with Z protons but only one electron such as a singly ionized helium He^+, a twice ionized lithium Li^{++}, etc. This form of the theory deviated, however, more and more from the experimental data as the number of electrons increased. Despite several attempts to adjust the theory to more complex cases it became evident that the theory had insurmountable limits and soon it was forced to leave the way clear for the development of quantum mechanics.

5.4 The Electronic Theories of the Chemical Bond

The Bohr–Sommerfeld theory, even if far from supplying a correct description of the structure of matter at the microscopic level, still allowed one to establish a convenient basis for a description in electronic terms of a useful theory of the chemical bonds. At the beginning of the twentieth century, the concept of chemical bond, although of central interest for the chemical community, was still unclear until two German physicists, Richard Abegg and Walther Kossel, thanks to their imaginative research, succeeded in clarifying its nature, contributing to its understanding in terms of electronic theory.

The German physicist Richard Wilhelm Heinrich Abegg (1869–1910), student of Nernst at Göttingen, was one of the pioneers of the chemical valence theory, a term coined by the former student of Kekulé, Carl Hermann Wichelhaus in 1867 (Wichelhaus 1867). In collaboration with Guido Bodländer (1855–1904), professor at the University of Braunschweig, Abegg developed an electronic theory of chemical affinity to describe the valence in terms of interactions of electrons (Abegg et al. 1899). In 1904, he realized that the noble gases, showing a particular chemical inertia to react, possess a complete external electron shell of eight electrons and defined the concept of normal valence and of positive and negative contra-valence for the atoms, maintaining that every element always had eight available "valence places." The positive valence corresponded to the number of valence places occupied by electrons and the contra- or negative valence to the number of free valence places in the atom. Every element therefore possesses a maximum positive and a maximum negative valence such that their difference was always equal to eight (Abegg's rule) (Abegg 1904). For example, sulfur has valence +6 in H_2SO_4 and valence −2 in H_2S. In the same way, nitrogen has valence +5 in HNO_3 and valence −3 in NH_3 while chlorine has valence −1 in HCl and +7 in $HClO_4$.

Abegg and his wife Line were both avid enthusiasts of hot-air-balloon trips. He was the founder and chairperson of the Silesian club for aeronautics in Breslau. This hobby ultimately caused his death at the age of 41 when he flew his balloon together with an engineer and female relative. When the balloon encountered strong winds, they decided to descend. The balloon reached the ground safely and the other two were able to exit the basket. Abegg hesitated to exit and was caught by a gust of wind and crushed by the basket.

In 1913, in the period in which Bohr elaborated his theory, the young German physicist Walther Kossel (1888–1956), son of the Nobel Prize winner for medicine Albrecht Kossel (1888–1956) and assistant of Lénárd at the University of Berlin, went to work with Sommerfeld at Munich, where he dedicated himself to the theoretical study of the emission and absorption of x-rays with atoms. On the basis of Abegg's ideas, Kossel developed the famous octet theory (Kossel 1916), independently proposed in the same year by Gilbert Lewis in the USA, and succeeded in explaining the formation of positive and negative ions such as those encountered in electrochemical processes (Kossel 1920). According to Kossel, the electrons of all element atoms, except hydrogen and helium, were divided into internal and external shells. The internal shells possessed a number of electrons equal to that of the closest noble gas. The start of any new period in the periodic system of Mendeleev corresponded to the formation of a new electron shell. The electrons located in the external shells were those determining the reactivity of the elements. If the element had an incomplete external shell it had the tendency to acquire electrons up to a number that would fill it to reach the electronic structure of the following noble gas in the periodic table. If instead the element had an excess of electrons with respect to the noble gas situated before in the periodic system, it would prefer to lose them to reach a stable structure. External electrons could be therefore lost or gained, giving rise to positive and negative ions, respectively. The affinity of the chemist of the nineteenth century was finally interpreted on a theoretical basis as the tendency to fill the external electron shells.

The ideas of Abegg and Kossel explained perfectly the formation of ions in the framework of Bohr's theory, but could not justify the stability of the majority of molecules which are neutral and do not contain ions, in particular the existence of diatomic molecules formed by two identical atoms such as Cl_2, N_2, O_2, etc.

After the discovery of the electron, J.J. Thomson, who undoubtedly possessed a very creative mind, tried to explain the existence of neutral molecules developing a valence theory of electrical "force tubes" binding the atoms through a polarization mechanism consisting in the transfer of a "corpuscle," the electron, from one donor to an acceptor atom (Thomson 1907a, pp. 138–139):

> For each valency bond established between two atoms the transference of one corpuscle from the one atom to the other has taken place, the atom receiving the corpuscle acquiring a unit charge of negative electricity, the other by the loss of a corpuscle acquiring a unit charge of positive.

In his book *The Corpuscular Theory of Matter* of 1907, Thomson represented this electron transfer process in molecules by means of arrow from the donor to the acceptor atom (Thomson 1907a). For example, in the H–C bond, the electron of the hydrogen atom migrated toward the carbon atom whereas in the case of the Cl–C bond, a carbon electron migrated toward the chlorine atom.

In this way he represented methane, carbon tetrachloride, and ethane with the formulas

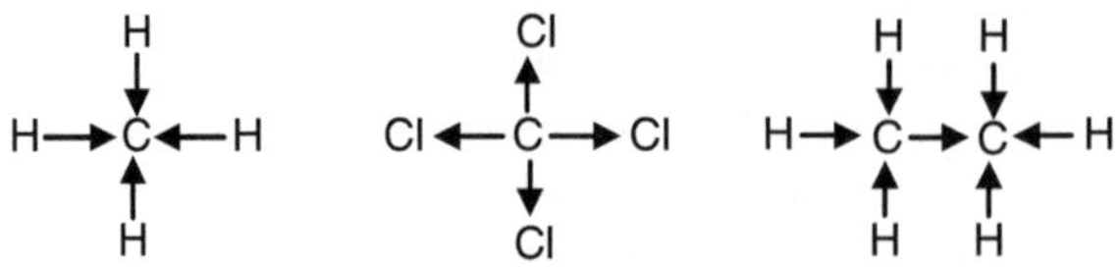

Thomson's idea was resumed taken up again in 1910 by Kaufman George Falk (1880–1953), professor at the MIT and by John Maurice Nelson (1876–1965) professor at the Columbia University in New York, who improved Thomson's electronic valence theory (Falk and Nelson 1910) assuming that the electrons shifted between two atoms in chemical bonds as a function of their relative position in the periodic system. Falk and Nelson also extended the arrow symbolism to represent double and triple bonds:

hinting even at the possibility of the existence of isomers.

This theory of polar bonds, although purely heuristic in nature, had some success among organic chemists interested in the effect of substituents in organic reactions and was later used by the English school in the theory of reaction mechanisms.

An important step toward the understanding of the basic nature of the chemical bond was made in 1916 by the American chemist Gilbert Lewis who, with his research, brought America to the forefront in chemistry. Son of a Bostonian lawyer, Gilbert Newton Lewis (1875–1946) had joined the University of Nebraska before moving at the age of 17 to Harvard, where in 1899 he obtained a Ph.D. in chemistry. After the Ph.D. he stayed for 1 year at Harvard before leaving, thanks to a fellowship, for Europe where he studied under the supervision of Ostwald at Leipzig and Nernst at Göttingen. Lewis did not get along too well with Nernst to the point where they developed a true and lifelong enmity which later retarded Lewis's nomination for the Nobel Prize in chemistry. Back in the USA, he worked for 3 years at Harvard as chemistry instructor under the direction of Theodor Richards who had studied also at Leipzig with Ostwald.

Owing to his strong character, Lewis's stay at Harvard did not last for long. Richards had adopted the same skeptical position of his former master Ostwald toward the concept of the atom, even if his scrupulous determinations of atomic weights would have provided him in 1914 with the Nobel Prize in chemistry. Lewis, who could not stand his very authoritative behavior toward his coworkers and his scorn for any form of theory of the chemical bond as interaction between the atoms, decided to leave Cambridge in 1904, accepting a position of Superintendent of Weights and Measures for the Bureau of Science of the Philippine Islands in Manila. Next year, however, he returned to the USA with a research position at the Massachusetts Institute of Technology (MIT) where he had the chance to work with a group of excellent physical chemists interested in the study of strong electrolytes directed by Arthur Amos Noyes, also a former pupil of Ostwald. In 1907 he became assistant, in 1908 associate, and in 1911 full professor. During this period his scientific activity was essentially devoted to the theory of relativity, a field in which he published a series of papers on the relationship between mass and energy, derived in a different fashion compared to Einstein. At the same time, he also published a physicochemical paper on osmotic pressure (Lewis 1908). In 1912, he moved to California where he later became professor of chemical physics and director of the College of Chemistry at Berkeley. The research activity of Noyes's research group pushed the young Lewis to get interested in thermodynamics (Lewis 1907), a field in which he would became known worldwide for his famous textbook *Thermodynamics and the Free Energy of Chemical Substances*, written in collaboration with Merle Randall (Lewis and Randall 1923). Even more influential was the stimulating MIT atmosphere that extended his interest to the study of the chemical bond both through the reading of Werner's papers on coordination complexes and through the study of the new theories of atomic structure such as that of J.J. Thomson in which the atom resembled a watermelon with electrons immersed at the side of the seeds in a pulp of positive electricity.

At the beginning of the century in the lectures that he used to give to first year chemistry students, Lewis used to represent the electrons in an atom with points or small circles at the vertex of a cube. At that time there started to circulate in the chemical community the idea that the stability of the rare gas elements depended upon the fact that they possessed eight electrons in the external Thomson's shell, as described by Abegg in 1904 and later reinforced by Kossel in 1916. Having realized that a cube possesses eight vertices, Lewis had the idea to represent, for purely teaching purposes, the atom as a cube with eight positions available to arrange the external electrons.

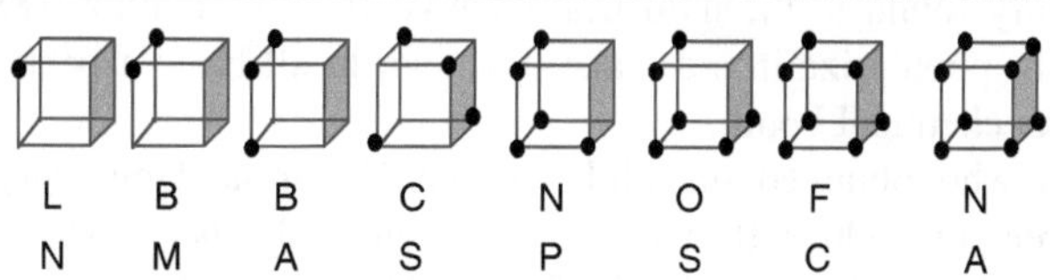

Lewis cubic atoms

By formally representing the atoms as cubes, Lewis arranged at the vertices a number of external electrons equal to the position of the elements in the eight columns of the periodic table. Starting from these formal structures he proposed the octet rule that bears his name. This strange theory of the cubic atom had clearly only didactical finality, but in 1916 he recovered it to formulate his theory of the chemical bond (Lewis 1916). In 1913, Lewis had the opportunity to read the manuscript of an English Ph.D. student, Alfred Lauck Parson (1889–1970), visiting Berkeley for a year, who suggested that the electron was not simply an electrical charge but also a magnet and that a bond could be formed between two atoms if two electrons were shared between the two atoms. Lewis represented this situation with two cubes with a side in common as shown in the figure. On this basis (Lewis 1916), Lewis introduced the concept of the covalent bond, a term actually coined only a few years later by his friend and coworker Irvin Langmuir. In the same way, ionic bonds were formed by transferral of one electron from one cube to another without sides in common. The main difference with previous theories was that in his formulation of the shared electron pair, Lewis allowed for partial transfers of electrons from one atom to another, in contrast with the positive–negative theory. He describes the partial transfer of two electrons, one from each of the two bonding atoms, so that there is a shared pair of electrons between them. This eliminates the need for the formation of oppositely charged atoms when there was no indication of individually charged atoms (ions) in a compound. This was the first true description of covalent bonding. Lewis theorized that electrons in an atom pair up around the nucleus, usually forming a tetrahedral arrangement. Although he never actually used the term "octet" for four pairs of electrons, the octet rule is often associated with his name. His main concern was with individual bonds between atoms rather than with all the electron pairs around each nucleus. Lewis's book *Valence and the Structure of Atoms and Molecules* is a classic, one of the greatest contributions to modern bonding theory (Lewis 1923).

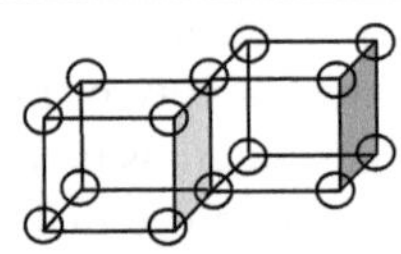

Two cubic atoms with one side in common

Lewis theory found in the industrial chemist Irvin Langmuir (1881–1957) the ideal person to popularize it and make it known in all basic aspects necessary to understand the chemical bond.

Langmuir, who obtained his Ph.D. at the Columbia University, started his scientific career in 1906 with a Ph.D. thesis entitled *Über partielle Wiedervereinigung dissociierter Gase im Verlauf einer Abkühlung*, developed in Göttingen under the supervision of Nernst (Langmuir 1906), who had just discovered his filament lamp. Langmuir, who was employed at the General Electric Company, soon became internationally known for his brilliant industrial applications of the kinetic theories and of surface adsorption as well as for the study of thermal effects in gases that led him to develop a filament lamp which was much more efficient than that of his master (Langmuir and Orange 1913). Impressed by Lewis's papers and by the importance that Lewis assigned to electrons as responsible for the "chemical forces" between atoms, Langmuir dedicated himself in the period 1916–1921 to the development, with great efficacy and ability, of the octet theory and to diffusion in the chemical milieu, thanks also to the creation of well-selected words such as *octet* and *covalent bond*, easily introduced into everyday slang in the laboratory. Starting from the Lewis theory he divided the electronic shells in cells, each able to host two electrons. In filling, the external shell around the nucleus could not host electrons until the internal ones were completely filled. An innovative aspect of his formulation of the octet theory was that two electrons could fill the same cell only if all others possessed at least one electron, a condition that anticipated the Hund rule.

According to the octet theory, enlarged and completed by Irving Langmuir, two atoms were held together by covalent bonds when each of the two atoms shared with the other one or more electrons to complete its external octet of electrons (Langmuir 1919b, c). For example, the chlorine atom possesses seven electrons in its external shell L. If two chlorine atoms share one electron each, both atoms complete their L shell with eight electrons, giving rise to a stable Cl_2 molecule as schematically shown in the figure below, where the electrons are represented by black and white dots.

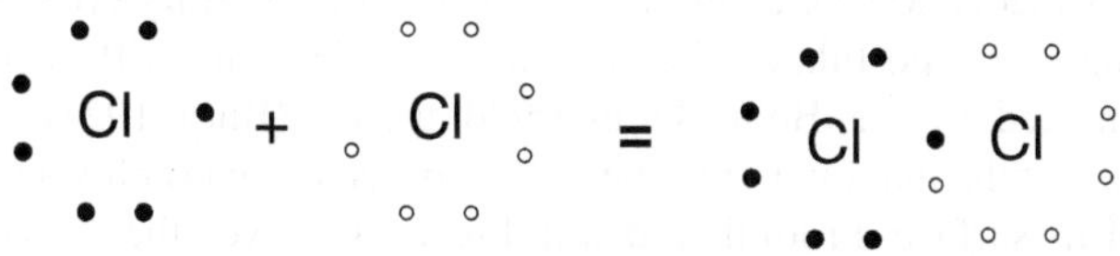

. Formation of a chlorine molecule from the separated atoms

The ethylene molecule $CH_2 = CH_2$ is made of two carbon atoms each with four electrons in the L shell and of four hydrogen atoms each with an electron in the K shell. By shearing the electrons as shown in the figure, the hydrogen atoms complete their K shell with two electrons whereas the two carbon atoms complete their L shells, with eight electrons. The four shared electrons between the two carbon atoms form a double bond. In the case of acetylene the two carbon atoms share six electrons forming a triple bond.

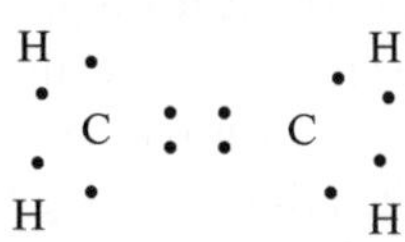

The Lewis ethylene molecule

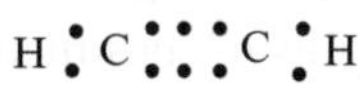

The *Lewis* acethylene molecule

Langmuir also invented in 1919 the concept of *isosterism* (Langmuir 1919a) to associate molecules like nitrogen N_2 and carbon oxide CO with the same number of electrons to which he attributed similar chemical properties (Langmuir 1920). The concept was later extended to biological systems by Friedrich Erlenmeyer in 1932.

The Lewis–Langmuir theory was soon found to be a simple but extremely effective structure that clarified the importance of the electrons in chemistry. More than that, it transformed the chemistry of the elements into the chemistry of the electrons. The valence electrons became in fact the true protagonists of the molecular structure and of chemical reactions, whereas the nuclei and the inner electrons remained passive bystanders in this new world of chemistry that replaced that invented a century before by Dumas. The theory that possessed enormous didactic power was largely used by the synthetic chemists since it provided not only a convenient instrument to represent the molecular structures correctly, but also a practical formalism to explain many reaction mechanisms and to predict substitution of chemical groups.

Its static nature, the absence of any connection with first principles, and the disconnection from physical theories, in particular the Bohr–Sommerfeld one, made it, rather than a theory, a simplifying model endowed with, however, an enormous heuristic power and of a very strong predictive capacity. It represented the concrete performance of practical laboratory chemists with respect to the abstraction of theoretical physicists. It was, however, just this abstractness that, by assigning physical consistence to abstruse mathematical operators, finally succeeded, giving origin to that great revolution of physics at the microscopic level, quantum mechanics, that in few years would completely change our vision of the world.

Langmuir's theory was taken up in 1921 by Charles R. Bury (1890–1968), who replaced Langmuir's postulate of cells with two electrons with a different one directly connected to the Bohr–Sommerfeld theory (Bury 1921). Bury's new postulate was that the maximum number of electrons in each shell was proportional to the area of its surface. From this idea it directly followed that successive shells contained 2, 8, 18, and 32 electrons, respectively.

5.5 The Aufbau Principle

With Sommerfeld's help, and taking into account Abegg's and Kossel's ideas, Bohr developed in a series of papers from 1921 to 1923 the *Aufbau* (building) principle that established how the electrons are distributed in the atomic orbits of the elements of the periodic table (Bohr 1921). The Aufbau principle started from the hypothetical possibility of constructing the electronic structure of an atom of the periodic table adding one electron to the electronic distribution of the previous atom and applying the concept of quantization of the orbits. Starting from the hydrogen atom with only one electron, the energy levels of the following atoms were, one at a time, filled with electrons starting from the lowest energy levels up on the basis of essentially empirical rules.

The electronic orbits were thus distributed in the atoms in shells or "barks" that contained the nucleus like onion layers (Bohr 1922). The original form of the Aufbau principle, developed in the period 1921–1923, soon showed its limitations when Bohr tried to extend his idea to filling the electronic orbits of many-electron atoms. In 1924, a new and more efficient version was proposed separately by two English scientists, the chemist John David Main Smith of the University of Birmingham and the physicist Edmund Clifton Stoner who worked at the Cavendish Laboratory of Cambridge. In 1924, Main Smith published four short letters (Main Smith 1924a, 1925) and a book entitled *Chemistry and Atomic structure* (Main Smith 1924b) in which he corrected, on purely chemical grounds, the condition by which Bohr assumed that the second electronic shell included eight electrons distributed in two subgroups each with four electrons. Main Smith stated instead that each group necessarily had to include a subgroup with two electrons and that in each subgroup there could be located a maximum $2(2\ell + 1)$ electrons.

At the same time, Edmund Stoner (1899–1968) published in the Philosophical Magazine a paper (Stoner 1924) that reached the same results on the basis of magnetic and spectroscopic data and in addition established that the maximum number of electrons in each shell was equal to $2n^2$. Stoner's paper was much better known than that of Main Smith in the physicist community that almost completely ignored the existence of a journal of industrial chemistry. Wolfgang Pauli who by chance read Stoner's paper quickly succeeded in giving an axiomatic form to his conclusions, establishing the famous exclusion principle (Pauli 1925). The Aufbau principle was definitively completed in 1926 with the discovery of electron spin by Uhlenbeck and Goudsmit and with the introduction in the theory of the fourth spin quantum number. During a meeting at Leyden, after a discussion with Einstein, and after even longer discussions with Pauli, Bohr accepted – with some hesitation – the concept of spin, of which he later became a strong supporter.

In the final formulation of the Aufbau principle realized in the framework of the Bohr–Sommerfeld theory, each shell, indicated by a letter (K, L, M, N, etc.), was characterized by the principal quantum number n that could assume all integer values 1, 2, 3, etc. In each shell a maximum of $2n^2$ electrons were allocated. The K ($n = 1$) shell therefore included at maximum 2 electrons, the L ($n = 2$) shell at maximum 8, the M ($n = 3$) shell 18, and so on. Each shell was further divided into subshells (s, p, d, f, ...), characterized by the orbital quantum number ℓ which assumed all integer values from 0 up to $n - 1$. In each subshell a maximum of $2(2\ell + 1)$ electrons were positioned. The magnetic quantum number m assumed all integer values from $-\ell$ to $+\ell$, including the value 0. In this way, in the s subshell were located two electrons, in the p six, in the d ten electrons, etc., as summarized in the table for the first four values of n.

Bohr–Sommerfeld quantum numbers

Quantum numbers			Number of states	
n	ℓ	m	Subshell	Total
1	0	0	2	2
2	0 (s)	0	2	8
	1 (p)	−1, 0, +1	6	
3	0 (s)	0	2	18
	1 (p)	−1, 0, +1	6	
	2 (d)	−2, −1, 0, +1, +2	10	
4	0 (s)	0	2	32
	1 (p)	−1, 0, +1	6	
	2 (d)	−2, −1, 0, +1, +2	10	
	3 (f)	−3, −2, −1, 0, +1, +2, +3	14	

5.6 Electron Spin

The discovery of the Pauli exclusion principle is intimately entangled with that of electron spin and the complexity of their history reflects the difficulties that nineteenth-century physics had to face in the transition from classical to quantum physics. The first to suggest that a fourth quantum number could be connected with a spinning motion of the electron was a young American student of physics, Ralph de Laer Kronig (1904–1995), born and educated at Dresden in Germany, but then transferred to the Columbia University in New York where he obtained his Ph.D. in 1925. Kronig, upon the suggestion of Paul Ehrenfest who visited Columbia while he was studying there, moved to Europe at the end of 1924 to visit the most important centers of theoretical physics of the old continent. In particular he went to Leyden, where he collaborated with Samuel Goudsmit to compute the band intensity in the Zeeman effect (Goudsmit and Kronig 1925). At the Tubingen University, Kronig had the opportunity to listen to a lecture by Wolfgang Pauli on the need to introduce a new quantum number in the Sommerfeld quantum mechanical formalism. Back in America, the young Kronig proposed in 1925 the existence of a spin quantum number, assuming that the electron could whirl round on itself giving rise to a spinning angular momentum.

The idea that the electron could spin like a whipping top did not please Heisenberg or Pauli who suggested he give up insisting on this funny idea that he qualified as lacking any physical reality. Kronig, discouraged by these criticisms, gave up and decided not to publish his ideas. Unfortunately for him the same idea was published next year by Goudsmit. Kronig did not bear any grudge for Pauli, however; on the contrary, the two became excellent friends and Kronig had a brilliant career. In 1927, he found, in collaboration with Isidor Isaac Rabi (1898–1988), the exact solution for the Schrödinger equation in the case of the symmetric rotor (Kronig et al. 1927).

In reality, the need for a fourth quantum number had already been put forward by Sommerfeld who in 1920 proposed the existence of an internal quantum number associated with a "hidden" rotation (Sommerfeld 1920) to describe the anomalous reaction of many electron atoms to an external magnetic field (anomalous Zeeman effect).

In 1925, Pauli published his *Ausschliessungsprinzip* exclusion principle (Pauli 1925) that proposed the existence of a fourth quantum number. The Viennese Wolfgang Pauli (1900–1958) arrived at the exclusion principle starting from the impossibility of explaining with only three quantum numbers the anomalous Zeeman effect. While in the spectra of atoms like hydrogen, triplets of equidistant lines were observed (normal Zeeman effect) in the presence of a magnetic field, perfectly explained with three quantum numbers, in some atoms the spectral lines were split into four, five, and even more components, with a separation larger than predicted by theory (anomalous Zeeman effect). This anomalous effect remained unexplained for several years and only with the introduction of a fourth quantum number was it possible to create a reasonable interpretative scheme for atomic spectra.

Another factor that pushed Pauli to look for a new quantum number was Stoner's paper, previously discussed, further supported by the fact that the Swede Rydberg had noticed that the number series 2, 8, 18, 32, . . ., defining the length of the periods of the periodic system, was the series $2n^2$. Pauli realized that this factor 2 recurring in Rydberg's formula, as well as in Bohr's and in Langmuir's theory, had no theoretical justification and arose from an as yet undiscovered condition.

The Pauli exclusion principle dictates that two electrons cannot have the same set of four quantum numbers. When an electron occupies an energy state defined by four values of the quantum numbers, that state is filled and cannot host another electron. This rule is actually valid only for particles like the electrons obeying the Fermi–Dirac statistic (fermions).

For several years, the Pauli principle represented an important integration of the Bohr–Sommerfeld theory, although without a plausible explanation of its presence. In his papers, Pauli in fact never explained the theoretical reasons for the existence of a fourth quantum number. Only in 1945 when he received the Nobel Prize for physics did Pauli supply in his Stockholm speech a full explanation of the principle in terms of quantum mechanics and not of the old quantum theory. In quantum mechanics, for an atom with two electrons at the a and b levels, the correct wavefunction in which the two electrons are indistinguishable and the function is antisymmetric with respect to the exchange of the two electrons has the form

$$\psi = \psi_1(\mathrm{a})\psi_2(\mathrm{b}) - \psi_1(\mathrm{b})\psi_2(\mathrm{a})$$

and this function vanishes if the two electrons are at the same level, a = b.

In 1926, the Swedish physicists George Eugene Uhlenbeck (1900–1988) and Samuel Abraham Goudsmit (1902–1978), working at Leyden in Holland under the supervision of Ehrenfest, read the just published Pauli's paper where he mentioned a fourth quantum degree of freedom. Goudsmit was an experimentalist who knew well the anomalous Zeeman effect, whereas Uhlenbeck was a theoretician who had worked in Italy and met Enrico Fermi. It was actually Uhlenbeck who, once he understood that Pauli was looking for a fourth quantum number, realized that this meant the occurrence of another degree of freedom for the electron and that the only possible additional degree of freedom was a spinning motion around its own axis, an idea that Konig had already suggested a year before. The two friends published the spin theory right away (Uhlenbeck and Goudsmit 1925) in papers in which the electron was assimilated to a small sphere of negative electricity rotating around the nucleus and spinning like a small whip top. Being an electric charge in rotation in an orbit, it had to be associated with an intrinsic magnetic moment. The two Dutchmen then applied to the rotation of the electron the condition that the spin angular momentum could only have the value (½)h/2π and that the magnetic moment could be oriented in a magnetic field in only two ways, parallel or antiparallel to the direction of the field (Uhlenbeck et al. 1926). Application of the Pauli exclusion principle then decrees that if two electrons have the same values of the quantum numbers n, ℓ and m must have opposite spins (antiparallels), one with value +1/2 and the other with value −1/2. The fourth quantum number therefore became the spin quantum number s that could have only two values, +1/2 and −1/2. The spin theory was later formalized by Pauli in 1927. The introduction of the spin quantum number s for the electron turned out to be of extreme importance for the successive development of quantum chemistry.

At this point, the atomic structure was more or less understood, at least in outline. It was still obscure, however, why the mass of the nucleus did not correspond to the atomic number and that for the same atomic number different isotopes could exist. In 1921, Rutherford had already postulated the existence of a neutral particle of mass equal to that of the proton. The problem was definitively solved in 1932 with the discovery of the neutron by James Chadwick.

References

Abegg R (1904) Die Valenz und das periodische System. Versuch einer Theorie der Molekularvebindungen. Z Anorg Chem 39:330

Abegg R, Bodländer G (1899) Die Elektroaffinität, ein neues Prinzip der. chemischen Systematik. Z Anorg Chem 20:453–499

Aepinus FUT (1759) Tentamen theoriae electricitatis et magnetismi. Accedunt dissertationes duae, quarum prior, phaenomenon quoddam electricum, altera, magneticum, explicat, auctore F.V.T. Aepino. Typis Academiae Scientiarum, Petropolis

Balmer JJ (1885a) Notiz über die Spektrallinien des Wasserstoff. Ver Naturforsch Ges Basel Z 7:750–752

Balmer JJ (1885b) Notiz über die Spectrallinien des Wasserstoffs. Ann Phys Chem 25:80–87

Bohr N (1913a) On the constitution of atoms and molecules, Part I. Philos Mag 26:1–25

Bohr N (1913b) On the constitution of atoms and molecules. Part II. Systems containing only a single nucleus. Philos Mag 26:476–502

Bohr N (1913c) On the constitution of atoms and molecules. Part III. Systems containing several nuclei. Philos Mag 26:857–875

Bohr N (1921) Atomic structure. Nature 107:104–107

Bohr N (1922) The constitution of atoms and the physical and chemical properties of elements. Z Phys 9:1–67

Bury CR (1921) Langmuir's theory of the arrangement of electrons in atoms and molecules. J Am Chem Soc 43:1602–1609

Crookes W (1879a) Contributions to molecular physics in high vacua. Magnetic deflection of molecular trajectory. Laws of magnetic rotation in high and low vacua. Phosphorogenic properties of molecular discharge. Philos Trans R Soc Lond 170:641–662

Crookes W (1879b) The Bakerian lecture: on the illumination of lines of molecular pressure, and the trajectory of molecules. Philos Trans R Soc Lond 170:135–164

Darwin CG (1914a) Collision of ∝ particles with light atoms. Philos Mag 27:499

Darwin CG (1914b) The theory of X-ray reflexion. Philos Mag 27:675

de Kronig LR, Rabi JJ (1927) The symmetrical top in the undulatory mechanics. Phys Rev 29:262–269

Falk KG, Nelson JM (1910) The electron conception of valence. J Am Chem Soc 32:1637–1654

Geiger H, Marsden E (1909) On a diffuse reflection of the α-particles. Proc R Soc (Lond) A82:495–500

Geiger H, Marsden E (1913) The laws of deflexion of a particles through large angles. Philos Mag 25:604–623

Goldstein E (1876) Vorläufige Mittheilungen über Elektrische Entladungen in verdünnten Gasen. Monatsbericht der Königlich Academie der Wissenschaften zu Berlin 279–95

Goldstein E (1880) Über die Entladung der Elektricität in verdünnten Gasen. Monatsbericht der Königlich Academie der Wissenschaften zu Berlin 82–124

Goldstein E (1886) Über eine noch nicht untersuchte Strahlunsform an der Kathode inducirter Entladungen. Sitzungsberichte der Königlichen Academie der Wissenschaften zu Berlin 693–97 Published on the Annalen der Physik Vol 300 Issue 1 pages 38–48, 1898

Goudsmit S, de Kronig RL (1925) Die Intensität der Zeemankomponenten. Naturwissenschaften 13:90

Haas AE (1910) Über eine neue theorische Methode zur Bestimmung des elektrischen Elementarquantums des Wasserstoff atoms. Phys Z 11:537–538

Hertz H (1892) Kathodenstrahlen durch dunne metalschichten. Annalen der Physik 45:28–32

Hittorf W (1869) Über die Elektricitatsleitung der Gase. Ann Phys 136:1–31, 136:197

Kossel W (1916) Über Molekülbildung als Frage des Atombaus. Ann Phys 49:229–362

Kossel W (1920) Bemerkungen über Atomkräfte. Z Phys der Physik und Chemie 1:395–415

Kronig de Laer R Rabi JJ (1927) The Symmetrical Top in the Undulatory Mechanics Phys Rev 29:262–269

Langmuir I (1906) Über partielle Wiedervereinigung dissociierter Gase im Verlauf einer Abkühlung. Thesis, "Göttingen" L. Hofer 1906

Langmuir I (1919a) Isomorphisme, isosterism and covalence. J Am Chem Soc 41:1543–1559

Langmuir I (1919b) The arrangement of electrons in atoms and molecules. J Am Chem Soc 41:868

Langmuir I (1919c) The structure of atoms and the Octet theory of valence. Proc Natl Acad Sci 5:252

Langmuir I (1920) The Octet theory of valence and its applications with special reference to organic nitrogen compounds. J Am Chem Soc 42:274

Langmuir I, Orange JA (1913) Tungsten lamps of high efficiency. Gen Elec Rev 16:956

Lénárd P (1893) Über Kathodenstrahlen in Gasen von atmosphärischem Druck und im äussersten Vacuum. Sitzungsberichte der Königlich Preussischen Akademie der Wissenschaften zu Berlin 1:3–7

Lénárd P (1894) Über Kathodenstrahlen in Gasen von atmosphärischem Druck und im äussersten Vacuum. Ann Phys 287:225–267

Lewis GN (1907) Outlines of a new system of thermodynamic chemistry. Proc Am Acad 43:259

Lewis GN (1908) The osmotic pressure of concentrated solutions, and the laws of the perfect solution. J Am Chem Soc 30:668–683

Lewis GN (1916) The atom and the molecule. J Am Chem Soc 38:762–786

Lewis GN (1923) Valence and the structure of atoms and molecules. The Chemical Catalogue Company, New York

Lewis GN, Randall M (1923) Thermodynamics and the free energy of chemical substances. McGraw-Hill, New York

Main Smith JD (1924a) Atomic structure. Chem Ind 43:323–325, 437, 548–549

Main Smith JD (1924b) Chemistry and atomic structure. E. Benn Ltd, London

Main Smith JD (1925) The distribution of electrons in atoms. Philos Mag 50:878–879

Mayer AM (1878) Floating magnets. Nature 17:487

Nagaoka H (1904) Kinetics of a system of particles illustrating the line and the band spectrum and the phenomena of radioactivity. Philos Mag 7:445

Nicholson J (1912) The constitution of the solar corona. Mon Not R Astron Soc 72:139, 677, 729

Pauli W (1925) Über den Zussammenhang des Abschusses der Elektronengruppen in Atom mit der Komplexstruktur der Spektren. Z Phys 31:373, 765

Perrin J (1897) Rayon cathodiques et rayons de Röntgen. Ann Chim Phys 7:496–555

Plucker J (1858) Auf die Einwirkung des Magneten auf die elektrischen Entladungen in verdünnten Gasen. Annalen der Physik Vol 179 Issue 1 151–157

Righi A (1890) Sulle traiettorie percorse nella convezione fotoelettrica. Atti Accad Naz Lincei 6, Rendiconti, p. 81–86

Righi A (1896) Sulla propagazione dell'elettricità nei gas attraversati dai raggi di Röntgens. Mem Accad Sci Bologna Serie V. Tomo VI., pp. 231–303

Rutherford E (1911) The scattering of α and ß particles by matter and the structure of the atom. Philos Mag 21:669–688

Rutherford E, Royds T (1909) The nature of the α particle from radioactive substances. Philos Mag 17:281

Rydberg JR (1886) Die Gesetze der Atomgewichtzahlen, vol. 11. Svenska Vetenskaps Academiens Handlungar, P.A. Norsted & Söner Ed., Stockholm pp. 13

Rydberg JR (1890) On the structure of the linespectra in the chemical elements. Philos Mag 29:331–337

Sommerfeld A (1916) Zur Quantentheorie der Spektrallinien. Ann Phys 51:1–94, 125–167

Sommerfeld A (1920) Allgemeine spektroskopische Gesetze, insbesondere ein magnetooptischer Zerlegungssatz. Ann Phys 63(2):221–263

Stoner EC (1924) The distribution of electrons among atomic levels. Philos Mag 48:719–736

Stoney GJ (1881) On the physical units of nature. Philos Mag 11:384

Tait PG (1877) On knots. Trans R Soc Edinburgh 28:145–190

Tait PG (1884) On knots, Part II. Trans R Soc Edinburgh 32:327–342

Tait PG (1885) On knots, Part III. Trans R Soc Edinburgh 32:493–506

Thomson W (1869) On vortex motion. Trans R Soc Edinburgh 25:217–260

Thomson W (1875) Vortex statics. Proc R Soc Edinburgh 9:58–78

Thomson JJ (1883) A treatise on the motion of vortex rings. Macmillan, London

Thomson JJ (1885) The vortex ring theory of gases. On the law of the distribution of energy among the molecules. Proc R Soc Lond 39:23–36

Thomson JJ (1897) Cathode rays. Proc R Inst 15:419

Thomson W (1901) Aepinus atomized. Archives néerlandaises des sciences exactes et naturelles 2:6

Thomson W (1902) Aepinus atomized. Philos Mag 3:257

Thomson JJ (1904) On the structure of the atom: an investigation of the stability and periods of oscillation of a number of corpuscles arranged at equal intervals around the circumference of a circle; with application of the results to the theory of atomic structure. Philos Mag 7:237

Thomson JJ (1907a) The corpuscular theory of matter. Constable & Co., London

Thomson JJ (1907b) On rays of positive electricity. Philos Mag (Series 6) 13:561

Uhlenbeck GE, Goudsmit S (1925) Ersetzung der Hypothese von unmechanischen Zwang durch eine Forderung bezügluch des inneren Verhaltens jedes ienzelnen Elektrons. Naturwissenschaften 13:953

Uhlenbeck GE, Goudsmit S (1926) Spinning electrons and the structure of spectra. Nature 117:264

von Helmholtz H (1858) Über Integrale der hydromechanischen Gleichung welche den Wirbelbewegungen entsprechen. J Reine Angew Math 55:25

von Helmholtz H (1881) On the modern development of Faraday's conceptions of electricity. J Chem Soc 39:277

Wichelhaus H (1867) Über Constitution und Zusammenhang† der organischen Säuren von 3 At. Kohlenstoff. Ann Chem 144:351–357

Wien W (1898) Untersuchungen über die elektrische Entladung in verdünnten Gasen. Ann Phys Chem 65:440

Chapter 6
Radioactivity

. . .und nun ging der Teufel los.. (Wilhelm Röntgen)

6.1 The Revival of Inorganic Chemistry

At the end of the nineteenth century, inorganic chemistry had lost most of its importance as ancestor of all other disciplines in which chemistry was evolving and was at the point of becoming more of an introductory teaching vehicle than a busy line of research as in the past. The study of the individual elements started to appear to the new generations of chemists as poor relative to the two great branches of chemistry, organic and physical chemistry. These instead were witnessing a period of great splendor and intense development, the first thanks to the formidable successes of organic synthesis and the second for the rigorous theoretical organization in which it had framed the structure of the molecules and the identification of their physical properties.

The work of the many "element hunters," who in the previous century had achieved fame and celebrity by digging in the northern regions of Sweden and Germany for strange ores from which new exotic elements were extracted through a patient and boring series of analytical procedures, seemed already bypassed by a growing interest in the interaction between electromagnetic fields and matter. This new field had found its peak of fame in the discovery of x-rays and of the fluorescence that this new source of radiation was able to induce in many materials. It was only thanks to the perseverance of a young Polish girl, Manya Sklodowska, who arrived in Paris to study physics at the Sorbonne, that atomic nuclei started to whisper again in their faint voice in an old laboratory of Rue Cuvier in the Latin Quarter of Paris; this was thanks to another kind of radiation, this time spontaneously emitted by some uranium compounds. From there this voice became a stronger and stronger roar, being heard throughout the world and giving rise to the new fascinating field of nuclear chemistry, later leading to the discovery of atomic energy and to its use for both good and evil.

S. Califano, *Pathways to Modern Chemical Physics*,
DOI 10.1007/978-3-642-28180-8_6, © Springer-Verlag Berlin Heidelberg 2012

6.2 The Curie Couple

The discovery of radioactivity started from the observation made in 1896 by Antoine Henri Becquerel that potassium uranyl sulfate $K_2UO_2(SO_4)_2$ spontaneously emitted a strange radiation capable of affecting a photographic plate. Becquerel came from a family of physicists; the grandfather, Antoine César (1788–1878), developed a method to extract metals from their minerals and the father, Alexander Edmond (1820–1891), studied solar radiation, the fluorescence and phosphorescence of minerals, and was a leading authority in Europe on the phosphorescence of solids. Just from his father, Antoine inherited a rich collection of fluorescent and phosphorescent minerals that later turned out to be precious for his research activities.

Antoine Henri Becquerel (1852–1908) obtained the title of engineer at the Ecole Nationale des Ponts et Chaussées in 1877 and a Ph.D. in physics in 1888. In a short while, he became professor of applied physics at the Museum of Natural History in 1892, at the Ecole Polytechnique in 1895, and later at the University of Paris. In a session of the Académie des Sciences, of which he was a member, he had the chance to listen to an enthusiastic lecture from the mathematician Henri Poincaré (1854–1912) on the discovery of x-rays, which occurred almost by accident by observing the fluorescence that they produced on striking the walls. Poincaré had received a letter from Wilhelm Conrad Röntgen, a scientist he had never met, with some astonishing photographs of bones made using the newly discovered radiation. In his lecture, Poincaré advanced the hypothesis that x-rays could somehow be related to the phenomenon of phosphorescence and Becquerel, impressed by this idea, decided to restart the studies of his father on induced fluorescence and phosphorescence, using solar light as the exciting radiation.

After having studied several compounds, he realized that fluorescence and phosphorescence were similar physical processes, both stimulated by external radiation. The main difference that he noticed between the two processes was that while fluorescence ceased if the exciting radiation was stopped, phosphorescence continued for a long time.

On 25 January 1896, Antoine Becquerel started the study of induced fluorescence in crystals and on 24 February reported to the Academy that several materials, in particular the crystals of potassium uranyl sulfate $K_2UO_2(SO_4)_2$, a compound known for its phosphorescence properties, emitted rays that penetrated thick sheets of black paper. Becquerel's approach to the study of the induced phosphorescence was to expose the sample to solar light and position it on a photographic plate wrapped in light-tight black paper, leaving the experiment on a windowsill where, according to his hypothesis, the sunlight would stimulate the mineral to glow, emitting the unknown x-rays. Wednesday, 26 February, and Thursday, 27 February, were days without sunshine in Paris and Becquerel, being unable to use solar energy for his experiment, decided to position the sample on a package of photographic plates wrapped in black paper in a drawer in his laboratory and left it there. On 1 March he developed the plates and found that these had been impressed by an invisible radiation spontaneously emitted by the crystals without exposition to solar light. It is even possible, as told in some records of Becquerel's experiment, that he simply forgot the sample positioned in the drawer. In any case, when he developed the plate and saw the shadow of the crystals impressed on it, he reached the wrong conclusion that the solar radiation, adsorbed by the crystals, was transformed in x-rays.

The first of March was actually a Sunday and it is hard to believe that he went back to his laboratory without a serious reason to carry out an operation of little interest that his colleagues of the Academy had practically ignored and that did not seem important compared to Röntgen's discovery. In reality, Becquerel knew that one of the pioneers of photography in France, Claude-Félix-Abel Niepce de Saint Victor (1805–1870), had already observed in 1867 that uranium salts weakly blackened photographic plates even when wrapped in several sheets of black paper. The observations of Niepce were reported in a book written by Becquerel's father that he undoubtedly read. It is therefore highly possible that, remembering Niepce's findings, he went back to the laboratory to examine the plates and to make sure that the phenomenon described by Niepce was real.

After a few days, Becquerel reported his discovery to the *Académie des Sciences* (Becquerel 1896a, b), but the news was almost completely neglected by the members of the Academy, excited by the discovery of x-rays to which Becquerel's rays did not seem comparable. In addition, after Röntgen's discovery, the hunt for all kind of invisible rays emitted by matter had started in the scientific community, leading to a plethora of phenomena, often purely imaginary, that gave rise to quite a bit of confusion and overall to much diffidence.

Among the several inventors and professed physicists impressed by Röntgen's discovery, the French Gustave Le Bon (1841–1931) succeeded more than the others in attracting the attention of the press and even the sympathy and benevolence of important scientists such as Poincaré. Le Bon, social psychologist, sociologist, and amateur physicist, reported in 1896 having discovered a "black light," a new type of invisible electromagnetic radiation different, but somehow correlated to, x-rays. Even if he did not receive too much attention in the physics community, his ideas on matter, radiations, and the ether stirred the interest of a wide audience starting to be aware of the imminent revolution against positivism. His book *L'Évolution de la matière* published in Paris by Flammarion in 1905, in which he maintained that the matter is unstable and transforms slowly but inexorably into a radiation, had as many as 12 editions in France.

Less lucky was the physicist René Blondlot (1849–1930), professor at Nancy, known for his research on electromagnetism, on Maxwell's theory, and for having realized in 1891 the first measurement of the propagation speed of radio waves. Blondlot added to Le Bon's fantasies a famous mistake of experimental physics, announcing to the Académie des Sciences in 1903 the discovery of a new type of rays that he named N rays in honor of Nancy. In a very short time, several confirmations of the existence of the N rays appeared in the scientific literature and the Académie even awarded him a 20,000 francs prize. Unfortunately for him in 1904 the American physicist Robert William Wood (1868–1955), professor of optics at the Johns Hopkins University, visiting Blondlot's laboratory at the behest of the journal *Nature*, revealed (Wood 1904) that Blondlot's discovery was a phenomenon of collective self-deception and imagination and not a physical process since he had verified that the N rays were detected by Blondlot's coworkers even when he had surreptitiously removed a prism from the optical path! Despite the indifference and disinterest of the members of the Académie, Becquerel continued to work for some time on the radiation emitted by uranium salts, proving with the aid of a magnetic field that it consisted of charged particles. Later he devoted himself to the study of the effect of magnetic fields on spectral lines, an effect discovered in 1896 by Pieter Zeeman (1865–1943) at Leyden, and presented a series of communications on the Zeeman and Faraday effects to the Académie. Even his son, Jean-Antoine-Edmond Becquerel (1878–1953) followed in the footsteps of his father although with less success.

If the Academicians did not play too much attention to Becquerel's discovery, his work was continued by a young Polish woman, Marie Sklodowska, and by her husband, Pierre Curie. With their research they created a new branch of chemistry, radiochemistry, procuring for Becquerel the honor of the Nobel Prize in physics in 1903.

Pierre Curie (1859–1906), who had the position of instructor at the Municipal School of Industrial Chemistry and Physics in Paris, was an already established physicist when he met his future wife. In 1880, together with his brother Jacques, he discovered piezoelectricity (Curie and Curie 1880) and was internationally known for having found that the magnetic susceptibility of paramagnetic materials was inversely proportional to the absolute temperature (Curie–Weiss law) as well as for having identified the temperature, the Curie point, above which the magnetic properties of a material disappear (Curie 1895). Despite these significant scientific results, he had not received the respect from his peers to match the level of his research, mostly because of his reserved and shy nature, alien to academic politics, but also because he had not yet written his Ph.D. thesis that he completed in March 1895, a few months before getting married.

Marya (afterward changed to Marie) Sklodowska (1867–1934) was born in Warsaw, Poland, at that time part of the Russian empire. In 1891, she followed her elder sister Bronisława to go to study in Paris, where she graduated at the Sorbonne in physics in 1893 and in mathematics in 1894. To support herself at the University she started, under the direction of the physicist Gabriel Lippmann (1845–1921), to study the magnetic properties of steel, thanks to a fellowship financed by the society for the support to the national industry.

Soon after their marriage in July 1895, the Curie husband and wife started to work together on the magnetism of metals in the small and badly equipped laboratory that Pierre had directed since 1882. In 1897, Marie completed her work on steel magnetism and started to look for a new field for her Ph.D. thesis. The discovery of the strange radiations emitted by uranium salts, made 1 year before by Becquerel, fascinated Marie who decided to select this completely new, virgin, and very promising topic for her Ph.D. thesis. Using an electrometer based on the piezoelectric effect, invented by Pierre and by his brother Paul-Jacques (1856–1941), Marie dedicated herself to the determination of the conductibility of air exposed to uranium salt radiation, a technique that Becquerel had already used with success to evaluate radiation intensity. The apparatus employed by Marie consisted essentially of a plate condenser, on which the finely powdered active material was spread, making the air between the plates a conductor of electricity. In order to measure the conductivity, the plate was raised to a high potential by connecting it to the pole of a battery of small accumulators. Marie found that the emission of rays was proportional to the uranium content of the material, constant over time, and independent of temperature and the illumination of the sample. She therefore started to test all available compounds to understand which of them emitted the unknown rays.

In this research, she was helped by several chemistry colleagues who supplied samples of rare earths, and in April 1898 she discovered that the same radiation was also emitted by thorium compounds, an element identified by Berzelius in 1828. By systematically investigating different compounds, she soon reached the conclusion that the intensity of the emitted radiation was independent of the chemical composition and depended only on the quantity of uranium or thorium present in the samples and understood that the capability of the emitted rays was a property specific to the atoms of uranium and thorium. The immutable atoms of the Greek philosophers had finally started to speak and to reveal their complex internal structure.

On 17 February 1898, the Curie couple examined samples of some uranium ores: pitchblende, a black pitch like mineral from the Joachimsthal mines in Bohemia, and chalcolite (also known as torbernite). Both minerals contained uranium, thorium, and rare earth oxides, and produced an ionization of the air greater than that of uranium oxide. Pitchblende was four times as active as uranium itself, and chalcolite twice as active.

After repeated measurements, yielding always the same result, Marie Skłodowska-Curie, as she liked to be called to underline her Polish origins, reached the conclusion that unknown compounds, more radioactive than uranium, were present in the ores. Pierre, once he realized that what she observed was not a spurious effect and that his wife had in her hands an important discovery, abandoned his own research on crystal symmetry and magnetism to dedicate himself to identifying, with her, these new unknown compounds.

Marie Curie collected in her memoirs a detailed description of this famous period of her life, of the squalid laboratory with the floor in a precarious condition, of the poor available equipment, of the cold that they suffered in winter and the asphyxiating heat in summer, and, above all, of the back-breaking work of purification and extraction of the radioactive fractions from tons of raw material. Her daughter Ève Curie (1904–2007) reconstructed with passion the fascinating atmosphere of this romanced and romantic period of her mother's life in a well-known book (Curie 1938).

In this period, the Curie couple changed to analytical chemists and, using separation techniques involving an infinite series of interminable sequences of fractional crystallizations, succeeded in isolating from the pitchblende ores several fractions, one of which, rich in bismuth, and another, rich in barium, were both strongly "radioactive."

From the bismuth-rich fraction they obtained, at the end of June 1898, a new substance, chemically similar to bismuth, and about 400 times more radioactive than uranium. On 18 July 1898, they published in Comptes Rendus a paper entitled *Sur une substance nouvelle radioactive, contenue dans la pechblende* (Curie and Curie 1898) in which they suggested for the new element the name of polonium, in honor of Marie's native country. In this paper, the term *radioactive* was mentioned for the first time, a term coined by Marie to indicate the emission of the unknown rays that Becquerel had named "U rays."

The true nature of these U rays was discovered in 1898 by Ernest Rutherford who showed that they consisted of two kinds of particles that, with a minimum effort of imagination, were called α and β rays. In the same period of time, the U rays were studied in detail by Hans Mayer (1904) as well as by Stefan Meyer and Egon von Schweidler (1905).

After few months, the Curies identified in the barium rich fraction, in collaboration with the chemist Gustave Bemont (1857–1937), coworker of Pierre, the existence of a second radioactive element for which they suggested the name radium. The discovery of radium was published on 26 December with the title *Sur une nouvelle substance fortement radioactive, contenue dans la pechblende* (Curie et al. 1898).

At the beginning of 1900, Pierre brought to his research a young chemist André Debierne (1874–1949), assistant at the laboratory of Charles Friedel of the Ecole de Chimie et Physique. Debierne was a short, stocky, and bald introvert personality who had relatively little credit in the international physics community. Nevertheless, he became an important friend of the Curie family and intensively collaborated with them in their research on unknown radioactive elements.

Starting from 1900, Marie obtained a chair at the École Normale Supérieure of Sèvres. In 1902, she was able to announce success in isolating 0.1 g of pure radium chloride from more than a ton of pitchblende, obtained thanks to the courtesy of the geologist Eduard Suess (1831–1914) born in London but Austrian naturalized and a member of the Austrian Academy of Sciences. In 1902, in collaboration with Debierne, Marie isolated pure radium by electrolysis of the chloride using a mercury cathode and distilling off the mercury in a hydrogen atmosphere. From a measurement of the molecular weight and from the emission spectrum, it was easy to prove that it was really a new element. An important contribution to the preparation of appreciable amounts of pure radium came from the work of another Austrian scientist, the physicist Stefan Meyer (1872–1949), director of the Institute for Radium Research in Vienna. Meyer had started his scientific activity as assistant to Boltzmann at the Institute for Theoretical Physics, University of Vienna, working on the magnetic permeability of liquids. Influenced, however, by a lecture of Friedrich Oskar Giesel, a pioneer in the research and production of radium, he decided to work on radioactivity and, in collaboration with Egon von Schweidler (1873–1948), was able to show that the β rays could be deflected by magnetic fields. At the beginning of the twentieth century (1910), the Institute for Radium Research in Vienna was opened thanks to funds donated in 1908 by the Austrian industrialist

Karl Kupelwieser, and Meyer became its first acting director. By request of the Austrian Academy of Sciences, Meyer organized, with the assistance of technicians of the Auer von Welsbach chemical plant, the production of 4 g of radium by extraction from pitchblende obtained from the Sankt Joachimsthal mines, located in Austria-Hungary. The radium – which Meyer shared with the Curies in Paris, Rutherford in Manchester, and Ramsey in London – made him a key figure in the research on radium.

The fact that materials containing radium spontaneously emitted light, presented by the Curies at the First International Physics Conference held in Paris in 1900, excited the imagination of the public, making radioactivity the most popular branch of science of the time and giving rise to a series of commercial activities and of mass media interest, intimately connected to an increase of its public visibility.

In June 1903, Marie Skłodowska-Curie obtained a Ph.D. in physics and in the same year she and her husband Pierre Curie shared with Becquerel the Nobel Prize in physics. One half of the Prize was assigned to Becquerel for the discovery of radioactivity and the other half to Pierre and Marie for the discovery of polonium and radium. With the Nobel Prize, Marie and Pierre Curie suddenly became famous. The Sorbonne gave Pierre a professorship and permitted him to establish his own laboratory, in which Marie became the director of research.

In January 1904, the industrialist Armet de Lisle, who supported the Curies in the production of radium, opened new industrial installations for producing radium for medical applications. Thanks to De Lisle's help, the Curies were able to obtain larger samples of radioactive material than they would never be able to prepare on their own. In 1904, the journal *Le Radium* was founded, the first scientific journal devoted to radioactivity, edited by Jacques Danne, an assistant to Pierre Curie, and directed by Henri Farjas. *Le Radium* initially was intended to be an "instrument of popularization and research," but after only 6 months it became a strictly academic journal. In the same period, books for a broad public of nonexperts were published by Marie Curie, *Untersuchungen über die Radioactiven Substanzen* (Braunschweig, Vieweg, 1904), by Henri Becquerel, *On the radio-activity of matter* (from The Smithsonian Report for 1902, Washington, 1903), by Foveau de Courmelles, *Les applications médicales du radium* (Imprimerie P. Orsoni, Paris, 1904), and by Alphonse Berget, *Le radium et les nouvelles radiations: que faut-il en penser? Que faut-il en attendre?* (Paris, G. Charpentier 1904).

On 19 April 1906, a tragedy struck the Curie family. Pierre was killed in a street accident, struck by a horse-drawn car while walking in heavy rain across Rue Dauphine near the Pont Neuf. Marie, devastated by the death of her husband, was left alone with two daughters, Irène 9 and Eve 2 years old. Fortunately, the physics department decided to entrust her with the chair created for Pierre Curie and also to appoint her as director of the laboratory.

Marie was the first woman to obtain a Ph.D. in science in 1908, the first woman to become professor of physics at the Sorbonne, the first woman invited to a Solvay conference, and the first scientist to obtain two Nobel Prizes, the second for chemistry being awarded to her in 1911. In spite of these important recognitions and of the great international success of her researches, Marie had to face at the end of 1910 and for the whole 1911 a ferocious denigrating campaign in the French press. At the beginning of the nineteenth century, the nationalist and chauvinist French society, fueled by gloomy currents of xenophobia, anti-Semitism, and anti-scientism, would not easily accept that a Polish woman, with a last name like Sklodowska and an additional first name "Salomea," suggesting a Jewish origin, furthermore connected to a group of intellectual supporters of the cause of the traitor Dreyfus and suspected of antinationalistic political positions, could be considered a true representative of the great French culture. This disparaging campaign, directed by the journalist Léon Daudet in the journal *L'Action Française*, became more intense when a sentimental relationship of about a year's duration between Marie and the physicist Paul Langevin (1872–1946), a former student of Pierre Curie, was made public, portraying Marie as a home-wrecker. This resulted in a great press scandal that excited the "prudery" of affluent French society.

This "Langevin affair" rapidly became a national scandal that culminated in a pistol duel, fortunately without consequence, between Paul Langevin and Gustave Téry, editor of the journal *L'Oeuvre* who had published some extracts of compromising letters of Marie to Langevin, stolen from the Langevin house. The matter deteriorated further when Téry accused Langevin of being a cad and a scoundrel so that, to defend his honor, Langevin was forced to challenge him to a duel. The duel with pistols at a distance of 25 m was organized for the morning of 26 November 1911. However, it turned into a real farce. The inexperienced Langevin even had a difficult time finding seconds for the day of the duel, since his colleagues were afraid of being involved in this unpleasant story and only at the last moment did he succeed in convincing two friends, in particular the mathematician Paul Painlevé, who later would become prime minister of the French Republic, to assume this embarrassing job. The morning of the duel was gray and foggy. Langevin arrived first and Téry with his entourage arrived after a short while. Painlevé who had been chosen by lot to direct the procedure but had no practice of duels, after having asked the combatants if they were ready, to the general surprise, without waiting, counted with a loud voice one, two, three, fire, before the two adversaries could be ready. Langevin who was just as incompetent raised his arm with the pistol, but, seeing that Téry was not yet ready, lowered his arm again. At this point, Téry said that he had no intention of depriving France of one of her best physicists and he too lowered his arm. Painlevé then quickly decided that the duel was over and in a general silence the two combatants went back home.

However, the situation became more and more embroiled when, due to the discomfort created by a scandal that had assumed international dimensions, Svante Arrhenius, member of the Nobel committee, sent a letter to Marie suggesting that she should refuse to go to Stockholm to receive the prize until the situation could be fully settled. Marie obviously answered rather truculently that she had received the Nobel Prize for the discovery of plutonium and radium and that the evaluation of the Nobel Academy could not be influenced by gossip relating to her private life.

When Albert Einstein heard the story of the Langevin affair, he wrote to Marie:

> I will always be grateful that we have people like you and Langevin among us.... If the rabble continues to be occupied with you, simply stop reading that drivel. Leave it to the vipers it was fabricated for.

On 11 December, Marie, holding her head highly, pronounced the Nobel speech at Stockholm, and after the ceremony dined at the Royal Palace with the King of Sweden, Gustaf V. However, the stress provoked by this unhappy series of events gave rise to a strong nervous breakdown for which she was after hospitalized and for almost a year unable to go back to work.

For the rest of her life Marie dedicated herself to the development of the use of radium for the treatment of cancer, an important medical application that Pierre had predicted once he had realized that the radiations emitted by radioactive compounds could destroy organic molecules and in particular could be used to destroy malignant tumors (Quinn 1995).

6.3 Hunting for New Radio Elements

With her research Marie Curie opened a new chapter of inorganic chemistry that was soon found to be extremely rich for important developments, giving rise, soon after the discovery of radioactivity, to a large amount of new research that flourished in few years all over the world.

In France, André Debierne, who had dedicated himself to the research of an element that he suspected should have been present in an iron rich-fraction extracted from pitchblende, discovered in 1899 a new radioactive element, actinium, and presented his discovery in three papers (Debierne 1899, 1903, 1904). At Marie's death in 1934, Debierne became director of the Curie laboratory despite the fact that he did not have the international renown expected for the successor of a personality like Marie Curie.

In 1898, in Germany Gerhard Carl Schmidt (1864–1949), professor at the University of Erlangen, discovered slightly before Marie Curie the radioactivity of thorium. Schmidt reported his results to the *Deutsche Physikalische Gesellschaft* in Berlin on 4 February 1898 and published a paper on this subject on 24 March of the same year (Schmidt 1898), a few months before the communication presented by Marie Curie on 12 April at the Académie des Sciences of Paris. Unfortunately for him, Marie Curie's name had already reached such a high level of popularity in the physics community that it obscured his priority in the discovery of thorium radioactivity.

Always in Germany, Julius Elster (1854–1920) and Hans Friedrich Geitel (1855–1923), an inseparable couple of teachers at the Wolfenbüttel Schule, known essentially for having discovered the photoelectric cell, proved experimentally at the end of 1898 that the radiation emitted by uranium salts was the same at ambient pressure, under vacuum, and in a mine at a depth of 853 m. In 1901, they measured the radioactivity of the air and soon after that of the ground and studied the radioactive decay for long time. In the 1902–1905 period, Friedrich Oskar (Fritz) Giesel (1852–1927) from the Braunschweig University observed the presence of a new radioactive substance in minerals of lanthanum and cerium that he named *emanium* since it emitted a radioactive gas (Giesel 1900–1905). Soon, however, it became clear that the emanium was nothing other than actinium, previously discovered by Debierne.

In 1900, Becquerel, by precipitation of insoluble carbonates from a solution containing uranyl ions, discovered the strange fact that the uranium was left in solution in the form of a soluble uranyl carbonate but that most of the radioactivity disappeared from the solution, appearing instead in the precipitate that did not contain uranium. There was therefore some radioactive material that "emaned" from the uranium salts and that escaped from the solution with the precipitate (Becquerel 1901).

In 1900, while investigating the radiochemical properties of uranium, W. Crookes and Becquerel made an important discovery: by precipitation of a carbonate salt from a solution containing uranium, they discovered that while the uranium remained in the supernatant as a soluble uranyl carbonate complex, the radioactivity originally associated with the uranium was now present in the precipitate, which contained no uranium. Moreover, the radioactivity of the precipitate slowly decreased with time, while the supernatant showed a corresponding increase of radioactivity. Similar results were obtained by E. Rutherford and F. Soddy when investigating the radioactivity of thorium.

The existence of radioactive gases that seemed to be present in receivers "emanating" from radioactive materials added further complications to the research involved in arranging the newly discovered elements in the periodic table (Curie 1899). In 1899, Robert Bowie Owens (1870–1940) (Owens 1899) and Ernest Rutherford (1900a, b) identified an emanation of thorium, and in the same year Pierre and Marie Curie observed that even radium produced an emanation that, in contrast to that of thorium, remained radioactive for months (Curie and Curie 1899). The occurrence of thorium and uranium emanations was further confirmed by Friedrich Ernst Dorn (1848–1916), professor at the Friedrichs Universität of Halle (Dorn 1900).

In 1900, William Crookes, who in 1861 had isolated thallium, succeeded in isolating at the same time as Becquerel a uranium fraction that seemed to possess the full emission. He named this fraction of uranium as uranium-X (later identified as protactinium). He also observed that the radioactivity of uranium-X decayed very quickly while in the meantime the radioactivity of the original uranium was restored. In 1903, A. Debierne discovered the same emanation in actinium (Debierne 1903). In 1909, Daniel Strömholm (1871–1961), professor of chemistry at the University of Uppsala, and Theodor Svedberg (1884–1971), at that time still associate professor, used isomorphism to identify the chemical character of the new radio elements. By crystallization of different salts in solutions containing radio elements, they found that thorium X (^{208}Pb) crystallized together with the lead and barium salts, but not with other salts, and concluded that it was an alkaline earth substance in contrast to previous erroneous conclusions that it was a univalent element.

In 1904, William Ramsay and John Norman Collie (1859–1942) at University College London suggested eliminating the term emanation from the terminology of the radioactive compounds and using the prefix eka, initiated by Mendeleev, for the unknown elements of the periodic table. In this way the emanations would be called eka-thorium, eka-radium, eka-actinium, etc. Their proposal, however, fell on deaf ears. In 1910, Ramsay went back to the problem together with Robert Whytlaw-Gray (1877–1958), suggesting (Ramsay and Whytlaw-Gray 1910) the name *niton* for the radium emanation. Despite these repeated attempts, the term emanation survived until 1923 when the international commission for the chemical elements chose the name radon, for all emanations (Aston et al. 1923). The different emanations, in fact, all possessed the same absorption spectrum that resembled that of the rare gases, argon, krypton, and xenon, a fact that suggested that it was an element of the family of the rare gases. This new element was isolated and studied by Rutherford and Soddy in 1902 and by Ramsay and Whytlaw Gray (1877–1958) in 1908. In few years it was clear that the different emanations were all produced by the radioactive decay of various isotopes of the elements. Radon was found very useful in medicine, especially in cancer research and in industrial tests, and was even utilized in the control of industrial welding processes.

6.4 Transmutation of the Elements

At the beginning of the twentieth century, radioactivity appeared as a new chapter of inorganic chemistry, relative to a small group of rather exotic elements, rarely found in nature, and barely utilizable in industry, with at best some useful applications for medical treatments and for dating archeological objects. For the physicists, radioactivity continued to be a mysterious topic since the nature of the emitted rays was still unknown and nobody knew its relationship with atomic structure. The scientist that did realize the connection between atomic structure and radioactivity within few years, and paved the way to the study of the nuclear structure, was a New Zealander, Ernest Rutherford, another great representative of the school of physics that, from Lord Kelvin to John Thomson, had made Cambridge the center of the physical world of that period. Ernest Rutherford, whose research on the atomic structure are discussed in Chap. 5, started his studies on radioactivity when in 1896, as a young researcher at Cambridge in the laboratory directed by J. J. Thomson, he assisted in the spread of the news of the discovery of Becquerel's U rays. The unknown nature of these rays stimulated his interest and pushed him to investigate their composition. In 1898, he succeeded in showing that the U rays were made of two types of particles that he named α and β rays, different for their penetrating power into matter, as well as for their charge and for their mass. The α rays were absorbed by thin metallic sheets of a few millimeters thickness while the β rays were able to cross metallic sheets 100 times thicker. The paper, published in 1899 when Rutherford was at the McGill University in Canada (Rutherford 1899), was sent by J.J. Thomson to the editor of the "Philosophical

Magazine" on 1 September 1898, almost contemporaneously to that of Marie Curie on radioactivity. In 1899, Fritz Geisel, Antoine-Henri Becquerel, and Marie Curie proved that the β rays were very fast electrons. The nature of the α particles, however, was understood only after some years. In the period 1906–1912, the German Hans Geiger (1882–1945) moved to Manchester to work with Rutherford who was now professor of physics there. In 1908, Rutherford and Geiger built an instrument to measure and count α particles and proposed that these were He nuclei (Rutherford and Geiger 1908). In 1909, Rutherford and Thomas Royds (1884–1955) definitively confirmed this hypothesis (Rutherford and Royds 1909).

The emission of new rays from radioactive materials continued to present new surprises; in 1900 the French chemical-physicist Paul Villard (1860–1934), working at the École Normale Superieure of Paris, discovered that uranium compounds emitted a third type of radiation even more penetrating than x-rays, able to get through a lead plate of several centimeters thickness (Villard 1900). Villard proved that these rays, that he named γ rays, were not deviated by an electric or magnetic field and therefore concluded that they were of the same nature as x-rays.

Rutherford's research activity was recognized as being of the highest level by everybody, and in particular by Thomson, his principal supporter. The rules of the University of Cambridge were, however, extremely rigid and offered few hopes to obtain in a short time a stable position for him. An interesting opportunity occurred, however, when a professorship opened at the McDonald Laboratory of the McGill University of Montreal in Canada. Rutherford decided to try this adventure and in 1898 left England for the new seat. The laboratory had been founded in 1891 thanks to the financial support of a philanthropist, Sir William MacDonald, who also supplied the funds to establish there the chair of physics that Rutherford occupied.

The McGill laboratories were well equipped and had one of the best collections in the world of radioactive materials. Rutherford therefore had no difficulties in resuming the work that he had started at Cambridge. Once arrived at the McGill in 1899, he started to collaborate with Robert Bowie Owens (1870–1940), professor of electro-technics at the same university, who studied the ionizing power of the thorium radiation. Owens had started his research with a normal sample of thorium, but the results of his measurements seemed completely erratic, subject to inexplicable sudden variations depending on the strangest factors such as the opening of a door of the laboratory or the displacement of a person around the working table.

After some time, Owens was convinced that the measurements were completely untreatable and that the results were random, due to air currents. For instance, if the system was placed in a closed receiver in which air circulated, the radioactivity decreased but if it was left quiet for about a quarter of an hour, the radioactivity went back to the initial value. Owens, bored and annoyed, abandoned the problem in the hands of Rutherford who, in a relatively short time, was able to prove that the radioactivity was not induced in the air around the sample, but was an emanation escaping from the thorium oxide in the form of a radioactive gas capable of making radioactive all materials that it contacted (Rutherford 1900a, b).

> The phenomena exhibited by thorium compounds receives a complete explanation if we suppose that, in addition to the ordinary radiation, a large number of radio-active particles are given out from the mass of the active substance. This "emanation" can pass through considerable thicknesses of paper. The radio-active particles emitted by the thorium compounds gradually diffuse through the gas in its neighbourhood and become centres of ionization throughout the gas.

Furthermore, by plotting in a diagram the variation of the radioactivity intensity as a function of time, he obtained an exponential curve that allowed him to define the concept of half-life, i.e., the time necessary to reduce to one half the radioactivity of a sample. His first Canadian student, Harriet Brookes (1876–1933), who became the first female nuclear physicist in Canada, collaborated with him on the series of research on the thorium emanation.

In the years 1900–1901, Rutherford continued to study thorium and radium (^{226}Ra) emanations. In 1900, a young English chemist, Frederick Soddy (1877–1956), coming from the University of Oxford, where he had worked as researcher from 1898 to 1900, joined Rutherford's research group at the McGill University to work with him on thorium radioactivity. The collaboration with Soddy gave rise to an intense activity in analytical chemistry to identify the unknown atoms which appeared and disappeared in connection with the thorium emanation. With reference to this period, Rutherford used to tell his friends that the fastest transformation he knew was his own from a physicist to a chemist in these days!

Their work with microscopic quantities of substances which changed over time, without supplying the smallest hint of what was happening, led them to look for models in order to find the right direction to follow. They continued formulating new conjectures that, after a short while, were found to be wrong. For 2 years they collected a large amount of data and invented an equivalent number of fanciful interpretations of the thorium emission, trying to understand how they collected the energy necessary to cross successive layers of aluminum sheets.

In 1902, Rutherford and Soddy started also a research on the emission of thorium using a new sample of pure thorium nitrate arrived from Germany. Soddy decided to prepare thorium hydroxide from the nitrate by precipitation with sodium carbonate or ammonium hydroxide. By filtration he collected in the filter solid thorium hydroxide that had completely lost its radioactivity. By evaporation of the filtrate he found, however, a solid residue which was strongly radioactive. They named thorium-X this new radioactive material as done by Crookes who named uranium-X the emission of uranium. In January 1902, after the Christmas holidays, they discovered in a new experiment that thorium-X was produced from thorium at a constant speed and that it decayed following an exponential law. Rutherford and Soddy hurried to publish these results in two papers in which they clearly pointed out that a new type of matter was produced in a kind of chemical reaction that occurred inside the nucleus (Rutherford and Soddy 1902a, b):

> The position is thus reached that radioactivity is at once an atomic phenomenon and the accompaniment of a chemical change in which new kinds of matter are produced. The two considerations force us to the conclusion that radioactivity is a manifestation of subatomic

> chemical change. [...]The idea of the chemical atom in certain cases spontaneously breaking up with the evolution of energy is not of itself contrary to anything that is known of the properties of the atoms, for the causes that bring about the disruption are not among those that are yet under our control.

In order to spread this information throughout the physics community, they published the same article in the *Philosophical Magazine*, a typical physics journal, repeating with different words the same concepts (Rutherford and Soddy 1902b); after 2,000 years Rutherford had discovered that the alchemist's dream, the transmutation of the elements, was a reality and existed in nature! These conclusions were so important and innovative that Rutherford hurriedly wrote to Crookes, editor of the *Journal of the Chemical Society*, requesting him to ensure that publication would not suffer any delays.

Soddy and Rutherford continued their collaboration trying to measure the energy released in the radioactive decay. In February 1903, Rutherford found that the α particles emitted in the decay were heavy particles carrying a positive charge and obtained a ratio e/m between the charge and the mass of 6,000 with respect to the value 10,000 of e/m for hydrogen, concluding that about 99% of the energy was carried by the α particles. In three successive 1902 papers, they repeatedly pointed out that in all minerals containing uranium, helium always appears as an inclusion, a fact that encourages the supposition that it is a final product of the radioactive decay.

In 1903, Soddy left Canada to go to work with William Ramsay at University College London, where he continued his research on radium emanation. In the same year, Soddy and Ramsay published a paper in which they announced that, using emission spectroscopy, they had discovered that the helium was produced in the radioactive decay of radium bromide as well as in the decay of the emanation (Ramsay and Soddy 1903).

In 1904, Soddy took up an appointment at Glasgow's University where he stayed for 10 years until in 1914 he became professor of chemistry at the University of Aberdeen. At Glasgow, Soddy discovered the existence of the isotopes and developed his displacement law that establishes the rule governing the transmutation of an element during radioactive decay. According to this law, better known as the Fayans and Soddy law, a radioactive element emitting an α particle shifts back two places in the periodic table while it moves forward one place by emitting a β particle or back one place by capturing an electron. This law represented the rule for the construction of the radioactive families of elements (Soddy 1913).

The term "isotope" was suggested to Soddy in 1913 by Margaret Todd, a Scottish doctor and writer to whom he was distantly related. Talking to her, Soddy tried to explain that, according to his research, it did look like different elements occupied the same place in the periodic table. The learned Todd noticed immediately that in Greek "the same place" is called ισο-τόπος (*iso-topos*) and suggested that he use it. Soddy accepted the suggestion right away and since then the term isotope has been in the public domain.

For his investigations of the disintegration of the elements and of the chemistry of radioactive substances, and in particular for the discovery of the displacement law and for the concept of isotope, Soddy received the Nobel Prize for chemistry in 1921.

Almost at the same time the Pole Kazimierz Fajans (1887–1975), in collaboration with Otto Göhrling, studied the sequence of the radioactive transformations of uranium ^{238}U decay (Fajans 1913a, b) and also arrived at the idea of the existence of isotopes. Fajans and Göhrling actually realized, using the idea of the radioactive displacements, that by emission of an α particle followed by emission of β and γ rays, uranium 238 transformed into uranium 234 according to the nuclear reaction

$$^{238}\mathrm{U} \xrightarrow{\alpha} {}^{234}\mathrm{Th} \xrightarrow{\beta,\gamma} {}^{234}\mathrm{Pa} \xrightarrow{\beta,\gamma} {}^{238}\mathrm{U}$$

They called the unstable nucleotide ^{234}Pa of protactinium of very short half-life (77 s) brevium (from the Latin *brevis*). The following year, 1914, a paper appeared written by Theodore William Richards, professor at Harvard and future Nobel laureate for his very accurate atomic weights measurements, and by Max Lembert, a student of Fajans who had gone to work with Richards. They reported several different atomic weights for lead extracted from different samples (Richards and Lembert 1914). In this paper, the authors pointed out the agreement of their results with the hypothesis of the existence of isotopes of Fajans and Soddy:

> This amazing outcome is contrary to Harvard experience with several other elements, notably copper, silver, iron, sodium, and chlorine, each of which seems to give a constant atomic weight, no matter what the geographical source may have been. No attempt is made here to discuss the theoretical aspects of the facts presented, but attention is called to their qualitative agreement with the hypothesis of Fajans and of Soddy.

The existence of protactinium was confirmed after few years in 1918 when Otto Hahn and Lise Meitner in Germany and Frederick Soddy and John Cranston (1891–1972) in England independently identified the isotope ^{231}Pa. In 1934, Aristide von Grosse (1905–1985) from the University of Chicago succeeded in preparing 2 mg of the oxide and 7 years later he obtained the pure metal (von Grosse 1934).

Kazimierz Fajans, after having studied at Leipzig, Heidelberg, and Zurich, went to Manchester in 1910 to the laboratory directed by Rutherford where he worked in collaboration with Moseley. Later he returned to Germany, first to the university of Karlsruhe and then as professor of chemical physics to Munich. In 1935, with the growth of the Nazi persecution of the Jews, he left Germany and immigrated to the United States as professor at the Ann Arbor University in Michigan. In collaboration with Otto Hahn (1936) he discovered the conditions of precipitation and absorption of radioactive substances, an important technique of separation and cleaning of small traces of radioactive substances. In 1924, a solid candidature of Fajans was submitted to the Nobel Prize committee. The great majority of the scientific community involved in radioactive activities was almost certain that he would get it, up to the point that the Swedish magazine "Svenska Dagbladet" announced his victory the day before the final decision of the Nobel committee. Unfortunately for him, the next day the Committee announced that no prize, either in chemistry or in physics, was to be awarded that year. After this he was again proposed twice but without success.

The proof of the existence of isotopes came later from the J.J. Thomson group, in particular from Thomson's collaboration with Francis William Aston which presented experimental evidence for the existence of two isotopes of neon (Thomson 1913a, b).

At the beginning of the twentieth century, about 40 radioactive elements were already known and among them several had exactly the same chemical properties although a different atomic weight. For example, ten different varieties of thorium with atomic weights ranging from 232 to 212, two of uranium, and six of actinium had already been identified and there was no place in the periodic table to allocate this multitude of elements. This fact made several chemists suspicious that one should completely reexamine the periodic table, one of the milestone of the structure of chemical theory, since in the table there were only six places available to locate a much larger number of elements, between the position of hydrogen and that of uranium, corresponding to positions 43, 61, 72, 75, 85, and 87.

In 1907, Thomson started his research on the canal rays, made of positively charged particles. Letting a jet of neon atoms cross an electric and a magnetic field, he registered on a photographic plate two traces, corresponding to two types of particles, one with atomic weight 20 and the other 22. To explain this strange result Thomson proposed the existence of either an unknown neon compound, for instance, a neon hydride, or of a new element, meta-neon. The occurrence of two traces did not seem, however, too convincing for the chemistry community without a true chemical separation of the two types of particles and was considered by many as an experimental artifact.

In 1909, Francis William Aston (1887–1945), who started to be interested in cathode rays at Birmingham under the guide of the chemists William Augustus Tilden (1842–1926) and Percy Frankland (1858–1946) and of the physicist John Poynting (1852–1914), went to work at the Cavendish Laboratory upon the invitation of Thomson to assist him in his experiments on the canal rays. Thomson committed Aston first to the problem of improving the equipment that had been built and then to the study of the existence of this strange meta-neon.

Aston improved Thomson's apparatus and in 1913 with the new instrument Thomson and Aston again studied neon and announced that the element of mass 22 had the same properties as the neon of mass 20. After having identified spectroscopically the two isotopes of neon, Aston dedicated himself to their separation. First he tried without success the fractioning on coal cooled in liquid air. In 1913, he used with more success gas diffusion through tubes of porous clay, using a quartz microbalance to control the different stages of the separation procedure of the two species.

With the start of the First World War in 1914, Aston was forced to stop his research and, because of his competence in mechanics and combustion processes, was assigned as technical assistant to the Royal Air Force factory at Farnborough to study the effects of atmospheric conditions on materials for aeronautics.

Back at the Cavendish Laboratory in 1919, he built a new instrument much more efficient than the first. In Thomson's original apparatus, the electric and magnetic fields were superposed so that the particles crossed the two fields simultaneously. However, Aston separated them, letting the magnetic field act on the particles already separated by the electric field. Using several ingenious devices, Aston succeeded in focusing at the same point of the photographic plate all particles possessing the same charge/mass ratio. With this new instrument, Aston definitively proved the existence of two neon isotopes. Soon after, he discovered that even chlorine had two isotopes ^{35}Cl and ^{37}Cl (Aston 1920b) and, having understood that the existence of the isotopes was a general feature of matter, started a systematic study of the isotopic effect, discovering that bromine, lithium, boron, and argon also had two isotopes, that magnesium had three, krypton and sulfur six, and xenon seven. In the following 3 years, he analyzed more than 30 elements and in total in his full research activity he identified 212 isotopes of different elements. In 1921, he decided to build a new instrument with much better performances than the previous one, overall with a better resolving power, at the point that he succeeded in separating 6 isotopes of mercury in the range from 198 to 204 atomic masses. Aston summarized the results of his research in two famous papers (Aston 1919, 1920a) and in the book *Isotopes* of 1922 (Aston 1922).

In the introduction to the book he wrote

> The importance, from purely practical and technical points of view, of the theory of isotopes would have been insignificant had its application been confined to the radioactive elements and their products, which are present in infinitesimal quantities on the Earth. But now that the isotopic nature of many elements in everyday use has been demonstrated, the possibility of their separation to any reasonable extent raises questions of the most profound importance to applied science.

For this research, he was awarded the Nobel Prize for chemistry in 1922 "for his investigations into the disintegration of the elements, and the chemistry of radioactive substances." In 1937, Aston built a third prototype, far more advanced than the previous one, with a resolving power about 20 times better and a precision in the determination of mass about 100 times greater (Aston 1937).

The apparatus built by Aston was called a mass spectrograph since it allowed one to separate ions or neutral atoms of different atomic mass. With this technique, it was easy to characterize the isotopes of even well-known elements. For instance, William Francis Giauque (1895–1982) proved the existence of three isotopes of oxygen of atomic mass 16, 17, and 18 (Giauque and Johnston 1929a, b) and Harold Clayton Urey discovered deuterium (Urey et al. 1932a, b), the isotope of hydrogen of atomic mass 2.

Aston's mass spectrograph has been perfected over the years and represents today an analytical tool of fundamental importance in all fields of chemical, biological, and pharmaceutical research.

6.5 Completion of the Periodic Table

The discovery of so many new elements made it necessary to modernize the periodic table, introducing new classificatory techniques more in line with the recent theoretical developments. By now it was clear that atomic weight was too coarse a variable for correct positioning of the elements in the table. This concept had been already expressed in 1886 by Johannes Rydberg (1854–1919), the first to mention that the order number of the elements was the only independent variable that could be expressed by integer numbers (Rydberg 1886). In addition, new elements, such as polonium, actinium, radium, radon, and the whole series of different isotopes, had been discovered and many of them were not easily inserted into the Mendeleev table.

In 1911, a Dutch lawyer and amateur physicist, Antonius Johannes Van den Broek (1870–1926), suggested that the number representing the position of an element in the periodic table, later the atomic number Z, corresponded to the total number of electrons in the atom. This idea was presented in a paper that Van den Broek published in the journal *Nature* (Van den Broek 1911) in July 1911, only 1 month after the publication of the famous article of Rutherford on the atomic nucleus. In two successive papers, (Van den Broek 1913a, b) van den Broek maintained that it was an error to compare the nuclear charge to the atomic weight of an element, suggesting that instead one should use the atomic number which represents the number of positive charges on the nucleus as proved by the experiments of diffusion of the α particles.

Rutherford had limited his discussion in his paper to the observation that the nuclear charge was an integer number of the order of half of the atomic weight, starting from the assumption that the atomic nuclei were made of helium nuclei, each with a nuclear charge equal to one half of its atomic weight. This implicitly corresponded to supposing that the nuclear charge was equal to the atomic number.

Henry Gwyn Jeffreys Moseley (1887–1915) strongly contributed to the final arrangement of the periodic table. He was one of the most brilliant pupils of Rutherford, prematurely killed in action during the First World War at the age of 27 years in the battle of Gallipoli in Turkey on 10 August 1915, being hit in the head by a sniper while, as signal officer of the 38th brigade of the Royal Engineers, he was calling headquarters on the phone. Son of a professor of anatomy at Oxford and a student of Trinity College of the same university, Moseley went to work with Rutherford in 1910 when the latter was professor at Manchester. In 1913 he discovered the law that bears his name, a systematic relationship between the frequency of the X-rays and the position of the elements in the periodic table. In particular, he found that a perfect linear relation is found by plotting the square root of the x-ray frequency as a function of atomic number (Moseley 1913, 1914).

Up to that time the concept of atomic number defining the position of the element in the periodic table was considered only representative of the sequence of the atomic mass and possessed no direct structural meaning. Moseley's law showed instead that the atomic number was a perfectly measurable quantity, confirming the hypothesis of Van den Broek in that it represented the number of electrons in the atom and thus the number of positive charges in the nucleus.

Moseley's measurements allowed one to define the exact position of each element in the periodic table, solving many still unanswered questions. For example, in building his table on the basis of chemical properties, Mendeleev had inverted the position of cobalt with that of nickel, locating cobalt at position 27 and nickel at position 28, despite the fact that the atomic weight of cobalt was slightly higher than that of nickel. Moseley's data perfectly confirmed this genial intuition of Mendeleev. Furthermore, Moseley showed that in the atomic numbers sequence, the positions 43, 61, 72, 75, 85, and 87 were still empty, confirming on one hand Mendeleev's prediction and supplying on the other the key to localize exactly the missing elements.

The existence of the radioactive element 61, *promethium*, had been foreseen in 1902 by the Czeck chemist Bohuslav Brauner (1855–1935), professor of inorganic chemistry at the Carl IV University of Prague. Brauner was one of the principal experts on rare earths. He anticipated the existence of the isotopes and suggested the use of oxygen as a basis for the atomic weights of the elements. A good friend of Mendeleev, he investigated the position of the rare earths in the periodic table (Brauner 1901) that he extended alongside lanthanum, suggesting that all rare earths should be placed in a single position of the periodic table between lanthanum (57) and tantalum (73). Brauner's proposal was not accepted due to the opposition of Mendeleev, until Moseley confirmed beyond any doubt that only 14 elements existed after lanthanum. This gave rise to the birth of the lanthanide series, from lanthanum (57) to lutetium (71), forming a small separate periodic table. Actually the number of lanthanides was an open problem for chemistry at the beginning of the twentieth century, since for several of them the pure products did not exist in appreciable quantity, being difficult to separate chemically mixtures of very similar elements. It was only thanks to Moseley's data that the problem was definitively solved.

While working on the rare earths, Brauner realized that the difference between neodymium and samarium was too big in comparison to the other lanthanides and made the hypothesis of the existence of a new element at position 61 between them (Brauner 1926). Different research groups all over the world declared within a few years to have somehow identified element 61, without, however, being able to prove it. The discovery of new elements had actually become at the end of the nineteenth and the beginning of the twentieth century quite a mass phenomenon. A certain Kosmann, unknown as a scientist, even reported in 1896 in a chemical journal a kind of April Fool joke, saying that he had succeeded in detecting two new elements at once, with the pompous names of kosmium and neokosmium. It soon became apparent that the "discoverer" of kosmium and neokosmium had just been making fun of this discovery's popularity.

The two most famous cases of imaginative self-deception were those of the hypothetical elements *florentium* and *illinium*. In 1924, Luigi Rolla (1882–1960), professor at the University of Florence, Italy, insisted that he had discovered, in collaboration with his student Lorenzo Fernandes (1902–1977), element 61 and sent a letter to the Accademia Nazionale dei Lincei containing the result of his spectroscopic analysis (Rolla and Fernandes 1926). Two years later in 1926 at the

University of Illinois, Urbana, B. Smith Hopkins (1873–1952), an expert on the chemistry of the rare earths, and his coworkers Leonard Yntema and J. Allen Harris, maintained that they had also identified element 61 and proposed for it the name illinium (Hopkins et al. 1926). All attempts made by experts in this field, such as Wilhelm Prandtl as well as the Noddack husband and wife, to confirm the validity of the experimental arguments supplied by the two group for florentium and illinium, respectively, resulted in inconsistencies because it was soon clear that it should be a radioactive element. The proof of the existence of element 61 was only obtained in 1945 by Charles D. Coryell (1912–1971), chief of the section for the study of the products of nuclear fission of the Manhattan project and his coworkers Jacob Akiba Marinsky (1918–2005) and Lawrence E. Glendenin (1918–2008) during the analysis of some by- products of uranium fission, made using an ion exchange column. The presence of the isotope of mass 147 of element 61 was confirmed using mass spectroscopy.

Due to the secrecy that surrounded all research connected with the war, the discovery was only announced in 1947. Coryell, acting upon the suggestion of his wife, proposed the name *prometheus*. In 1949, the International Union for Chemistry decided on the present name of *promethium*.

Mendeleev had already predicted the existence of element 75 with chemical properties similar to those of manganese that he named eka-manganese. In 1925, two researchers from Berlin, Walter Noddack and Ida Tackle, who married in 1926, after having searched for long time for eka-manganese in different minerals such as columbite, gadolinite, molybdenite, and several minerals of platinum, published a paper (Noddack et al. 1925) declaring that they had identified, with the help of the x-ray expert Otto Berg from the Siemens laboratories, element 75 that they called *rhenium* in honor of the Rhine river. Only in 1927, however, did they succeed in obtaining 1 g of rhenium from more than 600 kg of molybdenite (Noddack and Noddack 1927). The couple also claimed to have identified the element of atomic number 43 that they called *masurium* in honor of the Masuria, region of Eastern Prussia where Walter Noddack was born, but they never succeeded in proving experimentally that they really had found it. Element 43 was instead identified in 1937 at the Physics Institute of the University of Palermo, Italy, by the mineralogist Carlo Perrier (1886–1948) and the physicist Emilio Segrè (1905–1989). It was called *technetium* from the Greek τεχνητος, artificial, because it was artificially produced (Perrier and Segrè 1937). Segrè had brought from Berkeley to Palermo a sample of molybdenum obtained from Ernest Lawrence with whom he collaborated. The molybdenum was bombarded with deuterium nuclei at the Berkeley cyclotron, creating the technetium isotopes ^{95}Tc and ^{97}Tc. Technetium was thus the first artificially created element.

The element 72, hafnium (from *Hafnia*, the Latin name of Copenhagen) was discovered in 1923 by the Hungarian George Charles de Hevesy and by the Dane Dirk Coster at Copenhagen, the city where both worked with Niels Bohr.

The most credited version of the discovery of hafnium is based on the belief that the chemists of the time thought that element 72 had to be a rare earth. Apparently Niels Bohr, analyzing in 1921 the electron distribution of the electrons in the atomic levels, realized instead that with lutetium the fourth electronic shell was completed and therefore that element 72 could not be a rare earth. It is therefore said that Bohr asked his coworkers de Hevesy and Coster to look for it in the zirconium minerals, a choice that allowed them to find it in a relatively short time, and that Carl Popper utilized it to confirm his theory of the reduction of chemistry to the quantum theory. This version, however, has been recently disputed by Eric R. Scerri of the department of chemistry and biochemistry of the University of California, Los Angeles, and editor of the journal *Foundations of Chemistry* (Springer) (Scerri 1994). Scerri bases his different version on a series of documents obtained from the son of Friedrich (Fritz) Paneth, showing that it was not Bohr but the same Paneth, with whom de Hevesy had collaborated at the radium institute at Vienna, who suggested looking for hafnium in zirconium minerals. Friedrich Adolf Paneth (1887–1958), in addition to being an expert in radiochemistry, was also competent with epistemological problems of chemistry and published several papers on the topic, in particular a manuscript of more than 200 pages of a lecture that he gave at Konigsberg in honor of Kant, titled *Die erkenntnistheoretische Stellung des chemischen Elementbegriffs*. This lecture, published in German in the records of the *Königsberger Gelehrten Gesellschaft*, was later translated into English by his son with the title *The epistemological status of the chemical concept of element* (Paneth 1962). Paneth was an antireductionist, convinced that the network of empirical relationships created by the great chemistry of the nineteenth century could not be replaced by mathematical techniques, as he pointed out in his lecture:

> Even if the character of chemistry should change essentially in the future owing to penetration by mathematico-physical methods, its history during the nineteenth century, in which it achieved such successes without mathematics, must never be ignored in its philosophic evaluation.

This could explain why the suggestion of Paneth to de Hevesy was based on chemical knowledge and not on electronic configuration.

Element 87 was the last to be discovered, even if there were clear hints that it should be chemically similar to the alkaline metals. Mendeleev had in fact called it eka-cesium, and, in addition, it was expected by experts that it would be radioactive, being intermediate between two radioactive elements, radon and radium.

Despite the fact that few doubts existed on its radioactive nature, several attempts were made to identify a stable eka-cesium, obviously without results, except that of filling the literature with fanciful names such as *russium* (1925), *alkalinium* (1929), *virginium* (1932), *moldavium* (1937), *mosandium* (named after the Swedish chemist, Mosander), etc. With the publication of the law of the radioactive displacement of Moseley it was clear that element 87 could either be derived from actinium 89 by emission of an α particle or from radon by emission of a β electron:

$$^{227}_{89}\mathrm{Ac} \longrightarrow {}^{223}_{87}\mathrm{Ac} + {}^{4}_{2}\mathrm{He} \qquad {}^{222}_{86}\mathrm{Rn} \xrightarrow{\beta} {}^{222}_{87}\mathrm{eka-Cs}$$

and therefore research on element 87 was essentially limited to these two possible nuclear reactions, preferentially relative to the actinium decay since there were serious doubts about the possibility of a β emission from radon.

At the end of the 1930s, the Institute du Radium of Paris was strongly involved in the study of actinium, discovered in 1900 by André Louis Debierne, the director of the institute, where also worked Irène, daughter of Marie Curie and wife of Jean Joliot. Debierne had a bad character and was by no means a good director. He did not get along too well with Irène Joliot-Curie to the point where they worked on the same problem without even informing each other of their progress. At their service worked a young technician, Marguerite Perey (1909–1975), appointed by Marie Curie in 1929 who had learned from her the techniques of purification and handling of radioactive materials.

Debierne and Irène Curie both separately asked Marguerite Perey to purify actinium samples. Irène Joliot-Curie was interested in determining the half-life of the 227 (^{227}Ac) nucleotide whereas Debierne on his side was interested in looking for new, not better identified, radio elements (Adloff and Kauffman 2005).

Working on the purification of actinium, Perey discovered that the 227 actinium could decay either by emission of a β electron, thereby giving rise to thorium 227, or by emitting an α particle and producing a new element according to the scheme

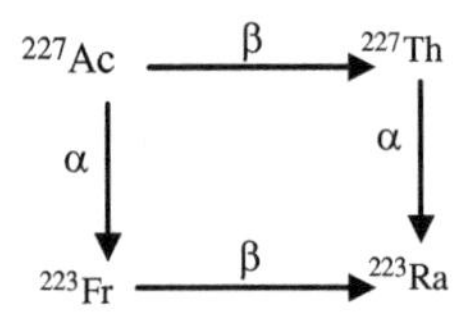

This new element, temporarily called *catium*, produced by emission of a β electron the 223 isotope of radium.

Marguerite Perey separately informed Debierne and Irène Curie of her experiment with the expected result that each of the two contenders considered themselves as the true inspirer of Perey's research. In particular, when Irène Curie went to see Debierne and told him that the technician Perey discovered a new element during research that she, Irene Curie, had directed, a violent fight started between them so that at the end they both decided not to be associated with the research that was therefore published under the name of only Marguerite Perey (1939). The name *catium* selected by Perey did not, however, please Irène Curie or the other members of the institute, and was replaced by the name *francium*, obviously acceptable to everybody.

The development of nuclear chemistry has strongly benefited from the availability of projectiles such as α particles and from the discovery of a new nuclear particle, the neutron. In 1919, Rutherford, striking nitrogen atoms with α particles, realized for the first time in the laboratory the alchemist's dream of transforming one chemical element into another (Rutherford 1920):

$$^{4}_{2}He^{++} + ^{14}_{7}N \rightarrow ^{17}_{8}O^{+} + ^{1}_{1}H^{+}$$

In this nuclear reaction, he obtained the emission of ionized hydrogen atoms (protons) and realized that the protons were part of the structure of the atomic nucleus:

> We must conclude that the nitrogen atom is disintegrated under the intense forces developed in a close collision with a swift alpha particle, and that the hydrogen atom which is liberated formed a constituent part of the nitrogen nucleus.

The term proton was probably coined just by Rutherford since, according to Pais (1991), it appeared for the first time in Rutherford's paper *Nuclear Constitution of the Atoms* (Rutherford 1920).

In 1920, Rutherford knew only two types of particles, protons and electrons, and with only these two particles it was problematic to build a theory of nuclear structure that could satisfy the laws of physics and at the same time account for the fact that the mass of the atom was roughly double that of the nuclear charge. There was another problem to explain, that of the emission of the β rays, electrons that undoubtedly came from the nucleus. If electrons were present in the nucleus as true particles, they had to travel around the protons on orbits of extremely small dimensions, this implying that they possessed monstrously huge energies. For example, the α particles, that were helium nuclei, should have been made of four protons bound together by two electrons. The impossibility of this kind of nuclear structure was theoretically proved in the 1930s by the Russian Armenian astrophysicist Viktor Ambarzumian (Hambardzumyan) and by the Ukrainian theoretical physicist Dmitri Iwanenko (Ambarzumian and Iwanenko 1930a, b; Iwanenko 1932).

Even more difficult to explain was the nuclear structure of heavy elements like uranium, for which hundreds of electrons had to be contained in the extremely small volume of the nucleus. Rutherford therefore made the hypothesis of the existence of a third type of elementary particle, electrically neutral, in which in an unknown way a proton and an electron were fused together. He called this unknown particle a *neutron* from the Latin root *neutral* with the Greek ending *on* in order to find the correct assonance with the names of the known elementary particles proton and electron. Of course all students of the Cavendish Laboratory were immediately addressed to look for this new particle without, however, any significant result.

A new series of research opened the road to the discovery of the neutron. In 1930, Walther Bothe, one of the great German nuclear physicists, Nobel Prize winner for physics in 1954, and his coworker Herbert Becker discovered that beryllium ($_4{}^9Be$) bombarded with α particles emitted a new type of highly penetrative radiation able to cross up to 20 cm of lead, 200 times more efficient than protons, and concluded that it was made of very high-energy γ rays.

In 1932, Irène Joliot-Curie and her husband Jean Joliot published a paper entitled *Émission de protons de grande vitesse par les substances hydrogénées sous l'influence des rayons γ trés pénétrants* (Curie and Joliot 1932) in which they maintained that by bombarding paraffin or other substances containing hydrogen with these γ rays emitted from beryllium, one obtained the emission of very high-speed protons (De Gregorio 2006):

> Nous avons étudié ces rayonnements par l'ionisation qu'ils produisent dans une chambre montée sur un èlectromètre Hoffmann.[...] Le courant augmente notablement quand on interpose des écrans de substances contenant de l'hydrogène comme la paraffine, l'eau, le cellophane. L'effet le plus intense a été observé avec la paraffine[...] Ces rayons γ de grande énergie sont capables de projeter des protons de grande vitesse quand ils traversent une substance hydrogénée.[...] Cet effet s'accompagne d'une absorption considérable du rayonnement par les noyaux d'hydrogène.

In the meantime, an old pupil of Rutherford at Manchester, James Chadwick, who in 1913 had left England to work in Berlin with Hans Geiger, went back in 1919, at the end of the war, to work again with his old master at the Cavendish Laboratory where in 1923 became research vice-director. At the outbreak of war in 1914, as an English citizen, he had been interned in the *Zivilgefangenenlager* of Ruhleben. When Chadwick read the paper of the Joliot Curie couple, he went to speak with Rutherford who immediately answered "I do not believe it!" As a matter of fact, in order to expel high-speed protons from paraffin, the γ rays should have had energies of the order of 50 million eV. At this point, Chadwick repeated the Bothe–Becker experiment, bombarding the beryllium with α particles, and succeeded in proving that the emitted rays were made of particles. By measuring (indirectly) their mass, he showed that the mysterious rays, emitted according to the nuclear reaction

$$ {}^{4}_{2}\mathrm{He} + {}^{9}_{4}\mathrm{Be} \rightarrow {}^{12}_{6}\mathrm{C} + {}^{1}_{0}\mathrm{n} $$

were not γ rays but the neutrons predicted by Rutherford.

Soon afterward Chadwick also discovered how to obtain neutrons by bombarding boron with α particles:

$$ {}^{4}_{2}\mathrm{He} + {}^{11}_{5}\mathrm{B} \rightarrow {}^{14}_{7}\mathrm{N} + {}^{1}_{0}\mathrm{n} $$

The discovery of the neutrons, sent on 27 February 1932, to *Nature* and successively to other scientific journals (Chadwick 1932) procured for Chadwick the 1935 Nobel Prize for physics and offered the nuclear physicists a kind of projectile much more convenient than α particles or protons, since, being neutral, they were not repelled by the nuclear charge. After the appearance of neutrons on the physics scene, it became common practice to utilize the mass number, i.e., the sum of the number of protons and neutrons in place of the atomic mass of the isotopes.

6.6 Transuranium Elements

When the periodic table had been completed from hydrogen to uranium on the basis of Moseley's law, the problem arose of the existence of elements of atomic weight greater than that of uranium, i.e., of transuranium elements. Once it was understood that unstable elements existed, able to spit pieces off their nucleus in the form of α or β particles, releasing great amounts of energy as γ rays, it was natural to pose the question whether it was possible to profit from this instability to create new elements, bombarding atoms with projectiles able to smash their structure or even possibly to be swallowed in their nucleus, thereby increasing the atomic weight. At Rome in 1934 the Italian physicist Enrico Fermi (1901–1954) decided to profit from the fact that the neutrons had no electrical charge and to bombard uranium atoms, hoping to obtain new transuranium elements. Fermi and coworkers found that at least four new radioactive substances were formed (Fermi et al. 1934; Amaldi et al. 1935) and assumed, perhaps too hastily, that among them new transuranic compounds could be present. Fermi's idea that bombarding uranium with neutrons could produce transuranium elements was soon criticized by Ida Noddack. She insisted that it was easier for a nucleus filled with protons and neutrons to split into smaller pieces rather than to absorb new neutrons, namely, that the fission of the nucleus was more probable than the formation of transuranic elements. Noddack's objection was, however, completely ignored by the physicists, not only because of the scientific prestige of Fermi but also because of the doubts arising from her false discovery of masurium.

As we shall see, both Fermi and Noddack were actually right. In 1939, Otto Hahn, Lise Meitner, and Fritz Strassmann realized at Berlin the fission of uranium, bombarded with neutrons. Later it was understood that the fission was the consequence of the formation of an unstable transuranic element that decayed into smaller fragments.

Fermi's idea was soon developed by Otto Hahn (1879–1968) in Berlin and led to the development of nuclear fission from which came both the atomic pile and the atomic bomb.

Otto Hahn, son of a rich, well-to-do glazier of Frankfurt am Main, studied chemistry at Munich and Marburg where in 1901 he received his Ph.D. in organic chemistry. He was not destined to study radioactive elements, being bent on a career as an industrial chemist. After the Ph.D., he went to England to improve his English and to visit Ramsay's laboratory in London where he found himself fully immersed in the world of nuclear physics and in direct contact with the principal representatives of this new branch of science.

Ramsay, who had at his disposal an impure sample of radium, asked him to purify it, and during this work Hahn discovered a new radioactive substance that he named radium-thorium. Excited by the discovery and encouraged by Ramsay, Hahn decided to forget his plans to join the chemical industry and decided to continue instead with the study of radioactive substances. To improve his competencies in this new field, he went to work from the autumn of 1905 to the summer of the following year with Rutherford at the McGill University in Canada. In this very profitable year, he discovered radioactinium, and learned how to study the emission of α particles from radioactive compounds, publishing a paper on their mass (Hahn 1908). Back in Germany he obtained a position at the Emil Fischer's Institute in Berlin, where he started his academic career. At the end of 1907, he met an Austrian theoretical physicist, Lise Meitner, Ph.D., from the University of Vienna, who had visited Berlin to follow Planck's lectures. She started a collaboration with him that lasted more than 30 years. For some years they worked together in temporary rooms, since the university did not accept that a woman could have an official position, until in 1912 they moved to the new Institute of Chemistry of Kaiser Wilhelm Gesellschaft (today Max Planck Gesellschaft) at Berlin-Dahlem where Fritz Haber was the director of the Chemical Physics Institute and where Hahn became director of the Institute of Radiochemistry. At the outbreak of war, Hahn was recalled to arms and sent, together with James Franck and Gustav Hertz, to a special unit of the German army directed by Fritz Haber, specializing in the production of chlorine and mustard gases, while Lise Meitner became a Red Cross nurse, in charge of the x-ray service of the Austrian army.

At the end of the war they restarted their collaboration in Berlin and in 1918 discovered protactinium 231, the mother element of the actinides series (Hahn and Meitner 1918). In 1921, Hahn discovered uranium Z, the first example of nuclear isomerism, i.e., the first example of the existence of a metastable nucleus theoretically explained in 1936 by the physicist and philosopher Carl Friedrich von Weizsäcker (1912–2007), a former assistant of Lise Meitner who was later involved in the uranium project for the construction of a German atomic bomb.

For more than 12 years Hahn dedicated himself to the study of the application of radioactive techniques to chemical problems until, at the beginning of 1938, with Lise Meitner and his assistant Fritz Strassmann (1902–1980), he decided to continue the kind of research started by Fermi in Italy, bombarding uranium with neutrons. Unfortunately with the annexation of Austria to Nazi Germany, Lise Meitner, of Jewish origin, was forced to leave Germany, and with the help of Niels Bohr was able to immigrate clandestinely to Sweden where she met her nephew, Otto Frisch, also a theoretical physicist. Before leaving Berlin, she published her last paper in collaboration with Hahn and Strassmann, maintaining that she had produced a new transuranic element with a half-life of 60 days, probably an isotope of iridium (Meitner et al. 1937).

After Meitner's departure, Hahn and Strassmann continued the experiments of bombardment with neutrons while still keeping in touch with Lise, the only true theoretician of the group. Lise Meitner and her nephew Otto Frisch understood Hahn's problem and wrote a letter to him suggesting that he should determine whether barium had been formed as a consequence of the bombardment of uranium with neutrons, since in this case the uranium had been broken into pieces. The research on barium, made using an organic salt of barium obtained from the chemist Wilhelm Traube (1866–1942), gave positive results and thus, at the end of 1938, Hahn and Strassmann communicated that as consequence of bombardment with neutrons the uranium atom had broken into two pieces (Hahn and Strassmann 1938). In further papers published in 1939 they announced that the other product of the fission was krypton (Hahn and Strassmann 1939a) and also that they had realized the fission of thorium (Hahn and Strassmann 1939b). At the same time, the theoretical interpretation of the nuclear fission process developed in Sweden by Lise Meitner and her nephew Otto Frisch appeared in *Nature* (Meitner and Frisch 1939).

For the discovery of nuclear fission, Hahn obtained the 1944 Nobel Prize in chemistry, a prize that he could collect only in 1946 when he went back to Germany, since at the end of the war he had been interned at Farm Hall, Godmanchester, close to Cambridge in England, being suspected to having contributed to the development of the German atomic bomb.

After the demonstration that the bombardment of the nucleus with neutrons gave rise to fission, as predicted by Noddack, the proof also arrived that transuranic elements could be formed as maintained by Fermi. In 1931, at Berkeley in California, Ernest Orlando Lawrence (1901–1958) constructed the first cyclotron (Lawrence and Livingston 1931), able to accelerate protons at energies of one million eV and in 1932, with the creation of the Radiation Laboratory, he succeeded in producing protons with energies of about four million V. With this new instrument, Lawrence discovered the transmutation of sodium through bombardment with deuterium atoms. In 1940, Edwin Mattison McMillan (1907–1991), in collaboration with Philip H. Abelson, discovered the first transuranic element, neptunium 239, bombarding uranium 238 with slow neutrons produced by the Berkeley cyclotron (McMillan and Abelson 1940):

$$^{238}_{92}U + ^{1}_{0}n \longrightarrow ^{239}_{92}U \xrightarrow{\beta} ^{239}_{93}Np$$

and in 1940 Glenn T. Seaborg, Edwin M. McMillan, J.W. Kennedy, and A.C. Wahl discovered that neptunium 239 has a half-life of 2 days, one half decaying into plutonium by the emission of β electrons:

$$^{239}_{93}Np \xrightarrow{\beta} ^{239}_{94}Pu$$

The discovery of plutonium was kept secret up to the publication of the paper at the end of the war (Seaborg et al. 1946).

The discovery of plutonium, for which McMillan and Seaborg received the 1951 Nobel Prize for physics, went far beyond their imagination and expectations. It was a discovery that has changed the course of history with the creation of the most terrible instrument of death that the human mind has devised – the atomic bomb.

In 1942, Enrico Fermi and his group at Chicago transformed uranium 238 into plutonium through a chain reaction that they achieved inside the first nuclear reactor. With the development in 1944 of the Manhattan project for the construction of the atomic bomb, research on nuclear fission increased very rapidly. Among the chemists involved in the Manhattan project, two researchers, Glen Seaborg and Albert Ghiorso, assumed a leading position in the research of transuranic elements and succeeded in giving back to inorganic chemistry the fundamental role that it had in the past.

Glen Seaborg (1912–1999) started his academic career at Berkeley under the supervision of Gilbert Newton Lewis with a Ph.D. thesis on the inelastic scattering of fast neutrons and afterward worked in inorganic chemistry and in particular in the search for the isotopes of the elements and for new transuranic elements.

He followed closely the developments that Enrico Fermi's group realized in Italy by bombarding uranium with neutrons, and the research that Otto Hahn performed in Berlin on the transuranic elements.

At Berkeley, he performed important research on artificial radioactivity, collaborating with the physicist John J. Livingood to use the newly completed Lawrence Radiation Laboratory 37-inch cyclotron to produce and discover several dozen new isotopes. During his lifetime he identified more than 100 isotopes of different elements. In 1937, in collaboration with John Livingood and Fred Fairbrother, he created isotope 59 of iron (^{59}Fe), later widely utilized in research on hemoglobin and in 1938 the isotope 131 of iodine (^{131}I), a very important nucleotide in the cure of thyroid diseases (Livingood and Seaborg 1938).

Together with his coworkers at the Lawrence Laboratory that he directed, and in particular with the electro-technical engineer Albert Ghiorso (1915–2010) who later succeeded him as director of the group, he prepared as many as ten transuranic elements (Ghiorso et al. 1950).

In 1940, Seaborg prepared plutonium 239 (^{239}Pu) by bombardment of uranium with deuterons. The result was officially published only in 1946 (Seaborg et al. 1946) due to the limitations on publication of material of military interest during the war. In 1944, he isolated americium 241 (^{241}Am), bombarding plutonium 239 with α particles in a nuclear reactor, obtaining in succession the isotopes ^{240}Pu and ^{241}Pu. Plutonium 241 transformed into ^{241}Am by β decay. In 1944, Seaborg prepared curium (^{242}Cm) by bombarding plutonium with α particles according to the nuclear reaction

$$^{239}\mathrm{Pu} + {}^{4}\mathrm{He} \rightarrow {}^{242}\mathrm{Cm} + {}^{1}\mathrm{n}.$$

In 1949, he isolated berkelium (^{243}Bk) by bombarding americium with α particles and producing ^{241}Am plus two neutrons. In 1950, he synthesized californium (^{245}Cf) by bombarding curium (^{242}Cm) with α particles and producing ^{245}Cf plus one neutron (Seaborg et al. 1950). In 1952, he identified fermium (^{255}Fm) and einsteinium (^{253}Es) in the debris of the explosion of the first atomic bomb and accounted for a complex nuclear reaction in which 15 neutrons were absorbed by uranium 238, giving rise to a chain of 7 beta decays (Ghiorso et al. 1955b). In 1955, Berkeley's group produced mendelevium (^{256}Md) plus one neutron, bombarding a sample of einsteinium with α particles (Ghiorso et al. 1955a). In 1958, with the aid of a linear accelerator Berkeley bombarded curium with carbon ions, obtaining the isotope 254 of element 102 named nobelium (^{102}No) (Ghiorso et al. 1958). Finally, in 1974, Berkeley's group and independently a group of soviet researchers of the Joint Institute for Nuclear Research at Dubno, directed by the physicist Georgij Nikolaevič Flerov (1913–1990), discovered the isotopes of mass 263 and 259, respectively, of element 106 (Ghiorso et al. 1974; Oganesian et al. 1974).

An important theoretical contribution of Seaborg was the identification of the series of the actinides separated from the rest of the periodic table as that of the lanthanides. Seaborg reached this conclusion after a series of vain attempts to synthesize americium and curium.

The synthesis of mendelevium, nobelium, fermium, and einsteinium gave rise to long nationalistic disputes between the United States and the Soviet Union where the group of nuclear physicists working at Dubno had obtained the same results and boasted priority.

After the Second World War, Seaborg became one of the most prominent American physicists, deeply involved in the nuclear energy policy and in the production of nuclear weapons as adviser of several presidents from Truman to Clinton. From 1961 to 1971 he was president of the *United States Atomic Energy Commission.*

References

Adloff J-P, Kauffman GB (2005) Francium (Atomic Number 87), the last discovered natural element. Chem Educ 10(5):387–394

Ambarzumian V, Iwanenko D (1930a) Les électrons inobservables et les rayons. Compt Rend Acad Sci Paris 190:582

Ambarzumian V, Iwanenko D (1930b) Über eine Folgerung der Diracschen Theorie der Protonen und Elektronen. Dokl Akad Nauk SSSR-A 6:153–155

Amaldi E, d'Agostino O, Fermi E, Pontecorvo B, Rasetti F, Segre E (1935) Artificial radioactivity produced by neutron bombardment. II. Proc R Soc A 149:522

Aston FW (1919) Isotopes and atomic weights. Nature 104:393

Aston FW (1920a) The constitution of the elements. Nature 105:617

Aston FW (1920b) The mass spectra of chemical elements. Philos Mag 39:611

Aston FW (1922) Isotopes. Edward Arnold & Co., London

Aston FW, Baxter GP, Brauner B, Debierne A, Leduc AF, Richards TW, Soddy F, Urbain G (1923) Report of the international committee on chemical elements. J Am Chem Soc 45:867

Aston FW (1937) A second-order focusing mass spectrograph and isotopic weights by the doublet method. Proc R Soc Lond A 163:391–404

Becquerel H (1896a) Sur les radiations emises par phosphorescence. Compt Rend 122:420–421

Becquerel H (1896b) Sur les radiations invisibles emises par les corps phosphorescents. Compt Rend 122:501–503

Becquerel H (1901) Sur la radioactivité de l'uranium. CR Acad Sci 83:977–978

Brauner B (1901) Die basis der Atomgewichte. (IV. Abhandlung.). Z Anorg Chem 26:186–202

Brauner B (1926) The new element of atomic number 61: illinium. Nature 118:84–85

Chadwick J (1932) Possible existence of a neutron. Nature 129:312, The existence of a neutron. Proc R Soc Ser A 136:692

Curie E (1938) Madame Curie. Gallimard, Paris

Curie J, Curie P (1880) Développement, par pression, de l'électricité polaire dans les cristaux hémièdres à faces inclinées. CR Acad Sci 91:294–295

Curie M (1899) Sur le poids atomique du métal dans le chlorure de baryum radifère. Compt Rend 129:760–762

Curie P (1895) Lois expérimentales du magnétisme. Propriétés magnétiques des corps à diverses temperatures. Ann Chim Phys 5:289–405

Curie P, Curie M (1898) Sur une nouvelle substance radioactive, contenue dans la pechblende. CR Acad Sci 127:175–178

Curie P, Curie M, Bèmont G (1898) Sur une nouvelle substance fortement radioactive contenue dans la pechblende. CR Acad Sci 127:1215–1217

Curie P, Curie M (1899) Sur la radioactivite provoquee par les rayons de Becquerel. CR Hebd Acad Sci 129:714–716

Curie I, Joliot F (1932) Émission de protons de grande vitesse par les substances hydrogénées sous l'influence des rayons γ trés pénétrants. Compt Rend 194:273
Debierne AL (1899) Sur une nouvelle matiére radio-active. Compt Rend 129:593
Debierne AL (1903) Sur la radioactivité induite provoquée par les sels d'actinium. Compt Rend 136:446
Debierne AL (1904) Sur l'actinium. Compt Rend 139:538–540
De Gregorio A (2006) Enrico Fermi and the discovery of neutron-induced radioactivity: a project being crowned. Stud Hist Philos Mod Phys 37:330
Dorn FE (1900) Die von radioactiven Substanzen ausgesandte Emanation. Abhandlungen Naturforschenden Gesellschaft (Halle) 23:1–15
Fajans K (1913a) Die radioaktiven Umwandlungen und das periodische System der Elemente. Berichte 46:422–439
Fajans K (1913b) A relationship between the nature of a radioactive transformation and the electrochemical behavior of the respective elements. Phys Z 14:131
Fermi E, Amaldi F, d'Agostino O, Rasetti F, Segrè E (1934) Artificial radioactivity produced by neutron bombardment. I. Proc R Soc A 146:483
Ghiorso A, James RA, Morgan LO, Seaborg GT (1950) Preparation of transplutonium isotopes by neutron irradiation. Phys Rev 78:472
Ghiorso A, Harvey B, Choppin G, Thompson S, Seaborg G (1955a) New element mendelevium, atomic number 101. Phys Rev 98:1518
Ghiorso A, Thompson SG, Higgins GH, Seaborg GT, Studier MH, Fields PR, Fried SM, Diamond H, Mech JF, Pyle GL, Huizenga JR, Hirsch A, Manning WM, Browne CI, Smith HL, Spence RW (1955b) New elements einsteinium and fermium, atomic numbers 99 and 100. Phys Rev 99:1048–1049
Ghiorso A, Sikkeland T, Walton JR, Seaborg GT (1958) Element 102. Phys Rev Lett 1:17–18
Ghiorso A, Nitschke JM, Alfonso JR, Alonso CT, Nurmia M, Seaborg GT, Hulet EK, Lougheed RW (1974) Element 106. Phys Rev Lett 33:1490–1493
Giauque WF, MacDougall DP (1933) *Attainment of temperatures below 1° absolute by demagnetization of* $Gd_2(SO_4)_3{\cdot}8H_2O$. Phys. Rev. 43:768
Giauque WF, MacDougall DP (1935) The Thermodynamic Temperature Scale in the Region Below 1° Absolute, Phys. Rev. 47, 885–886
Giauque WF, Johnston HL (1929a) An isotope of oxygen of mass 17 in the Earth's atmosphere. Nature 123:831
Giauque WF, Johnston HL (1929b) An isotope of oxygen, mass 18. J Am Chem Soc 51:1436–1441
Giesel FO (1900) Über radioactive Stoffe. Ber Deut Chem Ges 33:3569
Giesel FO (1902) Über radium und radioactive Stoffe. Ber Deut Chem Ges 35:3608–3611
Giesel FO (1903) Über den Emanationskörper aus Pechblende und über Radium. Chem Ber 36:342
Giesel FO (1904) Über emanium. Ber Deut Chem Ges 37:1696–1699
Giesel FO (1905) Über Emanium. Ber Deut Chem Ges 38:775–778
von Grosse A (1934) Element 91. Science 80:512–516
Hahn O (1908) Ein kurzlebiges Zwischenprodukt zwischen Mesothor und Radiothor. Phys Z 9:246
Hahn O, Meitner L (1918) Die Muttersubstanz des Actiniums, ein Neues Radioaktives Element von Langer Lebensdauer. Phys Z 19:208
Hahn O (1936) Applied radiochemistry. Cornell University Press, Ithaca
Hahn O, Strassmann F (1938) Die chemische Abscheidung der bei der Spaltung des Uran. Naturwissenschaften 26:756
Hahn O, Strassmann F (1939a) Über den Nachweis und das Verhalten der bei der Bestrahlung des Urans mittels Neutronen entstehenden Erdalkalimetalle. Naturwissenschaften 27:11
Hahn O, Strassmann F (1939b) Nachweis der Entstehung activer Bariumisotope aus Uran und Thorium durch Neutronenbestrahlung; Nachweis weiterer aktiver Bruchtucke bei der Uranspaltung. Naturwissenschaften 27:89–95

Hopkins BS, Yntema L, Harris JA (1926) The element of atomic number 61; illinium. Nature 117:792

Iwanenko DD (1932) The neutron hypothesis. Nature 129:798

Lawrence EO, Livingston MS (1931) The production of high speed protons without the use of high voltages. Phys Rev 38:834

Livingood JJ, Seaborg GT (1938) Radioactive iodine isotopes. Phys Rev 53:1015

Mayer H (1904) Die neueren Strahlungen; Kathoden-, Kanal-, Roentgen-Strahlen und die radioaktive Selbststrahlung (Becquerelstrahlen), 1st edn. Mährisch-Ostrau R. Papauschek

McMillan EM, Abelson PH (1940) Radioactive element 93. Phys Rev 57:1185

Meitner L, Frisch OR (1939) Disintegration of uranium by neutrons: a new type of nuclear reaction. Nature 143:239–240

Meitner L, Hahn O, Strassmann F (1937) Uber die umwandlungs Reihen des Urans die durch Neutronen Bestrahlung erzeugt verden. Z Phys 106:249

Meyer, Stefan and Schweidler, Egon von (1904) "Influence of Alternations of Temperature on Radioactive Substances." Pysicalische Zeitschrift, 5:319–320

Meyer S, von Schweidler E (1905) Sitzungberichte der Akademie der Wissenschaften zu Wien, Mathematisch-Naturwissenschaftliche Classe, p 1202 (Table 5)

Moseley HGJ (1913) The high frequency spectra of the elements, part 1. Philos Mag 26:1024

Moseley HGJ (1914) The high frequency spectra of the elements, part II. Philos Mag 27:703–713

Noddack W, Tacke I, Berg O (1925) Die Ekamangane. Naturwissenshaften 13:567

Noddack W, Noddack I (1927) Über den Nachweis der Ekamangane. Z Angew Chem 40:250

Oganesian YT, Tret'yakov YP, Il'inov AS, Demin AG, Pleve AA, Tret'yakova SP, Plotko VM, Ivanov MP, Danilov NA, Korotkin YS, Flerov GN (1974) Synthesis of neutron deficient isotopes of fermium,kurchatovium, and element 106. ZhETF Pis Red 20:580–585

Owens RB (1899) Thorium radiation. Philos Mag 48:360

Pais A (1991) Niels Bohr's times: in physics, philosophy, and polity. Oxford University Press, Oxford

Paneth FA (1962) The epistemological status of the chemical concept of element. Br J Philos Sci 13:1

Perey M (1939) Sur un élément 87, dérivé de l'actinium. Compt Rend 208:97

Perrier C, Segrè E (1937) Alcune proprietà chimiche dell'elemento 43. Rend Lincei 25:723, serie 6

Quinn S (1995) Marie Curie a life. Heinemann, London

Ramsay W, Soddy F (1903) Experiments in radioactivity, and the production of helium from radium. Proc R Soc 72:204–207

Ramsay W, Whytlaw-Gray RW (1910) Density of the emanation of radium. Compt Rend 151:126

Richards TW, Lembert ME (1914) The atomic weight of lead of radioactive origin. J Am Chem Soc 36:1329–1344

Rolla L, Fernandes L (1926) Über das Element der Atomnummer. Z Anorg Allg Chem 157:371

Rutherford E (1899) Uranium radiation and the electrical conduction produced by it. Philos Mag 47:109–163

Rutherford E (1900a) A radioactive substance emitted from thorium compounds. Philos Mag 49:1–14

Rutherford E (1900b) Radioactivity produced in substances by the action of thorium compounds. Philos Mag 49:161

Hahn O (1907) Mass of the α-particles from thorium. Philos Mag 12:371–378

Rutherford E, Soddy F (1902a) The radioactivity of thorium compounds I. Trans Chem Soc 81:321–350

Rutherford E II (1902) The radioactivity of thorium compounds II. The cause and nature of radioactivity. Trans Chem Soc 81:837–860

Rutherford E, Soddy F (1902b) On the cause and nature of radioactivity. PhilosMag 4 (19–24):370–396

Rutherford E, Geiger H (1908) The charge and nature of the alpha particle. Proc R Soc A81:162

Rutherford E, Royds T (1909) The nature of the α particle from radioactive substances. Philos Mag 17:281

Rutherford E (1920) Nuclear constitution of atoms. Proc R Soc Ser A97:374

Rydberg JR (1886) Die Gesetze der Atomgewichtzahlen. Svenska Vetenkaps Academiens Handlungar 11:13

Schmidt GC (1898) Über die vom Thorium und den Thorverbindungen ausgehende Strahlung. Ann Phys 65:141

Scerri ER (1994) Prediction of the nature of hafnium from chemistry, bohr's theory and quantum theory. Ann Sci 51:137

Seaborg GT, McMillan EM, Kennedy JW, Wahl AC (1946) Radioactive element 94 from deuterons on uranium. Phys Rev 69:366–367

Seaborg GT, Thompson SG, Street K Jr, Ghiorso A (1950) Element 98. Phys Rev 78:298–299

Soddy F (1913) The radio-elements and the periodic law. Chem News 107:97–99

Thomson JJ (1913a) On the appearance of helium and neon in vacuum tubes. Nature 90:645

Thomson JJ (1913b) Rays of positive electricity. Proc R Soc A 89:1–20

Urey HC, Brickwedde FG, Murphy GM (1932a) A hydrogen isotope of mass 2. Phys Rev 39:164–165

Urey HC, Brickwedde FG, Murphy GM (1932b) A hydrogen isotope of mass 2 and its concentration. Phys Rev 40:1–15

van den Broek A (1911) The number of possible elements and Mendeléeff's "Cubic" periodic system. Nature 87:78

van den Broek A (1913a) Die Radioelemente, das periodische System und die Konstituenten der Atome. Phys Z 14:32–41

van den Broek A (1913b) Intra-atomic charge. Nature 92:372–373

Villard P (1900) Sur la transparence de l'aluminium pour le rayonnement du radium. Compt Rend 130:1154–1157

Wood RW (1904) The N-rays. Nature 70:530–531

Chapter 7
From Quantum Mechanics to Quantum Chemistry

... the electron and the atom do not possess any degree of physical reality as objects of daily experience. ... Investigation of the type of physical reality which is proper to electrons and atoms is precisely the subject of atomic physics and thus also of quantum mechanics.

(Werner Heisenberg)

7.1 Planck's Quanta

The birth of quantum mechanics is rooted in the body of research that, at the end of the nineteenth century, dealt with the problem of electromagnetic radiation absorption and emission by an ideal absorbing-emitting system, the black body. The term "black body," coined in 1860 by Gustav Kirchhoff (1824–1887), designates an ideal body able to emit and absorb all possible frequencies of the electromagnetic spectrum, without reflections. Kirchhoff's experiments (Kirchhoff 1860) show that the radiation intensity emitted by a black body as a function of frequency depends on its temperature. In 1893, Wilhelm Wien (1864–1928) proved (Wien 1893) that the wavelength of the emission peak is a function of the inverse of the absolute temperature T (dotted curve in the figure below). In 1879, the Slovene physicist and poet Jožef Stefan (1835–1893) started also a research on black body emission. His experiments showed that the energy emitted by a black body per unit time and surface is proportional to the fourth power of the absolute temperature (Stefan 1879, 1881). His pupil, Ludwig Boltzmann, reached in 1884 the same conclusion on a purely theoretical basis, derived from thermodynamics principles. For this reason, the law that he proposed is known as the Stefan–Boltzmann law (Boltzmann 1884). In 1896, Wilhelm Wien computed the spectral density $\rho(\nu,T)$ (energy per unit of volume) emitted by a black body in the form

S. Califano, *Pathways to Modern Chemical Physics*,
DOI 10.1007/978-3-642-28180-8_7,

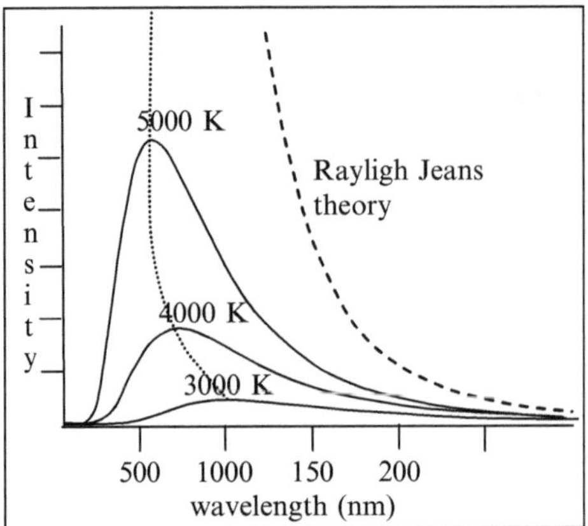

. Black body radiation curves

$$\rho(\upsilon, T) = a\upsilon^3 e^{-b\upsilon/T}$$

where v is the frequency and a and b are empirical constants (Wien 1896). Wien's law provides a relatively accurate model of the black-body emission in the visible region. However, it strongly deviates from the experimental curve in the far-infrared region. A different theoretical model, providing better agreement in the far infrared but a worse one at higher frequencies, was derived by Lord Rayleigh, based on the energy equipartition theorem (Rayleigh 1900). According to this model, the spectral energy distribution should increase with the square of the frequency, resulting in a very poor fit with experimental data in the ultraviolet region of the spectrum. This phenomenon, called "ultraviolet catastrophe," contrasted a very rapid increase in the theoretical curve with a decrease in the experimental one, which goes to zero asymptotically (sketched curve in figure). The "ultraviolet catastrophe," an expression coined by Paul Ehrenfest in 1911, is the direct consequence of the classical energy equipartition principle, which assigns the same average energy, $KT/2$, to all degrees of freedom of a physical system. The difficulties of the classical approach were also highlighted by Lord Rayleigh (1905) and by James Jeans (1905), who independently arrived in 1905 at the expression

$$\rho(v, T) = \frac{2v^2 KT}{c^2}$$

confirming the absurd result obtained by Wien.

This period of uncertainty and confusion in the field found an unexpected solution on 19 October 1900 when Max Planck announced that he had solved the dilemma of the Kirchhoff function. From that day, physics was not the same anymore.

Max Planck (1858–1947) received his Ph.D. in physics at the University of Kiel in 1879 with a thesis on the second law of thermodynamics. In 1889, he moved to the University of Berlin, where later, in 1892, he succeeded Kirchhoff as professor of physics. At that time, owing to Kirchhoff's influence, several experiments on black body emission were in progress at Berlin. This stimulated Planck to investigate further why the classical theory had failed. Planck was convinced that the second law of thermodynamics was an absolute truth. As such, he would not accept Boltzmann's probabilistic interpretation according to which an increase in entropy in the spontaneous evolution of physical systems is justified because it is by far more probable than a decrease. His belief in the absolute truth of the second principle is well represented by a famous quotation of the English astrophysicist Sir Arthur Eddington who in his book *The Nature of the Physical World* wrote:

> If someone points out to you that your pet theory of the universe is in disagreement with Maxwell's equations – then so much the worse for Maxwell's equations. If it is found to be contradicted by observation – well, these experimentalists do bungle things sometimes. But if your theory is found to be against the second law of thermodynamics I can give you no hope; there is nothing for it but to collapse in deepest humiliation.

The possibility of an entropy decrease in a physical process became a war horse in the debate between Boltzmann and the French mathematician Henry Poincaré. The mathematician Henry Poincaré (1854–1912) demonstrated in 1890 a famous theorem, the recurrence paradox (Poincaré 1889), asserting that any physical system evolving from a given state must unavoidably revisit that state given sufficient time. The waiting time did not represent a critical problem in the debate on the foundation of physics, which hinged instead on the relationship between Newtonian mechanics and statistical thermodynamics. The German mathematician Ernst Zermelo (1871–1953) used in 1896 (Zermelo 1896a, b) the Poincaré result to attack the mechanistic approach, arguing that any theory inconsistent with the second law of thermodynamics must be wrong. Boltzmann, however, showed that the recurrence theorem was consistent with the statistical viewpoint, and that physical processes associated with negative entropy variations are not in principle forbidden but only highly improbable (Boltzmann 1897). In particular, the waiting time for a physical system to go back to the initial state could even be greater than the duration of the existence of the universe.

At that time, Planck was convinced of the general validity of Wien's law, which seemed to model correctly the black body emission at low frequencies. Soon enough, however, he had to revise his position, as new measurements made by a Berlin group showed that, even at low frequencies, the experimental curves deviated considerably from Wien's law, showing an almost linear dependency instead of an exponential one.

To model the black-body experimental curve correctly, Planck considered a cavity containing a large number of independent oscillators emitting and absorbing radiation at thermal equilibrium, namely, at a constant entropy value. To fit the experimental data, he needed to satisfy two conditions: at low frequencies, the theoretical curve should display an almost linear behavior, while, at high frequencies, it should depart radically from the exponential trend of Wien's equation. Since the high-frequency behavior of the curve was determined by the radiation entropy, S, whereas in the low frequency regime the dominant term was the oscillators' average energy, U, Planck tried to interpolate between these two contributions, finally obtaining the expression

$$\rho(v,T) = \frac{8\pi hv^3}{c^3}\frac{1}{e^{hv/KT}-1}$$

which accurately modeled the experimental data. He presented this equation, known as Planck's law, to a meeting of the *Deutsche Physikalische Gesellschaft* on 19 October 1900, and published it on 14 December in the acts of the society and then in the *Annalen der Physik* (Planck 1900).

To reach this conclusion, Planck was forced to abandon his blind confidence in the absolute truth of the second law and to accept Boltzmann's probabilistic interpretation. In addition, to avoid divergence of his model from the experimental data at high frequencies, similar to Wien's model, he had to make a key hypothesis, which seemed completely foolish at the time but soon proved to be of fundamental importance for the birth of quantum mechanics. Planck's hypothesis was that the energy U_N of the oscillators of the black body cavity was not continuous but rather the sum of discrete quantities that he called energy quanta (Planck 1901):

> Hierzu ist es notwendig U_N nicht als eine stetige, unbeschränkt teilbare, sondern als eine discrete, aus einer ganzen Zahl von endlichen gleichen Teilen zusammengesetzte Grösse aufzufassen.[1]

The energy of a quantum was proportional to the frequency of the radiation and was given by the relation $E = hv$ where the constant h had the dimensions of an impulse (energy·time, erg·second). Initially, Planck had some difficulty in justifying his formula on a theoretical basis. Indeed, for a long time, he considered his hypothesis more as a mathematical trick than a true physical theory.

In 1931, remembering that period, Planck wrote (Planck and Wood 1931):

> I can characterize the whole procedure as an act of desperation, since, by nature I am peaceable and opposed to doubtful adventures. However, I had already fought for six years with the problem of equilibrium between radiation and matter without arriving at any successful result. I was aware that this problem was of fundamental importance in physics, and I knew the formula describing the energy distribution . . . hence a theoretical interpretation had to be found at any price, except for the inviolability of the two laws of thermodynamics.

For a while, around the beginning of the twentieth century, Planck's work remained virtually ignored in the physics community. The idea that energy could

[1] Moreover, it is necessary to interpret U_N [the total energy of a blackbody radiator] not as a continuous, infinitely divisible quantity, but as a discrete quantity composed of an integral number of finite equal parts.

be emitted or absorbed in discrete quantities was too new and strange to be easily accepted within the scheme of classical physics. Even more difficult to digest was the appearance of a new universal constant, h, Planck's constant, that established the ratio between energy and frequency. It was only thanks to Albert Einstein's talent that the quantum theory finally prevailed. In the period between 1905 and 1907, in order to explain the specific heat of solids, Einstein also suggested a similar quantum nature for the photoelectric effect, introducing the concept of a light quantum, the photon, associated with a discrete impulse $h\nu/c$ (Einstein 1906a, b, 1907).

Einstein could thus explain the photoelectric effect in quantum terms by assuming that metals emit electrons only if hit by radiation of a frequency ν higher than a predefined threshold frequency ν_0 specific for each metal. He hypothesized that in the interaction with the metal, the electromagnetic radiation behaves as if it were composed of individual particles, the photons, each one having energy $h\nu$, where h is Planck's constant. Only if this energy is higher than the threshold energy $h\nu_0$, the electrons may escape from the metal.

Einstein's idea was revolutionary, since it associated the energy of a light particle (light packet) to a frequency, a physical quantity characteristic of waves and not of particles. Conversely, it extended the concept of momentum, previously relegated only to the particle domain, to light waves.

It took an additional 20 years, after Plank's and Einstein's initial hypotheses, for the wavelike nature of elementary particles and the corpuscular nature of electromagnetic radiation to achieve their final interlinked status in the new physics. This was accomplished thanks to the definitive experimental data obtained from electron diffraction measurements and from the Compton Effect.

Nonetheless, even before experimental confirmation, Planck's and Einstein's ideas were used by Niels Bohr in 1913 to develop the theory of the electronic structure of the atom, described in Chap. 5, a theory that reached its pinnacle with the final organization of the periodic table in terms of the Aufbau principle. Bohr's theory represented an attempt by a great physicist to save the marvelous construction of classical physics, the triumph of the previous century. However, its conceptual structure contained an intrinsic defect that was impossible to eliminate. It was becoming increasingly difficult to explain the physical data at the elementary particles level, since a large number of crucial experiments conducted over the years showed without doubt a double nature, sometimes corpuscular, sometimes wavelike, of both the matter and the radiation.

In 1923, Arthur H. Compton (1892–1962), 1927 Nobel laureate in physics, in a famous paper (Compton 1923) on x-ray scattering, discussed in detail the difference between the wave and the particle behavior. If light is made of photons, each of them, when hitting an electron, should behave like a billiard ball hitting another ball. In other words, the energy of the photon cannot be distributed throughout the electrons of the metal but only to one of them. As a consequence, the photon must transfer some momentum to the electron that will thus be deviated from its initial trajectory, scattering the photon at an angle with that of the incident radiation, as shown in the next figure.

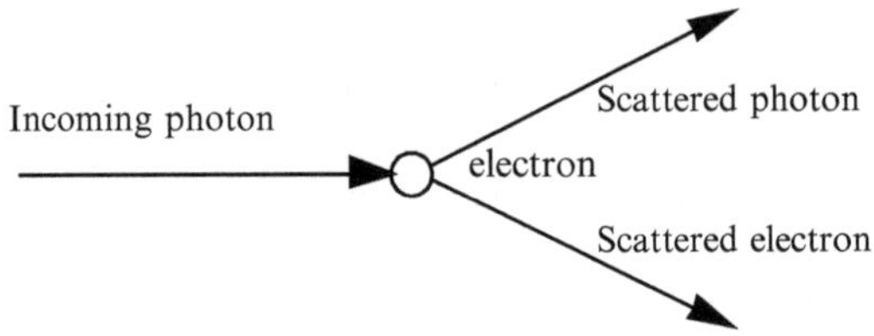

. Schematic representation of the Compton Effect.

In 1922, the French physicist Louis de Broglie (1892–1987), carrying Einstein's hypothesis to extremes in his thesis entitled *Recherches sur la théorie des quanta*, concluded that if the radiation had a double nature of wave and particle, an electron could also have the same dualistic behavior:

> L'atome de lumière équivalent en raison de son énergie totale à une radiation de fréquence ν est le siège d'un phénomène périodique interne qui, vu par l'observateur fixe, a en chaque point de l'espace même phase qu'une onde de fréquence ν se propageant dans la même direction avec une vitesse sensiblement égale (quoique très légèrement supérieure) à la constante dite vitesse de la lumière.

De Broglie summarized the results of his doctorate thesis in a paper entitled *Ondes et quanta* (De Broglie 1924). According to de Broglie, a wave of energy $E = h\nu$ and of wavelength $\lambda = h/p$, where p was the electron momentum, was associated to the electron. In an atomic orbit, the wave associated with the electron was a stationary wave, as pointed out by Nicholson in 1912.

This hypothesis was finally verified (Davisson and Germer 1927) in 1927 at the Bell Telephone Laboratories in America by Clinton Joseph Davisson (1892–1962) and Lester Halbert Germer (1896–1971), as well as in England, only a month later, by George Paget Thomson, the son of J.J. Thomson, showing that electrons, just like x-rays, can be scattered by matter (Thomson and Reid 1927). For these works, Davisson and Thomson received the Nobel Prize for physics in 1937.

7.2 Quantum Mechanics

When the idea that both electromagnetic radiation and electrons possessed a double nature of wave and particle started to spread, another pillar of classical physics started to waver: the concept of orbit.

By the beginning of 1924, the Dutch physicist Hendrik Anthony Kramers (1894–1952), one of the coworkers with Bohr at Copenhagen, had developed the quantum theory of light dispersion, without speaking of electron orbits. Kramers's ideas were summarized in a famous paper by Bohr, Kramers himself, and a young American physicist, John Clark Slater (Bohr et al. 1924), who developed a general theory of light emission, absorption, and scattering. Even if not completely correct, this work represented the starting point for the birth of quantum electrodynamics.

J.C. Slater would soon become one of the fathers of quantum chemistry in the United States.

In 1924, there were two important centers of theoretical physics in Europe: the Niels Bohr Institute at Copenhagen and that of Max Born at Göttingen. In these laboratories, the suspicion that the concept of orbit was actually responsible for the difficulty of extending classical mechanics to the world of the electrons was already being voiced. Among the physicists involved in the discussion of this problem, the young Werner Heisenberg (1901–1976) was the one who, at the age of only 23 years, started to explore the possibility of eliminating orbits from particle dynamics. The basic idea of Heisenberg was that a correct dynamical treatment of the electron's motion should be based only on observable quantities. In classical dynamics, the orbits are determined by Newton equations and by the initial conditions. He realized that this deterministic description was correct for objects of the macroscopic world where orbits are directly observable, but was not easily transferable to the microscopic world, by arbitrarily assuming that electrons move as planets or satellites.

At the end of April 1925, Heisenberg, suffering from a violent attack of hay fever, asked his director Max Born for permission to go on holiday to the island of Heligoland in the North Sea to cure himself. During this holiday, Heisenberg went back to the work started at Göttingen before his illness, devoted only to the identification of physical observables in electron dynamics. For this purpose he started collecting tables of physical quantities that he considered true observables, such as spectral frequencies, moments, scattering amplitudes, etc. The mental process by which he arrived at a new mechanics has been well illustrated by Abraham Pais (1991) and testifies to the great conceptual labor in which not only Heisenberg but a full contingent of members of the new physics were involved during these years.

Classically, an orbit is described by time-dependent coordinates $q(t)$ and moments $p(t)$, continuously varying as a function of time. The classical solutions of the dynamics of an object such as an electron are obtained by solving the equations of motion. Here, the potential energy is normally written as a function of the squares of the coordinates q and the kinetic energy as a function of the squares of the moments p. In this way, however, one unavoidably arrives at a description of the object's motion in terms of trajectories or orbits just because coordinates and moments are continuous variables.

Heisenberg was well aware of the fact that the quantum mechanical theory of the electron arising from the works of Planck, Einstein, and Bohr required instead the existence of discrete stationary energy states (energy levels) for the atoms. He was equally well aware that atomic radiation absorption and emission processes were then defined as instantaneous transitions between any two of these energy levels. Thus, to discuss electron dynamics in terms of energy levels and of transitions among them, while continuing to use coordinates and moments as in the classical treatment, Heisenberg decided to define discrete quantum coordinates $q_{nn}(t)$ and $q_{nm}(t)$ to describe the electron in a stationary energy level n and in the transition from an energy level n to an energy level m, respectively. In the same way he defined discrete moments $p_{nn}(t)$ and $p_{nm}(t)$ of the electron in the n level and in the transition $n \rightarrow m$, respectively.

To calculate the energies E_n of the quantum levels, Heisenberg used the classical approach of computing the total energy $H = V + T$, where V is the potential and T the kinetic energy. In order to calculate V and T he needed the squares of the coordinates and moments and thus had to face the difficulty of squaring quantities with a double index, a problem never before encountered. After a long struggle, Heisenberg used the expressions

$$q_{nm}^2(t) = \sum_k q_{mk}(t) \cdot q_{kn}(t)$$

$$p_{nm}^2(t) = \sum_k p_{mk}(t) \cdot p_{kn}(t)$$

where the sum extends over all possible values of the index k, namely, over all possible quantum states for $m = n$ and over all possible transitions for $m \neq n$.

In the same way he wrote products between two different quantities $q(t)$ and $p(t)$ in the form

$$[p(t) \cdot q(t)]_{mn} = \sum_k p_{mk}(t) \cdot q_{kn}(t)$$

$$[q(t) \cdot p(t)]_{mn} = \sum_k q_{mk}(t) \cdot p_{kn}(t)$$

Heisenberg developed this formalism thanks to a true stroke of genius that only after some time was recognized as correct. Actually, at that time, Heisenberg did not know matrix algebra. It was his professor, Max Born (1882–1970), to whom he had given the manuscript to read, who realized that what Heisenberg was doing was nothing else but an application of this kind of algebra. Max Born, who had studied with great mathematicians such as Klein, Hilbert, and Minkowski, the "mandarins" of German mathematics, knew this branch of mathematics well and had no difficulty in translating and extending Heisenberg's paper into matrix language (Born and Jordan 1925, Born, Heisenberg, Jordan), with the help of his pupil Pascual Jordan (1858–1924). Jordan was also a good mathematician, who has made important contributions to quantum mechanics. Despite his political ideas, as a member of the Nazi party and of the "Brown Shirts Organization," Jordan always fought in defense of the scientific merits of his Jewish colleagues, even if the other members of the party tried to discredit them in every way.

Working with his strange mathematical expressions, Heisenberg was shocked to realize that in general the product $q(t)\cdot p(t)$ is not equal to the product $p(t)\cdot q(t)$, namely, as mathematicians use to say, in this case the product is not commutative. A more elegant and general result on this matter, of extreme importance for the axiomatic formulation of quantum mechanics, was presented by Born and Jordan in their work, showing that the $p\cdot q$ product obeys the relation

$$\sum_k (p_{mk} \cdot q_{kn} - q_{mk} \cdot p_{kn}) = \begin{cases} h/2\pi i & \text{if } m = n \\ 0 & \text{if } m \neq n \end{cases}$$

where $h/2\pi$ represents the unit of angular momentum. Later Dirac introduced the symbol $\hbar = h/2\pi$ to represent the unit of angular momentum in quantum mechanics.

Then Heisenberg computed the energies H_{nn} of the quantum levels and published in 1925 in the *Zeitschrift für Physik* (Heisenberg 1925) a paper that marks the date of birth of quantum mechanics, followed next year by a second paper in *Naturwissenschaften* (Heisenberg 1926). In a short while, having mastered matrix algebra, he wrote another paper in collaboration with Born and Jordan (Born et al. 1926), reformulating his ideas in matrix form.

At the same time, the English physicist Paul Adrien Maurice Dirac (1902–1984) published an article (Dirac 1925) in which the quantum mechanical equations were formulated in terms of operators. Dirac, the most elegant theoretician of quantum mechanics from a formal point of view, developed the relativistic formulation of quantum mechanics in 1928, obtaining from first principles the existence of the spin, and proposing the famous Dirac equations that many consider one of the real wonders of twentieth-century physics.

The year 1926 was unbelievably rich in ideas for the new physics. While Heisenberg, Born, and Jordan had improved the matrix formulation of quantum mechanics and Dirac had supplied a more elegant interpretation in terms of quantum operators, a formally completely different theory, wave mechanics, came into the limelight thanks to the work of a Viennese physicist Erwin Schrödinger, supporter of the physics of the continuum against that of the discrete. Erwin Schrödinger (1887–1961) started his scientific career at the University of Vienna where he studied under the leadership of Fritz Hasenöhrl (1875–1915). Hasenöhrl, who succeeded Boltzmann on the chair of physics, died during the war in Italy near Folgaria in 1915, hit by a grenade during an attack by the Italian army. In 1914, with the outbreak of the First World War, Schrödinger was first sent to the Italian border, then transferred to Hungary in 1915, and then sent back again to the Italian front.

After the war Schrödinger returned to Vienna and continued his career as assistant to Max Wien, until in 1922 he became professor at the University of Zurich. There he met Peter Debye and Hermann Weyl and struck a long-lasting friendship with them. At the beginning of his scientific activity he was interested in the theory of solids, in thermodynamics, in statistical mechanics, and even in physiology, but soon he shifted to quantum mechanics. In 1927, he joined the Institute of Physics of the University of Berlin where he met Albert Einstein and where he stayed until 1933, when he decided to leave Germany, disgusted by the Nazi policy and by the persecution of the Jews. In the same year, he was awarded the Nobel Prize in physics together with Paul Dirac. However, his strange family situation (he actually lived with two women) created difficulties with the European academic society that forced him to wander for 7 years between several

European Institutions in Austria, Great Britain, and Belgium. Finally in 1940 he settled at the Dublin Institute for Advanced Studies where he stayed for 15 years, until he retired to Vienna in 1956 as emeritus professor of physics.

During the period of his Viennese studies, Schrödinger acquired complete mastery of the mathematical techniques to solve equations with eigenvalues and eigenfunctions, typical of the continuum media. To clarify his approach, we remind the reader that in mathematics objects able to transform functions into other functions are called "operators." For example, the mathematical object d/dx is an operator that transforms the function sin(x) into the function cos(x). When application of an operator to a function yields the same function multiplied by a constant, the function is called an *eigenfunction* of the operator and the constant is called an *eigenvalue*. As an example, e^{kx} is an eigenfunction of the operator d/dx with eigenvalue k since $(d/dx)e^{kx} = ke^{kx}$.

Inspired by the ideas of de Broglie on the wave nature of matter, Schrödinger tried to develop a quantum theory of the continuum, in opposition to the quantum theory of the discrete of the German school. Owing to his theoretical background, he knew well that for a continuum medium, the solutions of the wave equation for simple systems, such as a vibrating cord fixed at the extremities, always led to a discrete number of waves. These include the fundamental wave ψ_1, of frequency ν, and the overtones $\psi_2, \psi_3, \ldots \psi_n$, of frequency $2\nu, 3\nu, \ldots, n\nu$, etc., as well as all their possible combinations:

$$\psi(q.t) = \sum_n c_n \psi_n(q.t)$$

In other words, even in the classical treatment, the wave equation leads to a quantized solution for the vibrations of the cord, without the need to introduce additional hypotheses.

Schrödinger spent the 1925 Christmas holidays at Arosa on the Swiss Alps, together with a Viennese girlfriend, while his wife stayed in Zurich. In this mountain hut he had the idea to insert the wavelength $\lambda = h/p$ and the energy $E = h\nu$, proposed by de Broglie, into the classical wave equation. To describe the dynamics of an electron in a hydrogen atom, he defined a set of wavefunctions ψ_n, whose temporal evolution satisfied the wave equation

$$i\left(\frac{h}{2\pi}\right)\frac{\partial\psi_n}{\partial t} = \left(-\frac{h^2}{8\pi^2 m}\nabla^2 + V\right)\psi_n$$

where i is the imaginary number $\sqrt{-1}$, V the potential in which the electron moves, and

$$\nabla^2 = \frac{\partial^2}{\partial x^2} + \frac{\partial^2}{\partial y^2} + \frac{\partial^2}{\partial z^2}$$

is the sum of the second derivatives with respect to the coordinates used.

From this equation, by introducing a simple hypothesis on the time dependence of the ψ_n functions, one obtains a time-independent equation whose eigenvalues E_n define the energy of the stationary states of the atom, namely, the quantum energy levels.

The time-independent Schrödinger equation has the form

$$\left(-\frac{h^2}{8\pi^2 m}\nabla^2 + V\right)\psi_n = E_n\psi_n$$

more conveniently written in the compact form

$$H\psi_n = E_n\psi_n$$

which shows that the Schrödinger equation is a typical differential equation in which the operator H

$$H = \left(-\frac{h^2}{8\pi^2 m}\nabla^2 + V\right)$$

known as the Hamiltonian operator, acts on the eigenfunctions ψ_n to extract from them the eigenvalues E_n.

By solving this equation for hydrogen-like atoms, Schrödinger obtained the correct energy eigenvalues and proved that it automatically generates the three quantum numbers n, ℓ, and m of the old quantum theory of Bohr as well as Balmer's expression for the hydrogen frequencies.

Schrödinger's paper on the new quantization theory (Schrödinger 1926a), published in January 1926, represented another fundamental achievement of the twentieth century and opened a new era for both physics and chemistry. A few weeks later he published a second paper (Schrödinger 1926b), presenting a new version of his equation and its application to the harmonic oscillator, the rigid rotor, and the diatomic molecule. In the same year he published a third paper to show the equivalence of his theory to that of Heisenberg (Schrödinger 1926c) and in 1927 a fourth paper containing the solution of the time-dependent equation (Schrödinger 1927).

This second version of the quantum mechanics, completely different from that of Heisenberg, pleased a large number of physicists and chemists since it used a kind of mathematics that they were already accustomed to, in contrast to that of Pauli and Dirac which was almost unknown, too formal, and too difficult to understand. Of course, supporters of the matrix theory despised the fans of the wave approach and, conversely, the latter repudiated the abstruse mathematical techniques of the matrix theory that deprived the reality of the electron of its physical meaning. In any case, from 1926 onward the physicists were forced, willingly or not, to admit that two different theories, at first glance irreconcilable, produced the same results.

At the beginning of July 1926, Schrödinger made the first move toward a combination of the two approaches (Schrödinger 1926c), maintaining that particles were nothing else than confined wave packets. This was a commendable attempt made in the spirit of saving classical physics and of deriving from it the new mechanics as a particular case. Of course, this attempt soon failed, since it is physically impossible for a wave packet to be confined forever.

What Schrödinger could not achieve was achieved instead by Max Born. On 10 July 1926, he published a paper (Born 1926a) on the diffusion of a beam of particles, in which the idea was first mentioned that the probability of diffusion of the electron could be related to Schrödinger's wavefunction. This concept was rediscussed and improved in a second paper published with the same title on 14 September (Born 1926b), in which the connection between the theory of Schrödinger and that of Heisenberg was realized by assigning the probability $|\psi|^2 d\tau$ of finding the electron inside a small element, of volume $d\tau$, to the square of the wavefunction multiplied by the volume $d\tau$. Born's paper was extremely important, but it did not enjoy the interest that it deserved, essentially because in the same year a paper on the equivalence of the two theories was published by Schrödinger. Born himself probably did not realize immediately how important his work was and how much it would influence the future of the interpretation of quantum mechanics, in just a few years.

The success of quantum mechanics continued in 1927 with the publication of the famous indeterminacy relationship of Heisenberg (1927), originated from an endless series of inflamed discussions between Bohr and Heisenberg at Copenhagen, which continued for months without either of the two scientists changing his ideas. While Bohr was looking for a formulation of quantum mechanics that allowed the existence of both waves and particles, Heisenberg insisted in completely excluding the wave formalism.

To convince Bohr, Heisenberg imagined a series of *Gedanken Experimenten*, expression invented by Einstein to define ideal experiments, practically unrealizable although conceptually justifiable. Once the concept of orbits was abandoned, Heisenberg was still left with the problem of what it meant to follow the motion of an electron. Orbits and trajectories are temporal sequences of positions of an object in motion and there was little sense in continuing to use this concept for objects of dimension of the order of 10^{-8} cm. In other words, without orbits the new mechanics could represent the position and the speed of electrons only within given limits. Heisenberg pointed out that in order to observe an electron it is necessary to illuminate it with radiation of wavelength shorter than the electron radius, e.g., with γ rays, and to observe the scattered radiation with a suitable microscope. In this case, however, due to the Compton effect, the γ photon that hits the electron transmits a momentum $h\nu/c$ to it that displaces it from its initial position and at the same time changes its speed.

With arguments of this kind, Heisenberg succeeded in proving that if one tries to measure the position q and the momentum p of an electron at the same time, one unavoidably finds that the product of the error Δq on the coordinate times the error Δp on the momentum is given by the relation

$$\Delta q \cdot \Delta p \geq h$$

In the same way the product of the error on the energy ΔE times the error Δt on the time measurement obeys the relation

$$\Delta E \cdot \Delta t \geq h$$

Heisenberg's argument, based on the use of classical instruments to study the properties of the quantum world, was strictly connected to the complementarity principle that Bohr developed during this time. This principle found its consecration at the physics meeting of Como on 16 September 1927, and at the Solvay meeting on 24–29 October of the same year.

Bohr considered the wave and the particle representations of quantum mechanics as complementary, but he maintained that their validity was always bound to the comparison with experiments, i.e., the results of measurements made with classical instruments. Therefore, according to Bohr, the behavior of an electron was not a property of the electron itself but rather of the kind of measurement made.

The indeterminacy principle had a profound influence on the philosophical concept of experiment, since it showed that the intervention of the observer always disturbs the system under observation. This disturbance is negligible in the macroscopic world but becomes important in the submicroscopic one. The Heisenberg principle also called into question the classical idea of causality. In classical physics, the instantaneous position and momentum of a particle in a given potential field completely define the corresponding position and momentum at a later point in time. Classical determinism, however, is lost at the electron level, its description becoming only probabilistic. This probabilistic description was already included in Born's paper on the meaning of a wavefunction (Born 1926c). The idea that in any infinitesimal volume element there is a finite probability of finding the electron led to a description of the space around the nucleus as the ensemble of points where it is most probable to find the atom's electrons. Higher probability also means higher charge density, and thus Born's interpretation corresponds to identifying the space regions around the nucleus with high density of negative charge. Thus, in Born's interpretative framework, the electron completely lost its identity as a particle and became a cloud of negative charge around the nucleus.

The electronic wavefunctions of the hydrogen atom, and by extension also the atomic and molecular wavefunctions in general, are today called orbitals, a term that appeared in the scientific literature only in 1932, thanks to Robert Mulliken. Similar to the old Bohr's orbits, each orbital is then characterized by four quantum numbers, n, ℓ, m, and s, with the great difference that they were now derived directly from the theory and not introduced as an ad hoc hypothesis, as in the old quantum theory. As an historical note, the same symbols s, p, d, etc., used by Bohr in his Aufbau principle for the electronic sub-shells, were used by Mulliken and are still used today.

There were thus s ($\ell = 0$), p ($\ell = 1$), d ($\ell = 2$), etc., orbitals according to the following scheme:

n	ℓ	m	Orbitals
1	0	0	1s
2	0	0	2s
2	1	−1, 0, 1	$2p_x$, $2p_y$, $2p_z$
3	0	0	3s
3	1	−1, 0, 1	$3p_x$, $3p_y$, $3p_z$
3	2	−2, −1, 0, 1, 2	$3d_{xx}$, $3d_{xy}$, $3d_{xz}$, $3d_{x^2-y^2}$, $3d_{z^2}$

Although atomic orbitals are mathematical functions that describe the wavelike behavior of electrons, they are most conveniently represented in polar coordinates as three-dimensional objects in space representing the electrons' charge cloud around an atomic nucleus. The representation of electrons as charge clouds around a nucleus started a new chapter in the fascinating history of the electron in chemistry. Simple pictures of the atomic orbitals in polar coordinates are shown in the next figure.

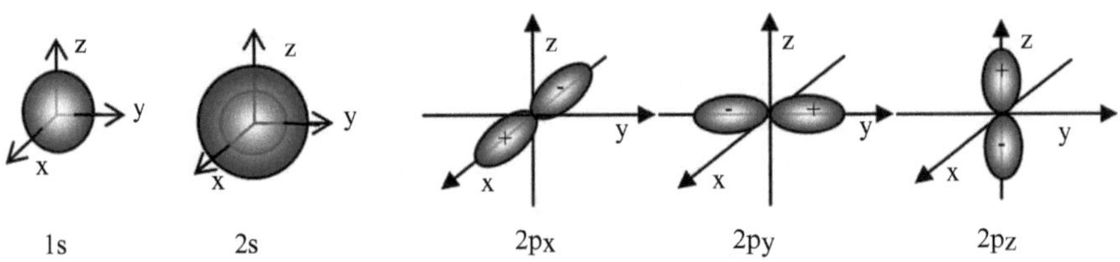

The simplest s and p atomic orbitals

The 1s and 2s orbitals have spherical shapes whereas the three $2p_x$, $2p_y$, and $2p_z$ orbitals have two lobes, with a node at the nucleus, oriented along the axes x, y, and z, respectively. The five 3d orbitals have more complex shapes with more lobes. The wavefunctions are positive on one side of the nodal planes and negative on the other.

These tri-dimensional orbital representations are also of great use to visualize the overlap between two atoms, building shared electron density areas corresponding to the electron pairs of Lewis.

The original form of Schrödinger's equation was only valid for the dynamics of a single electron. For atoms with many electrons, the mathematical formalism was too complex and intractable without the use of approximation techniques. The problem of many-electron atoms was debated through the contribution of several authors, including E. Fermi, P. Dirac, J.C. Slater, and above all of D.R. Hartree and V.A. Fock, and its solution gave birth to quantum chemistry and to the modern theory of the chemical bond.

In 1927, an important simplification of the mathematical technique to find a correct solution of these complex differential equations was developed by the English mathematician Douglas Rayner Hartree (1897–1958), expert in numerical analysis. Hartree studied the methods of numerical calculus at Cambridge, and during the Second World War he learned how to apply them to the study of anti-aircraft shell trajectories. At the end of the war he returned to Cambridge, where in 1921 he attended a series of lectures by Niels Bohr that convinced him of the importance of numerical analysis techniques for the solution of differential equations. In 1923, he visited Ehrenfest at Leyden and met both Goudsmit and Uhlenbeck, with whom he remained in correspondence, while becoming more and more oriented toward the study of physics.

Hartree's goal was to find a solution of Schrödinger's equation for a many-electron atom without using empirical parameters but rather starting from first principles. In 1927, after some preliminary attempts (Fues 1922; Hartree 1923) by himself, Erwin Richard Fues (1893–1970), and Robert Bruce Lindsay (1900–1985), he introduced a new procedure, called *self-consistent field method*, in the study of the solutions of differential equations with variable coefficients. This approach considers all electrons as independent from each other, thus allowing one to factorize the total wavefunction for an atom with n electrons as the product of n mono-electronic functions.

Hartree then applied an iterative technique that used the initial wavefunctions of $n-1$ electrons to compute the mean field acting on the nth electron. The calculation was repeated for all electrons to produce a mean potential used to compute new atomic wavefunctions. At the end of a refinement cycle, the atomic wavefunctions were used to compute the energy of the atom. The iterative process was then repeated as many times as necessary to converge on the final best value of the energy.

The Hartree method, presented at a meeting of the Cambridge Philosophical Society in November 1927 and then published in 1928 in two successive papers (Hartree 1928a, b), gave him notoriety in the circle of experts. He kept in touch with Heitler and London, with whom he had a long exchange of letters, especially with the latter.

In 1928, the Englishman John Arthur Gaunt and the American John Clark Slater (1928) independently proved that Hartree's method conceptually represents an application of the variation principle to a trial wavefunction, written as a product of mono-electronic wavefunctions.

John Arthur Gaunt (1904–1944), born in China from missionary parents, immigrated to Cambridge, England, in 1923 to specialize in applied mathematics. In 1926, he started to work in quantum mechanics under the leadership of Ralph H. Flower who suggested to him to verify the applicability of Hartree's method. Gaunt applied the method to compute wavefunctions and energies of atoms and showed that it was by far the most efficient one (Gaunt 1928). In 1929, he published two papers on the helium triplet state (Gaunt 1929), but immediately after this he left for China to teach at Hong Kong. He died in a Japanese concentration camp in 1944.

In 1927, Hartree met Slater for the first time at a meeting organized by Pauli in Zurich. In 1930, Slater published a short note, proving that Hartree's method did not fulfill the antisymmetry principle for wavefunctions, since it was based on an old formulation of the Pauli principle, which only prevented two electrons from occupying the same quantum state. It was easy for him to prove that using a determinant in place of the simple product of mono-electronic wavefunctions would be sufficient to satisfy automatically the conditions of antisymmetry for the poly-electronic wavefunction.

The same result (Fock 1930) was independently attained by one of the greatest Russian theoretical physicists of the first half of the twentieth century, Vladimir Aleksandrovich Fock (1898–1974), of St Petersburg University, known for the high quality of his school of mathematics that included scientists such as Chebyshev and Lyapunov. In the 1920s, he had already contributed to the rise of quantum physics by defining the Fock space (Fock 1932), an algebraic system corresponding to an enlarged Hilbert space, often utilized in quantum mechanics to describe quantum states with an unlimited number of particles.

Fock's method, which automatically accounted for antisymmetry and introduced the exchange energy, was based on group theory, a formal treatment of the symmetry relationships among physical or mathematical objects. Today, group theory is part of the normal curriculum of the students of physics and chemistry, but at that time it was too abstract for them and practically inaccessible except to a few experts.

In 1935, combining his own ideas with those of Fock, Hartree reformulated his approach more clearly and precisely into what was then called the Hartree–Fock method (Hartree & Hartree 1935). The method, which is also valid for molecules, is based on the definition of a Hamiltonian operator, the Fock operator, capable of acting on the wavefunctions of only one electron at a time. The Fock operator includes the kinetic energy of the electron, its interaction with the nuclei, the repulsion from all other electrons approximated by a uniform distribution of negative charge, and the exchange energy due to the antisymmetry.

Bertha Swirles (1903–1999), another pupil of Ralph Fowler at Cambridge (Swirles 1935, 1936), contributed to the final version of the Hartree–Fock method. She also contributed to the development of a more accurate way of introducing the repulsion among electrons by adding the configuration interaction to the Hartree method (Swirles et al. 1939).

Initially, due to its heavy computational requirements, the Hartree–Fock method was utilized only for many-electron atoms, thanks to the spherical symmetry that considerably simplifies the problem. With the arrival of electronic computers in 1950, however, the Hartree–Fock method was extended to molecules and, within a few years, became a standard technique of quantum chemistry.

The final quantum mechanics formalization was presented in 1930 by Paul Adrien Maurice Dirac in his book *The Principles of Quantum Mechanics* (Dirac 1930) that rapidly became the reference text for the theory. In this book, Dirac integrated both the wave mechanics and the matrix approach of Heisenberg within a unique formalism that used the operator technique, in vectorial space, to describe measurable quantities, thus proving that the two formulations were simply representations of the same quantities in different spaces. This book completed the formal framework of the basic quantum mechanics that became the fundamental tool for the study of the structure of matter.

7.3 Quantum Chemistry

Application of quantum mechanics in chemistry did not take long. In 1927, Schrödinger's equation for the H_2^+ molecule, the simplest molecule with one electron and two nuclei, was published. The calculation of the energy was made by Oyvind Burrau (1927), a Danish physicist working at Copenhagen in Bohr's laboratory. Burrau's manuscript was presented to the *Danish Royal Academy of Science* by Bohr himself on 7 December 1926. Burrau had solved Schrödinger's equation exactly for an electron interacting with two positive nuclei, thus defining the first form of binuclear molecular orbitals. Burrau succeeded in integrating Schrödinger's equation by variables separation, through a transformation to elliptical confocal coordinates. This approach, which involves complex mathematical transformations, was certainly correct, but it was applicable only to a relatively simple case and could not be extended to more complex systems.

Much more important for future developments was instead the application of wave mechanics to the hydrogen molecule H_2, realized by two young physicists, Walter Heitler and Fritz London, who, after leaving for Zurich to work with Schrödinger and soon realizing that he had no interest in collaborating with them, decided, after a month, to work together to calculate the "van der Waals forces," forces between two hydrogen atoms. There is no reason to believe that the subject had been suggested by Schrödinger himself or that they had discussed it with him. However, it is certain that Schrödinger knew of the work since he had talked about it with Mulliken, and he had actually introduced Heitler and London to him, specifying that he was very interested in their problem.

Walter Heitler (1904–1981) came from the University of Munich, where he had studied with Sommerfeld and Herzfeld and had obtained a doctorate in theoretical physics. In 1926, he was awarded a Rockefeller Foundation fellowship first to work with Bjerrum and Bohr at Copenhagen and then with Schrödinger at Zurich. Fritz London (1900–1954), who had instead started as a philosopher interested in the structure of theories, graduated at Bonn with a thesis entitled *Über die Bedingungen der Möglichkeit einer deduktiven Theorie. Ein Beitrag zu einer Mannigfaltigkeitslehre deduktiver Systeme*. With a professor of mathematics for a father and a famous physicist for a brother, it was unavoidable that sooner or later he would become interested in science. Actually, after graduation, he started working in quantum mechanics at the *Stuttgart Technische Hochschule*, under the supervision of Paul Peter Ewald (1888–1985). At Stuttgart, he collaborated with Helmut Hönl (1903–1981) on the calculation of the intensity of spectral lines. He then moved, first to Munich and then to Zurich, to study physics with Sommerfeld and Schrödinger.

Heitler and London attacked the problem of computing the energy of the H_2 molecule by trying to understand what happens when two hydrogen atoms that are far apart start to approach each other. When the two atoms are separated, the total energy is just double the energy E_0 of a single hydrogen atom. When the two atoms approach, however, new interactions take place between the nucleus and the electron of one atom and those of the other. It is easily seen from the figure below that this additional interaction is given by

$$H' = -\frac{e^2}{r_{b1}} - \frac{e^2}{r_{a2}} + \frac{e^2}{r_{12}}$$

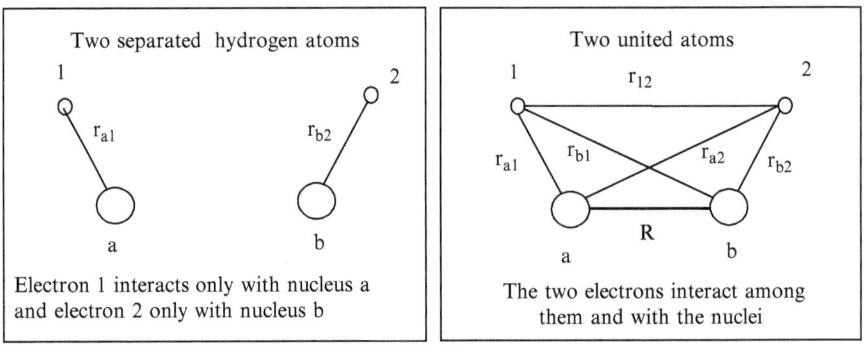

Scheme of the Heitler-London calculation

The crucial idea of Heitler and London, based on Heisenberg's suggestion that electrons are indistinguishable, was that when two atoms approach, the electron bound to one nucleus feels the attraction exerted by the other nucleus and vice versa. As a result, the probability that the electrons exchange between the two nuclei gradually increases as their distance decreases, until it becomes impossible to distinguish to which nucleus either electron belongs. It was thus possible to write either a wavefunction $\psi_a(1)\psi_b(2)$ describing the situation in which electron 1 is bound to atom a and electron 2 to atom b, or the function $\psi_a(2)\psi_b(1)$ in which the electrons have exchanged their position. Heitler and London struggled for long time on the choice between these two possible wavefunctions until Heitler suggested that, since the electrons are indistinguishable, the total wavefunction could only be written as a linear combination of both functions, in the form

$$\Psi_1 = \psi_a(1)\psi_b(2) + \psi_a(2)\psi_b(1)$$

or

$$\Psi_2 = \psi_a(1)\psi_b(2) - \psi_a(2)\psi_b(1)$$

To compute the energy in the two cases, Heitler and London utilized the variational theorem, a mathematical technique, often called variational principle, that Lord Rayleigh had developed long before to compute the energy minimum of an oscillating system. The variational principle is based on the fact that if one uses a test wavefunction, which is not a true eigenfunction of the Hamiltonian operator, the computed energy will always be higher than the true one. In order to apply the variational method, iterative techniques are therefore used to decrease the computed energy, until it becomes the best approximation to the experimental value.

Heitler and London used this approach to compute the energy of the two wavefunctions Ψ_1 and Ψ_2, obtaining two different expressions:

$$E_1 = 2E_0 + \frac{C+A}{1+S_{12}} \qquad E_2 = 2E_0 + \frac{C-A}{1-S_{12}}$$

where C, A, and S_{12} are integrals that they called Coulomb, exchange, and superposition integrals, respectively.

Since the Coulomb, C, and the exchange, A, integrals are both attractive terms (negative numbers) with $A > C$, the E_1 energy is smaller than that of the two separated hydrogen atoms $2E_0$ and therefore the wavefunction Ψ_1 corresponds to an attractive energy between the two atoms. Conversely, the E_2 energy is larger than $2E_0$ and thus the wavefunction Ψ_2 is repulsive. The exchange energy is thus responsible for the formation of a chemical bond. Obviously, the attraction due to the exchange energy was balanced by the electrostatic repulsion between the two positive nuclei, such that equilibrium was reached at a well-defined distance, that of the length of the chemical bond between the two atoms. This fact, however, led to the complication that only the wavefunction Ψ_2 is antisymmetric, satisfying the Pauli exclusion principle, since its sign is inverted if one changes electron 1 with electron 2. The wavefunction Ψ_1, which does not respect the Pauli principle, is attractive whereas the wavefunction Ψ_2, which respects it, is repulsive. This incongruent result is due to the fact that up to this point the electron spin had been neglected. Heitler and London removed this incongruence by multiplying the function Ψ_1 by an antisymmetric spin function, corresponding to a couple of electrons with antiparallel spins, and the function Ψ_2 by a symmetric spin function, corresponding to a couple of electrons with parallel spins.

Heitler and London did not expect that a new energy term, the exchange energy, binding the atoms together, would appear in the quantum calculations. Indeed, for a while, the exchange energy remained a true mystery, until it became clear that it was a true quantum effect with no counterpart in the classical treatment, as specified by Heitler:

> I think the only honest answer today is that the exchange is something typical for quantum mechanics, and should not be interpreted – or one should not try to interpret it – in terms of classical physics.

On 12 June 1927, the Heitler–London paper was presented to a meeting of the *Deutschen Physikalischen Gesellschaft* at Friburg im Bresgau and then published in *Zeitschrift für Physik* (Heitler and London 1927).

Calculations of energies and wavefunctions of homonuclear diatomic molecules started blooming in the second half of the 1930s. In 1933, an extremely precise calculation of both energy and bond length for the hydrogen molecule was made by Hubert M. James and Albert Sprague Coolidge at Harvard (James and Coolidge 1933) using a 13-term wavefunction expressed in terms of spheroidal electron coordinates and of the internuclear distance. They calculated an energy of 4.70 eV, very close to the experimental value of 4.72 eV, and a bond length of 0.74 Å, coinciding perfectly with the experiment.

Solving Schrödinger's equation to yield the energies and wavefunctions of complex molecules is a huge problem if one takes into account the full number of variables, i.e., the coordinates of all nuclei and of all electrons. For instance, in the relatively simple case of a triatomic molecule like water, H_2O, with only 3 nuclei and 10 electrons, one should solve a partial differential equation with 39 variables. In the slightly more complex case of benzene with 12 nuclei and 42 electrons, the equation would involve 162 variables.

In 1927, a significant simplification was introduced in the calculation by a paper of Max Born and of an American student of his, Robert Oppenheimer (1904–1967), the future father of the atomic bomb. Oppenheimer, once he had obtained his masters degree at Harvard, went to Cambridge, England, to work with Rutherford at the Cavendish Laboratory. As soon as he realized that he preferred theoretical to experimental physics, he left England to visit the Mecca of the theoretical physicists of the time, Göttingen's University, where he obtained his Ph.D. in physics in 1927 under the supervision of Max Born. At Göttingen, he collaborated with Born on several problems of quantum mechanics, and he published a paper with him entitled *Zur Quantentheorie der Molekeln* (Born and Oppenheimer 1927), which introduced an approximation (BO) that soon became crucial for the development of quantum chemistry. The Born–Oppenheimer idea was to separate the electronic from the nuclear motions, taking into account the fact that the nuclei are thousands of times heavier than electrons and thus move much more slowly, so that the electrons traveling very fast see the nuclei as practically at rest. In the Born–Oppenheimer approximation, the Schrödinger equation can be solved to compute the electronic energies holding the nuclei at their equilibrium positions. The nuclear dynamics is separately treated to compute the vibrational spectrum of the molecule, by solving the nuclear equation in which the electronic energy acts as a potential for the nuclear motions.

In the original Born–Oppenheimer paper, the nuclear motions were separated into small displacements of the nuclei from their equilibrium positions (internal motions) and larger-scale external motions, i.e., the translations of the molecular center of mass and the rotations around the molecular inertia axes. This separation was, however, treated in a complex and cumbersome fashion, which was difficult to interpret. The correct separation of the nuclear motions was realized, 8 years later, by Carl Eckart (1935) and afterward by Aaron Sayvetz in his famous paper *The Kinetic Energy of Polyatomic Molecules* (Sayvetz 1939).

7.4 Interatomic and Intermolecular Forces

Fritz London also made fundamental contributions to the theory of the forces giving rise to the condensation of gases to their liquid and solid phases, although still present in the gaseous phase and responsible for significant deviations from perfect gases behavior. The occurrence of weak attractive and repulsive forces between the atoms was already known to the sixteenth century atomists. Newton actually pointed out the existence of attractive and repulsive forces between particles and in 1758 the Jesuit Ruder Boscovich took up Newton's ideas, suggesting that the attractive force occurring between the constitutive particles of matter decreased as the distance increased, to become repulsive at very short distances.

Later, several authors of the eighteenth and nineteenth centuries accepted the existence of these forces not only to explain the nature of liquids and solids but also to account for several physical phenomena. For instance, Pier Simon Laplace (1749–1827) considered the existence of weak forces to account for capillary attraction, recalling a previous idea of Englishman Francis Hauksbee (1666–1713) relating to surface tension.

The concept of interatomic forces, however, become an integral part of chemical theory only after the publication of the Ph.D. thesis of the Dutchman Johannes Diderik van der Waals, entitled *Over de Continuïteit van den Gas-en Vloeistof-toestand* (on the continuity of the gas and liquid states). This work introduced an equation connecting these two states of matter, showing that they not only merge into each other but that they also have the same physical nature. Johannes Diderik van der Waals (1837–1923), born in Leyden, lacked a curriculum of classic studies. As such, according to existing Dutch law, he did not have the right to be enrolled at university and became a school teacher instead. Fortunately for him, new legislation was established whereby university students in science were exempt from the requirement of a prior classical education. This allowed van der Waals to join the University of Leyden officially and to obtain a Ph.D. in physics in 1873. The study of the papers of Clausius on the kinetic theory of gases and the experiments of Thomas Andrews (1813–1885) on the existence of the critical temperature of gases (Andrews 1869) stimulated the interest of the young van der Waals in the problem of the continuity of gaseous and liquid states. He realized that the deviations from the behavior predicted by the kinetic theory, observed for real gases, were due to the fact that their molecules have a nonnegligible volume and that, in addition, weak attractive–repulsive forces exist between them. Since then, these have been called van der Waals forces.

By taking into account the finite molecular volume and interaction forces, van der Waals deduced the new form of the equation of state for gases:

$$\left(P + a/v^2\right)(v - b) = RT$$

where a and b are constants that are specific to each gas. In 1910, he was awarded the Nobel Prize in physics for his studies on intermolecular interactions. van der Waals's work influenced the German theoretical physicist Gustav Adolf Feodor Wilhelm Ludwig Mie (1869–1957) who proposed (Mie 1903) the expression

$$\Phi_{12}(r) = \left(\frac{n}{n-m}\right)\left(\frac{n}{m}\right)^{m/(n-m)}\varepsilon\left[\left(\frac{\sigma}{r}\right)^n - \left(\frac{\sigma}{r}\right)^m\right]$$

where n and m are integers, r is the interatomic distance, and ε and σ are parameters that characterize the potential shape. Mie's expression was soon transformed by Lennard-Jones in a simpler semiempirical expression

$$\Phi_{12}(r) = 4\varepsilon\left[\left(\frac{\sigma}{r}\right)^{12} - \left(\frac{\sigma}{r}\right)^6\right]$$

Since then, this expression has become widely utilized in the scientific literature (Lennard-Jones 1924a, b) to describe not only atomic but also intermolecular interactions expressed as the sum of all possible interatomic forces between all atoms of two interacting molecules.

In addition to the interatomic van der Waals forces, other kinds of interactions are envisaged for atoms and molecules. These are normally separated into electrostatic, polarization, and induction forces. The electrostatic interactions and polarization forces are easily interpreted in terms of the classical theory, by which the charge distribution in the atoms or molecules is represented in terms of multipoles (charges, dipoles, quadrupoles, etc.) and polarizability. The basic formalization of the theory of intermolecular forces was mostly due to Dutch researchers, such as Peter Debye (1920, 1921) and Willem Hendrik Keesom (1876–1956), a coworker of Kamerlingh Onnes and the discoverer of solid helium (Keesom 1920, 1921a, b). Dispersion forces, however, have no classical equivalent term, and thus their contribution to interatomic and intermolecular interactions was understood only after the development of quantum mechanics. The existence of liquid helium, proven by the liquefaction experiments performed in 1908 by the Dutchman Heike Kamerlingh Onnes (1853–1926), did not find a reasonable explanation in terms of the classical and even of the earlier quantum theory until almost 1930. It was actually difficult to understand how it was possible that two inert and spherically symmetric atoms, lacking electrical charge, could attract each other giving rise to a condensed system. In 1927, Peter Debye who was visiting New York, in a discussion at Columbia with a young Chinese theoretical physicist, Shou Chin Wang (1904–1984), succeeded in convincing him to try using the new quantum mechanics to explain the existence of liquid helium. Wang, using a model of two hydrogen atoms simulated by two oscillating dipoles, succeeded in proving (Wang 1927a, b) the presence of an attractive interaction due to a long-range potential of the type $U(r) = -C_6\exp(r^{-6})$ that justified the occurrence of the liquid phase.

In 1930, the problem of the interaction between neutral atoms was reconsidered by London, who coined the expression "dispersion effect" (London 1930a, b, c) to account for contributions originating from the interaction of instantaneous dipoles occurring on the atoms due to their random charge fluctuations. In collaboration with the Austrian Robert Eisenschitz (1898–1968), who worked at the Kaiser Wilhelm Institut für Chemie in Berlin, he again faced the problem of the interaction between two hydrogen atoms with parallel spins. Using second-order perturbation theory, they (Eisenschitz and London 1930) validated Wang results and proved that the dispersion interaction was a pure quantum effect, due to the motion of the two electrons around the nuclei. Owing to this fast motion, every atom has an instantaneous dipole moment at each time instant, even if it rapidly averages to zero. The instantaneous dipole moment of an atom induces a field proportional to R^{-3}on a second atom, where R is the interatomic distance. The moment induced on the second atom now interacts with that of the first atom, giving rise to an energy variation that is also proportional to an R^{-3} factor. Thus, ultimately, this yields a total contribution to the energy proportional to R^{-6}. In the original London treatment, the interaction potential was expressed as a series of powers of R^{-1}, i.e., of the same series of the expansion of the electric charge in terms of electrical multipoles. However, for very short distances, an additional repulsive interaction is added to the attractive potential of medium and large distances. Eisenschitz and London proved that this was due to the need for the electronic wavefunction to be antisymmetric, with respect to the exchange of the two electrons. Later, however, it was shown that this expansion was not necessary and that it was possible to reformulate the dispersion energy in terms of polarization propagators (Oddershede et al. 1984). This new approach represents an important generalization of the treatment of interaction forces in terms of field theory(Casimir and Polder 1948). The perturbation method has also been successfully applied to the density matrix theory framework (Kohn and Sham 1965) for the calculation of the structure of dimers of helium, neon, water, and carbon dioxide.

The quantum treatment of interatomic forces was extended to the interaction between molecules by Slater and Kirkwood (1931) and by the same London (1937) who, in his previous papers of 1930, had already used the term "molecules." Important contributions to the study of intermolecular forces originated from the works of Joseph Oakland Hirschfelder (1911–1990) and of his coworkers. The treatise that Hirschfelder wrote in 1954, in collaboration with Charles F. Curtiss and Robert Byron Bird, entitled "Molecular Theory of Gases and Liquids," remains to this day a classic in this field (Hirschfelder et al. 1954). Other fundamental contributions were supplied by the English-Australian David Buckingham (1930–), mostly for quadrupole interactions (Buckingham et al. 1970, 1986).

For several years, and still today, in several simulations of the dynamics of molecules in their liquid and solid phases, the interaction between two molecules is expressed as the sum of all pair-wise interactions between all atoms of one and all atoms of the other molecule. These interaction potentials, often of a semiempirical character, have found broad applications in the study of very complex molecules, such as polymers and proteins, as well as in the study of the structure and dynamics of molecular crystals.

7.5 The Valence Bond Theory

Heitler and London's paper on the hydrogen molecule contained too much mathematics for the chemists of the time and in addition was published in a physics journal, *Zeitschrift für Physik*, that chemists were not used to reading. The paper would have probably had to wait for a long time before being finally accepted by the chemistry community, were it not for a young American chemist, from the California Institute of Technology (Caltech). In 1926, this young chemist, Linus Pauling, had been awarded a Guggenheim Foundation fellowship to go to Zurich, where he became acquainted with the work of Schrödinger and of his group. Pauling, who immediately realized the importance of quantum chemistry, became one of the founding fathers of quantum chemistry and twice a Nobel laureate, first for chemistry and then for peace, in recognition of his civil and political engagements.

Once he obtained a degree in chemical engineering, in 1922, Linus Pauling (1901–1994) became interested in the study of molecular structure by way of the papers of Lewis and Langmuir on the electronic structure of atoms and molecules. Luckily, he had a chance to pursue this interest, thanks to a fellowship that allowed him to enroll in the Ph.D. program of the School of Chemistry at Caltech.

At that time, under the presidency of the physics Nobel laureate Robert A. Millikan (1868–1953), Caltech was starting to become the best university of the West Coast of the United States. This was due to the presence of prestigious scientists, such as the astronomer George Hellery Hale, who built the great astronomical observatory of Mount Palomar, and the chemist Arthur Amos Noyes (1866–1936), who had previously directed the Research Laboratory of Physical Chemistry at MIT for 17 years, before being called to direct the Gates Chemical Laboratory at Caltech. Within a few years, Caltech also became the most important center for the study of genetics and molecular biology in the United States, due to the arrival in Pasadena of Thomas Hunt Morgan, the most famous American geneticist, who had discovered that genes were linearly arranged within chromosomes. The presence of an important school of genetics was seminal for the future of Linus Pauling, who later also made fundamental contributions to this field of research with his famous hypothesis on the helicoidal structure of proteins.

Linus Pauling, while working at Caltech under the supervision of the crystallographer Roscoe G. Dickinson on the applications of x-ray techniques to the study of crystal structures, also started to investigate the problem of the chemical bond. Once he obtained his Ph.D. in chemistry in 1925, he decided to visit Europe, at

Noyes suggestion, where the best centers existed for the theoretical study of the atomic structure within the framework of the rising quantum mechanics. Before leaving for Europe, Pauling followed a series of lectures on the old quantum theory that Sommerfeld held at Caltech. Based on these, he started to compute molecular energies and structures, applying the Lewis theory of shared electronic pairs to explain the stability of simple homonuclear diatomic molecules such as F_2, O_2, and N_2. The reading of the Burrau paper on the H_2^+ hydrogen molecule pushed him into adopting two-center orbits to create dynamical models of larger molecules such as benzene, still within the framework of the old quantum theory.

Thanks to a Guggenheim fellowship, Pauling left for Europe in 1926, planning to spend a year in Munich in Sommerfeld's laboratory, and then to visit other institutes, in particular those of Bohr in Copenhagen and of Schrödinger in Zurich. He thus had the luck to be in the right place at the right time to witness the birth and the growth of quantum mechanics, in direct contact with the principal authors of that revolution of physical thoughts.

When Pauling reached Munich, Sommerfeld's institute was still under the shock of the publication of Heisenberg's paper on quantum mechanics. He just had time to get settled with his young wife in a rented apartment before Schrödinger's papers on wave mechanics were published. For a chemist, used to thinking in terms of atoms and molecules, Heisenberg's theory, where orbits and electrons disappeared and all that was left were mathematical entities such as matrices, was too formal and abstract to be easily digested. Schrödinger's wave mechanics also eliminated the concept of orbits, but at least replaced it with something that possessed a minimum of physical reality, the stationary waves. Thus, Pauling had no doubts in choosing Schrödinger's approach over Heisenberg's to translate quantum mechanics to the chemical world.

During the 19 months spent in Europe, Pauling contacted several leading exponents of the new physics. This was made possible by the continuous flux of visitors to Sommerfeld's laboratory and to the large number of scientists he met while visiting Copenhagen and Göttingen for several weeks and Zurich for several months. While he was in California, in the chemistry milieu surrounding Noyes at Caltech, he was considered a high-level intellectual, with deep knowledge of mathematics and physics. It was only in Munich, Copenhagen, and Zurich, when he came in contact with great theoretical physicists, that he realized his limitations and recognized that he had to make great efforts to learn as much of the theory as quickly as possible.

Pauling stayed in Munich a little more than a year, while working to apply quantum mechanics to polyelectronic atoms, on the suggestion of Sommerfeld, a subject on which he published two papers. In the spring of 1927, while in Copenhagen visiting Bohr's institute, he met Samuel Goudsmit, who had recently developed the electron spin theory with Uhlenbeck, and he collaborated with him on the problem of the fine structure of the bismuth spectrum, although without great success. Nevertheless, he struck up a friendship with Goudsmit and committed himself to helping him in transforming his Ph.D. thesis into a textbook of atomic spectroscopy. The two friends actually went ahead with the project by mail and in

1930 the book was published with the title *The Structure of Line Spectra*. Pauling then moved to Zurich, where he remained for 2 months, but he only saw Schrödinger at seminars, without succeeding in getting him interested in his ideas. More important for his future was instead meeting Walter Heitler and Fritz London, who were developing their solution of the Schrödinger equation for the hydrogen molecule. Despite several discussions with them, mostly with Heitler, he was never informed of their work and he discovered it only later, in 1928, when he was back in the United States.

After returning to Pasadena, with a position of assistant professor of theoretical chemistry, he started teaching a course in quantum mechanics at the request of Noyes. Pauling was not a theoretical physicist of the caliber of those he had encountered in Europe, but had the great advantage over them of knowing the problems of chemistry when dealing with molecules. He understood right away that the Heitler–London paper offered great possibilities for the study of the structure of complex molecules and that the idea of exchange energy was the right way to understand the chemical bond.

The first step, however, was that of presenting quantum mechanics to the chemistry community in simple terms and, above all, in a form that could easily fit the chemists' approach to modeling the electron, framed in terms of the Lewis theory, easy to understand and capable of representing chemical bonds. In this context, he first prepared a short article for the Proceedings of the National Academy of Sciences, in the spring of 1928, where he pointed out the importance of Pauli's principle and of exchange energy for the theory of the chemical bond. He then followed with a broader review paper for *Chemical Reviews*, where he explained the scheme of the Heitler–London (HL) theory of the chemical bond in the hydrogen molecule, showing its equivalence to the Lewis theory of the shared electron pairs (Pauling 1928).

On the basis of the HL theory, Pauling developed the valence bond theory, an approach to chemical bonding used widely for more than 30 years, in which the electrons were assigned to single atomic orbitals and the chemical bonds were formed between atoms by sharing electron pairs with coupled spins. To account for the energy exchange, molecular wavefunctions were written as linear combinations of all possible permutations of electrons among the different atomic orbitals.

The electronic theory of the chemical bond, however, presented a series of problems in predicting the structure of even simple molecules. As discussed before, the 1s orbitals have a spherical shape, whereas the three 2p orbitals have a two-lobe shape, with a node at the nucleus, oriented along the x, y, and z axes, respectively. A simple combination of these orbitals, however, led to three-dimensional structures that were considerably different from the experimental ones. For instance, the oxygen atom has eight electrons, of which four with paired spins are located in the 1s and 2s orbitals, as shown in the scheme below.

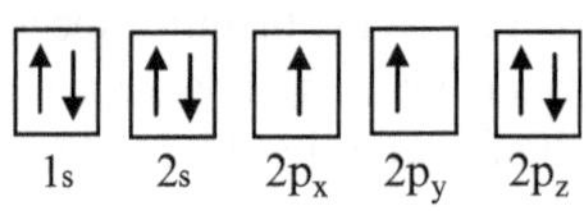

Four additional electrons are in the 2p orbitals, of which only two ($2p_x$ and $2p_y$) have unpaired spins available to form bonds with the 1s orbitals of the two hydrogen atoms. However, since the two 2p orbitals are oriented at 90° with one another, one would obtain a molecule with an angle of 90° degrees between the two O–H bonds, whereas the experimental value is 104.5°. To explain the difference it was necessary to introduce an ad hoc hypothesis, assuming that the discrepancy was due to the neglect of the repulsion between the electrons and between the nuclei. Still more difficult was to explain the different structures observed experimentally for molecules containing carbon atoms, i.e., the large majority of organic molecules from hydrocarbons to proteins. The electronic structure of the carbon atom, with only two unpaired electrons in the 2p levels, suggested that it should be bivalent, as actually occurs for the carbon monoxide molecule CO:

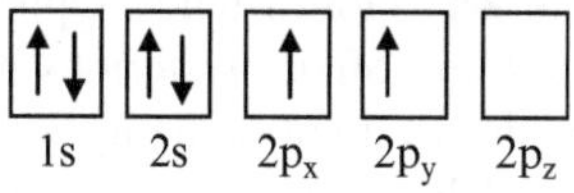

In all the other billions of molecules, carbon is tetravalent. In addition, its bonds are oriented either toward the vertices of a tetrahedron as in the saturated hydrocarbons, in a plane as in the case of ethylene, or even in a straight line as in acetylene:

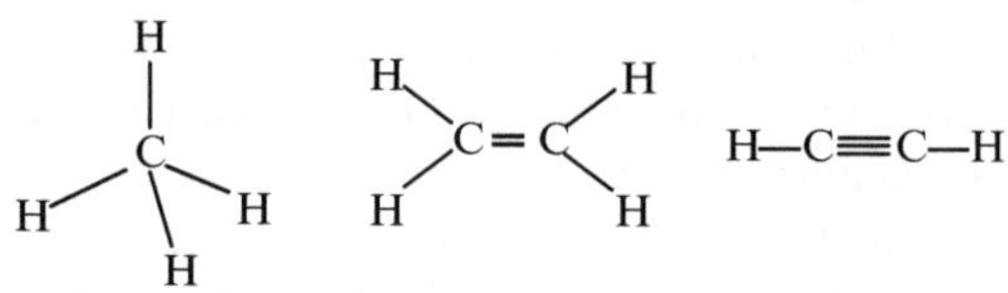

Methane Ethylene Acetylene

Pauling understood that, in order to have a tetravalent carbon atom, it was necessary to promote one electron from the 2s to the empty $2p_z$ orbital, giving rise to four unpaired electrons available to couple to the electrons of four monovalent atoms, to form four single chemical bonds.

Of course this operation required a given amount of energy to excite the 2s electron in the 2p orbital and it was necessary to understand where to find it. Pauling's idea was to resort to the exchange energy that he had encountered in the HL treatment of the hydrogen molecule. This idea was simple but brilliant: each time a C–H bond is formed, a given amount of exchange energy is produced and therefore four C–H bonds should be largely sufficient to compensate for the expense of energy necessary to promote the electron in the excited 2pz state:

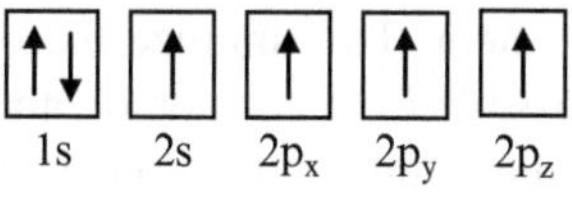

At the time, however, Pauling was unable to frame this idea in the correct mathematical form and, for the time being, preferred to put the problem aside and to concentrate his attention on crystallographic research.

In 1929, while in Pasadena, he met the physicist Robert Oppenheimer, whom he had already met in Munich, and in a short time they struck up an intense friendship and collaboration. With his competence in theoretical physics and mathematics, Oppenheimer perfectly complemented Pauling's expertise in chemistry in a project aimed at formalizing the chemical bond theory. Oppenheimer was a brilliant and fascinating personality, a cultivated, refined, and sensitive person who attracted many young students captivated by his speeches on the new physics and by his vision of a new world, free and without ties. Despite the gossip about his radical and even extreme political views on free love, and even on homosexuality, Pauling and Oppenheimer rapidly became very close friends. Oppenheimer used to make gifts and write poems with obscure and often mysterious meanings, on topics ranging from the readings of the classics to homosexuality, science, and art. In the boring and conformist environment of Pasadena, Pauling and his wife Ava Helen were bewitched by this unusual character and for a period used to see him very often. This friendship, however, came to an abrupt end in 1929, when one day Oppenheimer proposed to Ava Helen that they go to Mexico together for a short trip. Obviously, Ava Helen refused and that same evening she told the story to her husband who, from that moment, decided to cut his relationships with Oppenheimer completely, as well as the collaboration on chemical bond theory. After some years, Ava Helen told Pauling that, in her opinion, Oppenheimer was not in love with her but rather with him. Pauling agreed with her.

The separation from Oppenheimer deprived Pauling of valuable theoretical support to assure a rigorous mathematical formulation for his ideas on the four valences of the carbon atom. For a full year in 1929 he tried different approximations without significant results, until he decided to go back to Europe to see his friends in Munich, where he stayed for 3 months, and to visit new crystallography laboratories, especially in England.

Back in America, in October of 1930, Pauling concentrated again on the problem of the valences of the carbon atom. For 2 months he tried all possible combinations of the 2s and 2p atomic wavefunctions to obtain four bonds, correctly oriented and with the right energy. One evening in December 1930, he finally found the right way to simplify the atomic wavefunctions. In polar coordinates, the ideal basis to represent functions in a three-dimensional space, the wavefunctions depend on the distance (radial part) and on two angles (angular part). Pauling realized that, if one neglects the radial part, it becomes relatively easy to combine the four angular wavefunctions with suitable coefficients to obtain four new mixed functions, perfectly equivalent among them. To minimize the electrostatic repulsion, these orbitals must be as far as possible from each other, and thus they must be oriented toward the vertices of a regular tetrahedron with angles of $109^\circ\ 28'$. Pauling called this mixing of an atomic 2s with the three 2p wavefunctions an sp^3 hybridization. Each of the four new hybrid orbitals shared a quarter of the properties of the 2s and three quarters of those of the three 2p orbitals. Their shape was roughly similar to that of the 2p orbitals, but the two lobes were now of different dimensions, the negative one being very small and the positive one much larger.

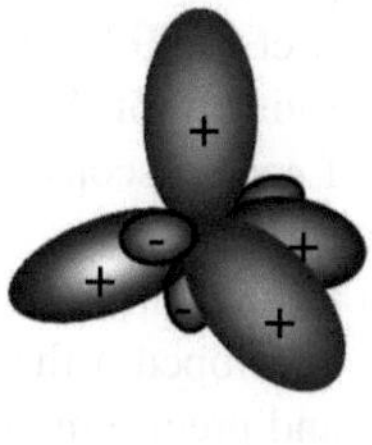

sp^3 hybridization

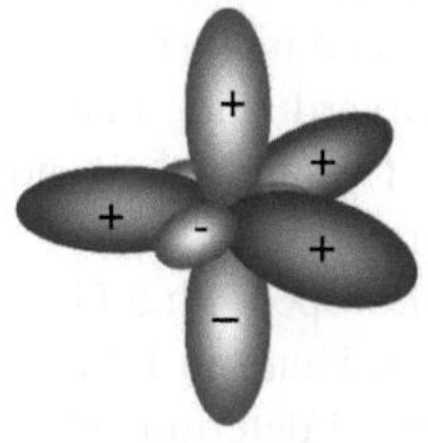

sp^2 hybridization

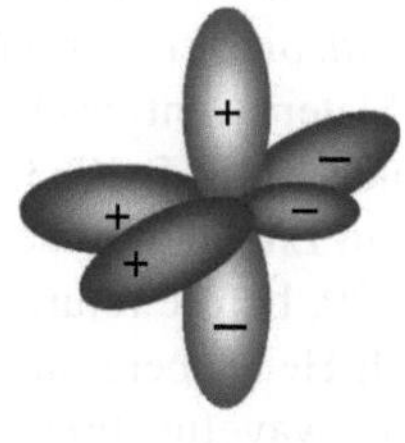

sp hybridization

Once he discovered the hybridization mechanism, it was easy for Pauling to explain the orientation of the carbon orbitals in ethylene. In this case, only the $2p_x$ and the $2p_y$ orbitals are combined with the 2s orbital. As a result, three identical hybrid orbitals oriented at 120° from each other are formed in the xy plane, whereas the $2p_z$ orbital remains perpendicular to that plane.

Pauling, thanks to his experience as a crystallographer and to a deep knowledge of the structure of molecules and complexes, was able to resolve rapidly several other hybridizations, such as the linear sp ($2s + 2p_x$), the square sp^2d ($s + 2p + d$), the octahedral sp^3d^2, and the tetrahedral sp^3d^3 hybrid structure (Pauling 1931a, b, c).

In the period 1931–1933, Pauling published a series of five famous articles entitled *The Nature of the Chemical Bond* (Pauling 1931a,b,c; 1932 a,b) that presented a general outline of the valence bond theory, based on the idea of electron pairs with antiparallel spins. The presentation was completed with the principle of maximum orbital overlap, to form stable chemical bonds, with the concept of resonance, and with the hybridization theory of the atomic orbitals.

On 1 March 1931, a month before the publication of Pauling's first paper in this series (6 April 1931), an article by the American physicist John Clark Slater, entitled *Directed Valence in Polyatomic Molecules* (Slater 1931), appeared in *Physical Review*. This article described a theory of the chemical bond that Slater had presented in April 1930 at a meeting of the National Academy of Sciences in Washington and again, in December of the same year, at a Cleveland meeting of the American Physical Society.

In this paper, Slater presented a more rigorous formulation of the sp^3 hybridization theory, as well as his theory of determinants, developed in 1929. Actually, Slater discussed the same concepts as Pauling, i.e., the idea of maximum orbital overlap, the relationships between ionic and covalent bonds, the formation of directional bonds, the importance of the exchange energy, and in particular the case of the carbon atom.

John Clark Slater (1900–1976) was a physicist born and educated in an academic milieu on the East Coast of the United States. In 1923, after obtaining a Ph.D. in physics at Harvard under the supervision of Percy William Bridgman (1882–1961), he left for Europe to study the principles of spectroscopy with Fowler at Cambridge. At the end of 1923, he moved to Copenhagen, where after no more than 6 months he had already contributed to developing a general theory of light emission, absorption, and scattering processes, in collaboration with Bohr and with his right-hand man Hendrik Kramers. This theory was published in a paper entitled *Über die Quantentheorie der Strahlung*, which made him famous (Bohr et al. 1924). In June 1924, Slater went back to Harvard and, under the supervision of Frederick A. Saunders (1875–1963), started to apply the fundaments of spectroscopy he had learned in England to the interpretation of the UV spectra of complexes.

In 1929, before returning to Europe with a Guggenheim fellowship to collaborate with Heisenberg and to meet Hund in Leipzig, Slater developed a theory to formulate wavefunctions in terms of determinants, a simple and precise method to write antisymmetric polyelectronic wavefunctions as linear combinations of monoelectronic wavefunctions, normally called spin-orbitals. A spin-orbital is the product of an orbital and of a spin function (Slater 1929). The idea that complete wavefunctions should be antisymmetric, with respect to a permutation of the orbital or of the spin component, had already been introduced in the theory by Heisenberg and Dirac, yet nobody before Slater had supplied a simple method to take such antisymmetry into account.

In addition, in 1930, Slater proposed a series of rules to write atomic orbitals in the form of exponential functions, known as Slater orbitals, with exponents representing the nuclear charges shielded by the presence of the electrons.

Pauling and Slater, although rivals, had an excellent relationship and great respect for each other. The fact that they developed the same theory independently was eventually proven by the name HLSP (Heitler–London–Slater–Pauling), used for several years to refer to their joint theory, until it finally became known as the valence bond theory, a theory with two separate fathers.

In 1931, Slater became professor of physics at MIT in Boston and shifted the focus of his interests to solid-state and microwave theory. Slater's research on the microwave electromagnetic theory, in collaboration with Julius Stratton and Nathaniel Frank, laid the basis for the development of radar. Slater was also a prolific writer of physics and chemical-physics textbooks. From 1933 to 1968, he wrote 14 books, including *Chemical Physics* (1939), *Quantum Theory of Matter* (1951), *Quantum Theory of Atomic Structure* (1960), and *Quantum Theory of Molecules and Solids* (1963–1966) that rapidly became classics that educated generations of physicists and chemists.

In 1931, Pauling published the second paper of the series *The Nature of the Chemical Bond*, describing the quantum-mechanical approach to specific bonds, such as those encountered in oxygen, boron, and nitrogen compounds.

At the beginning of 1932, Pauling discussed the transition from covalent to ionic bonds, showing the existence of all possible intermediate stages, and he resumed his studies on the concept of exchange energy, which he broadened into that of resonance. Pauling had already used exchange energy to explain the stability of the four carbon bonds in sp^3 hybridization. To this energy he now added that of resonance to explain molecular structures that could not be interpreted in terms of covalent bonds alone. The simplest case is that of hydrochloric acid which can be represented either by a covalent structure, in which the hydrogen and the chlorine atom share an electron pair with antiparallel spins, or by an ionic structure, in which the hydrogen has completely relinquished its electron to chlorine, thus becoming a positive ion. According to Pauling, this molecule was a hybrid, resonating between the two limiting structures and thus its wavefunction was the sum of the wavefunctions of the two resonating structures.

The concept of resonance explained well the occurrence of unsaturated double bonds and of several chemical situations, and was welcomed with enthusiasm by the chemical community, especially by the new generation of organic chemists, who saw the explanation of several reaction mechanisms in these resonant forms. In the 1920s, Thomas Martin Lowry had used the term *double semipolar* bond and Nevil Vincent Sidgwick the expression *coordinated bond* to describe the occurrence of partially ionic bonds.

The introduction of the concept of resonance in chemistry was certainly facilitated by the ability of Pauling and his coworker George Wheland to present it without mathematical equations, using a simple, clear style, supported by great explanatory abilities.

Resonance acquired special importance after the publication of the fourth paper in Pauling's series on the chemical bond, where he presented his method to evaluate the relative weight of the ionic versus the covalent part of a molecule as the difference between the energy computed assuming only covalent bonds and that computed assuming only ionic bonds. By comparing the energy for heteronuclear diatomic molecules, calculated within the covalent framework, with the experimental ones, Pauling could evaluate the stabilizing effect of the resonance with the ionic form. Since a high value for the ionic bond meant that one of the two atoms had a greater tendency to attract electrons, Pauling used these data to build up a new electronegativity scale that he used to evaluate the type and strength of chemical bonds.

In this way, he predicted that fluorine, located at the end of the scale, being strongly electronegative, should be able to extract electrons even from a rare gas atom, such as xenon. Initially, experiments made at Caltech produced negative results but, after 30 years, in 1962, his prediction was confirmed by the German chemist Rudolf Hoppe who synthesized the first rare gas compound, xenon difluoride XeF_2 (Hoppe 1964), and by the American Neil Bartlett, who synthesized xenon hexafluoroplatinate, $Xe^+[PtF_6]^-$ (Bartlett 1962). The resonance idea found further support in the application of the valence theory to the benzene molecule, which Pauling published in 1933 with his Ph.D. student George Wheland (Pauling and Wheland 1933). In this paper, the quantum-mechanical treatment of the benzene molecule was developed as a function of five resonant canonical structures, two called Kekulé and three Dewar structures. In 1865, Kekulé had hypothesized that the structure of the benzene molecule oscillated continuously between two chemical structures, with alternating double and single bonds, such that the double bonds of the first structure were shifted by one position in the second. In 1867, his pupil James Dewar, who studied under him at Gent, added another three structures to the first two, with a long bond connecting two opposite atoms in the hexagon, as also imagined by Carl Hermann Wichelhaus in 1869 (1869).

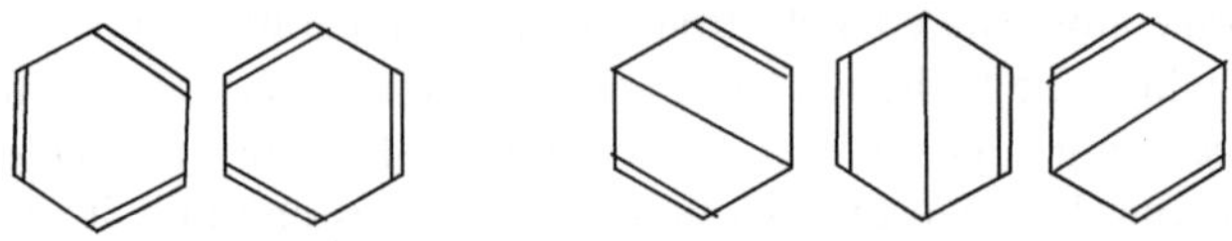

The two Kekulé structures The three Dewar's structures

Pauling maintained that the properties of the molecule could be represented as a mean value between those of the five canonical structures, i.e., that the true structure was the superposition of these rather than a physically unrealizable very fast interconversion between them. Then, in collaboration with Jack Sherman, Pauling extended his ideas of resonance to a series of conjugated double bonds, to complex aromatic systems with conjugated double bonds, and to molecules of biological interest such as lycopenes and carotenes.

An important contribution by Pauling to the vulgarization of quantum mechanics, and in particular of the valence bond theory, universally known since then as the VB method, was the publication in 1935 of the book *Introduction to Quantum Mechanics with Applications to Chemistry* (Pauling and Wilson 1935), followed in 1939 by the book *The Nature of the Chemical Bond* (Pauling 1939), both translated into all major languages. These rapidly became two of the "classics" of chemical literature.

In the mid 1930s, Pauling's research interests drastically changed, becoming oriented toward new fields, including genetics and biology, thanks to interaction with important biologists such as Thomas Hunt Morgan (1866–1945), Theodosius Dobzhanski (1900–1975), Calvin Bridges (1889–1938), and Alfred Sturtevant (1891–1970), who gave Caltech supremacy in genetics among United States institutions. Thanks to a grant from the Rockefeller Foundation, Pauling started to study the structure of hemoglobin. Thus, in collaboration with his Ph.D. student Charles D. Coryell (1912–1971), profiting from his expertise in crystallography and from his competence in the theory of chemical bonds, he proved that hemoglobin changes structure when bound to an oxygen atom with a covalent bond, discovering the diamagnetic properties of arterial blood with respect to the paramagnetism of venous blood. Around that time, Pauling's interests shifted again, this time toward the study of proteins, with the ambitious goal of theoretically predicting their structure.

With the birth of the molecular orbital, the valence bond theory started to lose ground and the concept of resonance, which initially seemed to represent the triumph of the valence bond theory, marked the road to its decline.

The word "resonance" was probably not the most well suited to represent Pauling's ideas to the chemistry community. In classical physics, this term is normally used to designate two bodies oscillating with the same frequency. For many chemists, this introduced the idea that an electron could jump between two positions or even that a molecule could continuously oscillate between two different structures. When Pauling first used this concept, a large majority of the chemists were convinced, right or wrong, that he meant that two or more forms of a molecule literally existed and that they transformed continuously into one another.

The debate on the meaning of the concept of resonance lasted for a long time. For some it was only a mathematical artifice that allowed the valence bond theory to achieve more degrees of freedom. For others it was a way of translating quantum mechanical concepts into chemical language. The discussion became even more inflamed in the period between 1940 and 1950 when the partners of the molecular orbital theory, mostly theoreticians from Lennard-Jones's school in England, started to prove the superiority of their simpler and less muddled theory, showing the absurdity of invoking more canonical forms for aromatic molecules when these could be derived in much simpler fashion with their theory.

7.6 The Molecular Orbital Theory

The molecular orbital theory was developed, practically at the same time in Germany by Friedrich Hund and in America by Robert Mulliken, with a series of closely related papers that followed the same theoretical path.

Both Hund and Mulliken were actually spectroscopists, and their aim was to interpret the electronic spectra of molecules. When their research projects were started, neither of them had any interest in creating a new theory of the chemical bond. In both cases, however, the close connection between the spectral patterns and the nature of the molecular electronic level drove them in an unexpected direction that ended in the molecular orbital theory.

In the presentation of the molecular orbital theory (MO), it is practically impossible to separate Hund's contribution from Mulliken's, even if they followed a different path. Mulliken, by cultural background and by university curriculum, was an experimental chemist who slowly shifted toward theoretical chemistry once he realized that he preferred theoretical rather than experimental activity. In contrast, Hund was trained as theoretical physicist and knew little about chemistry, even though in dealing with molecular spectra he was unavoidably forced to learn a minimum of this discipline to understand what he was doing. In an interview that he gave in 1963 to Thomas Kuhn, he actually confessed:

> Ich schemte mich etwas, weil ich furchtbar wenig Chemie konnte. Ich hab' nie Chemie studiert und fühlte mich darum allen chemischen Fragen gegenüber unsicher und war darum vorsichtig.

Robert Sanderson Mulliken (1896–1986) came from a Bostonian university milieu. His father, Samuel Parsons Mulliken, professor of organic chemistry at the Massachusetts Institute of Technology (MIT), had been a great friend of Arthur Amos Noyes, one of the most important chemists at the beginning of the twentieth century. Mulliken obtained his master's degree at MIT in 1917 and, around the beginning of the First World War, worked on the production of poisonous gases in the chemical service of the United States army. At the end of war, after a short period of activity as an applied chemist, he realized that his experimental practice in the laboratory was not his ideal and that he was rather interested in the more theoretical aspects of chemistry. He thus registered at the Ph.D. school of chemistry of the University of Chicago where, under the supervision of W.D. Harkins, he graduated in 1921 with a thesis on the isotopic separation of mercury by evaporation.

In Chicago, Mulliken became interested in the valence theory by reading the papers of Langmuir and Lewis, and he followed two physics courses given by the Nobel laureate Robert A. Millikan that introduced him to the quantum theory, a theory that he considered very confusing and described as disorganized chaos. After receiving his Ph.D., thanks to a 2-year fellowship from the British National Research Council, he moved to Harvard were he started to work on the quantum theory under the direction of Edwin Crawford Kemble (1889–1984), one of the first highly regarded American spectroscopists. At Harvard, he started reading spectroscopy books, such as *Atombau und Spektrallinien* by Sommerfeld, while working in the Jefferson Physical Laboratory directed by Theodore Lyman (1874–1954), where Frederick A. Saunders (1875–1963) was the head of a molecular spectroscopy team. Under the leadership of Saunders, Mulliken devoted himself to the study of the isotopic effect on vibro-rotational transitions, an effect in the infrared spectrum of gaseous hydrochloric acid that had been discovered only a few years before by Francis Wheeler Loomis (1889–1976) and by Adolf Kratzer (1893–1983). In the 1924–1925 period, Mulliken achieved a small personal success by correcting the assignment proposed by the Englishman Wilfred Jevons, who claimed to have identified the spectrum of boron nitride (BN). Instead, Mulliken showed that the identified spectrum was that of boron oxide and, after an exchange of letters and even a trip to England, he succeeded, in 1925, in convincing Jevons that his new assignment was correct.

The electronic spectra of a series of diatomic molecules (CO, N_2, NO, BO, CN, CO^+, and O_2) were already known at the beginning of the 1920s, and Mulliken focused his research on a general scheme for their assignment to the right energy level transitions. He started from the observation that these spectra showed evident similarities among them, which were even more evident in the spectra of isosteric molecules, i.e., molecules with the same number of electrons. Thus, borrowing criteria from the classification of atomic spectra, Mulliken started classifying molecular electronic spectra by grouping them into families and suggesting that similarities in the electron distribution corresponded to similarities in the energy level distribution.

These similarities led the American spectroscopist Raymond Birge (1887–1980), professor of physics at Berkeley, to suggest that the energy levels associated with the valence electrons in molecules corresponded to those of the valence electrons in the atoms (Birge 1926). Mulliken did not miss this opportunity and, having accepted Birge's suggestion at once, he rapidly found the way to interpret the known electronic spectra correctly and to assign new ones (Mulliken 1926). In his systematic investigation of the relationships between electronic spectra and molecular structures, Mulliken introduced for the first time the concept of atomic promotion, i.e., the idea that a rearrangement of electronic distribution takes place in the formation of a molecule, thus promoting some electrons to levels with a higher quantum number.

At Harvard, Mulliken met several first-rate scientists such as J. Robert Oppenheimer, John H. Van Vleck, Harold C. Urey, and John Clark Slater. Indeed, following their example, in 1925 he made the ritual pilgrimage to Europe, where he first visited several spectroscopists in England, then went to Copenhagen, and finally to the Institute of Theoretical Physics in Göttingen. In Max Born's institute, in addition to Born, he met his assistant Friedrich Hund with whom he struck up a fraternal friendship. Their collaboration became one of the most fruitful of the century and gave origin to the molecular orbital theory, even though they never published a joint paper. When in 1966 Mulliken was honored with the Nobel Prize, he openly declared that Hund equally deserved it.

Friedrich Hund (1896–1997), after having studied mathematics, physics, and geography at Marburg and Göttingen, was accepted at the Göttingen University, where he received his Ph.D. in physics in 1922, under Max Born's supervision. Hund became one of Born's closest coworkers and, as soon as he had completed his Ph.D. thesis, he helped him prepare the text of his lectures of atomic mechanics, *Vorlesungen über Atommechanik*, which Born gave in the 1923–1924 period. In 1925, after a period of interest in the relationship between molecular dimensions and crystal lattices, he shifted to atomic spectroscopy and in a short time he became a true expert of the spectral analysis of complex atoms using the Pauli exclusion principle and the basic criteria of the old quantum theory. Although Bohr's theory allowed establishing of the number and type of terms of the electronic configurations, to define the levels' energy it was necessary to analyze the experimental spectra. Within the framework of this research field, he developed two famous rules of great use in the identification of the spectral terms, known as Hund's rules. These allow prediction of the order in which electrons are organized in the different atomic levels. Hund's rules start from the idea that electrons always tend to occupy the lowest energy levels, following Bohr's aufbau principle, which states that no more than two electrons, with antiparallel spin can be arranged in each level. However, since electrons have negative charge, they tend to be as far as possible from each other and, therefore, to occupy all equivalent energy levels before occupying the same level with opposite spins. Hund's rules are very general and predict with great accuracy the electronic configurations of the majority of the elements, even though some exceptions exist. These occur when atoms with many electrons achieve greater stability by filling a new **s** level before completing the lower **d** or **f** levels.

From October 1925 to September 1926, Hund wrote the book *Linienspektrum und periodisches System der Elemente* that rapidly qualified him as the true expert on atomic spectroscopy. In 1926, he started a research project on molecular spectroscopy and published his first paper on the subject (Hund 1926) based on experimental data by Reinhard Mecke, Robert Mulliken, and Raymond Birge. In this paper, Hund put forward the idea that each electron in a molecule is defined by four quantum numbers that do not depend on the internuclear distance R. Starting from the knowledge of the electronic structure of atoms, he considered the two limiting cases: (1) two atoms separated by a distance $R = \infty$ and (2) two overlapping atoms separated by a distance $R = 0$. In this fashion, he built a series of correlation diagrams between the levels of the separated and united atoms as a function of R. He thus reached the conclusion that separated atom levels connected to empty overlapping atom levels are not favored to form a bond. Rather, those occupied for all values of R are favored.

He then started a systematic investigation of quantum mechanics that, thanks to a fellowship, drove him to work in Niels Bohr's laboratory. Hund was a typical product of Göttingen's theoretical school, with a deep knowledge of the theoretical aspects of physics and of mathematical techniques. This allowed him to master quickly the ideas and the algorithms of quantum mechanics. At the end of the 1920s, he was one of the first physicists to handle group theory, an abstract theory that has very important applications in quantum mechanics and molecular spectroscopy (Hund 1927b). At that time, the study of group theory was not part of the curriculum of physics and chemistry students. Thus, its appearance in the world of quantum mechanics was welcome with the name *Gruppenpest*, since it really panicked chemists and physicists, who found it unintelligible and so abstract as to represent true mathematical madness.

As soon as they met in 1925, Mulliken and Hund started discussing how to elaborate a principle of electron organization, analogous to the Aufbau principle of Bohr, thus assuming that electrons in molecules were also arranged in quantized orbits (Mulliken 1926). These orbits were organized as shells, similar to those of the atoms, with the significant difference that they were now extended over the whole molecule, including two or more nuclei.

Over the years, Mulliken and Hund exchanged abundant correspondence and they both published molecular spectroscopy papers in the period 1926–1927. In the meantime, however, the old quantum physics had merged into the new discipline of quantum mechanics and both understood that it was vital to reformulate their theories according to the new formalism.

Mulliken, owing to his relationships with the Göttingen's physicists, was exposed first to Heisenberg's matrix mechanics and only later to Schrödinger's. He soon realized, however, that he lacked the mathematical background to master Heisenberg's theory. Thus, he preferred to orient his approach toward Schrödinger's formalism, which was based on the solution of second-order differential equations that all students of chemistry were acquainted with at that time. In 1927, he went to Zurich to visit Schrödinger, who introduced him to Heitler and London. They showed him their recently published article, but Mulliken, who was too fixated on his own ideas, did not realize the importance of work that, in the hands of Pauling and John Slater, would soon become the valence bond theory, an important rival of molecular orbital theory.

In the summer of 1927, Hund and Mulliken simultaneously developed the interpretation of the electronic spectra of diatomic molecules and the transformation from atomic to molecular orbitals. Thus, in 1928, they had both written their first quantum mechanics paper.

From a methodological point of view, Hund's paper (Hund 1927a, b) was very different from Mulliken's (1926, 1928). Hund, who was a theoretical physicist, knew quantum mechanics well and was more interested in its abstract formalism than in the molecules themselves. Mulliken, instead, always maintained a pragmatic approach to the new theory, with respect to the old Bohr-Sommerfeld theory, that allowed him to visualize molecular orbitals analogously to electronic orbits. Hund, however, considered Mulliken's approach a crude application of the American's pragmatic attitude.

In 1928, Mulliken went back to the United States as associate professor in the department of physics of the University of Chicago. After a short time, he obtained a Guggenheim fellowship that he utilized in two stages: the first part, in 1930, to travel to Leipzig to meet Hund again and to get to know Heisenberg, Debye, Hückel, and Edward Teller, Heisenberg's assistant. The second, in the period 1932–1933,when the political atmosphere in Germany became oppressive and many were predicting Hitler's rise, to see his Leipzig friends again and to travel to Göttingen, Berlin, and Darmstadt, where he visited Herzberg.

The molecular orbital theory formulation of Mulliken and Hund in the 1928 was not yet competitive with that of the valence bond, being far from a true theory of the chemical bond and instead oriented essentially to the interpretation of the electronic molecular spectra (Mulliken 1928). It was only in the period from 1928 to 1931 that Hund and Mulliken slowly realized that they had in their hands a true valence theory, above all when other physicists like Gerhard Herzberg in Germany and Lennard-Jones in England started to use the concept of molecular orbitals in their works.

In 1929 a Discussions of the Faraday Society meeting entitled *Molecular Spectra and Molecular Structure* took place in Bristol, England, organized by W.E. Garner and J.E. Lennard-Jones. The best experts in the field took part in the meeting, among them Hund, Mulliken, Raman, Henri, and Herzberg. Lennard-Jones, professor at the University of Bristol, presented at the meeting his paper *The Electronic Structure of Some Diatomic Molecules* (Lennard-Jones 1929). This work, which was the first quantitative discussion of the MO method, included the approximation used to describe molecular orbitals as linear combinations of atomic orbitals (LCAO method). Lennard-Jones's paper, that derived the structure of the molecule of oxygen from first principles, represented the true launching pad of the molecular orbitals theory and convinced several chemists of the importance of quantum mechanics for the study of molecules. Gerhard Herzberg also studied the oxygen molecule in 1929 with similar methods and introduced into the Hund–Mulliken theory the concepts of bonding and antibonding orbitals, where the second energetically counterbalanced the first (Herzberg 1929). In addition, Herzberg, together with Heitler, showed, from the analysis of the rotational Raman spectrum of the N_2 molecule, that the nitrogen nucleus, which does not include electrons, obeys the Bose–Einstein statistics, contrary to what was believed at that time.

In the period 1930–1932, when the MO method was well established and had important representation in Europe, Mulliken wrote a series of papers entitled

The Interpretation of Band Spectra (Mulliken 1930), published in the prestigious journal *Reviews of Modern Physics*, and in the period 1933–1937 a second series of 13 papers entitled *Electronic Structures of Polyatomic Molecules*, published partially in *Physical Review* (Mulliken 1932) and partially in the *Journal of Chemical Physics* (Mulliken 1937). These papers, only a small part of his huge series of papers, gave a final structure to the MO theory and to his relationship to the electronic spectra.

Mulliken founded in Chicago the most important research center in the world for theoretical chemistry and electronic spectroscopy, the *Laboratory of Molecular Structure and Spectroscopy* (LMSS) in which several high-level American and European specialists worked, among them Charles C. Price, Hugh Christopher Longuet-Higgins, Michael Kasha, Klaus Ruedenberg, Robert Parr, Gerhard Herzberg, and Enrico Clementi.

Mulliken attracted a large number of pupils who ended up occupying prominent positions in the history of quantum chemistry. Among them, the scientific personality of the Dutchman Clemens C.J. Roothaan (1918) was particularly important. Born in Nijmegen in the Netherlands, he started his scientific career in 1935 by studying electro-technical engineering at Delft. During the German occupation of Holland in the Second World War, he was persecuted for his ethnic origin and ended up in a concentration camp, where he and other researchers and students were offered the opportunity to perform calculations for the Dutch company Philips. After a year in the camp, he was liberated and was awarded the master grade in recognition of the theoretical work done for Philips.

After immigrating to America with a fellowship, he was accepted at the University of Chicago, where Fermi, Teller, Goeppert-Mayer, and Robert Mulliken were teaching at the time. Thanks to the support of Maria Goeppert-Mayer, he then succeeded in securing a position in the research group of the Austrian physicist Karl F. Herzfeld, a pupil of Sommerfeld, who had immigrated to America in 1926 and was a professor at the catholic University of Washington. While working in Washington, under the supervision of Mulliken, he completed his Ph.D. thesis on the semiempirical calculation of molecular orbitals. In 1950, he was awarded a Ph.D. and was offered a job as professor in the department of physics of the University of Chicago.

In that period, Mulliken had realized the importance of electronic computers for the development of quantum chemistry, and he pushed the administration to equip the University of Chicago with a powerful electronic computer system, whose direction was entrusted to Roothaan. In 1951 Roothaan published his best known paper, *New Developments in Molecular Orbital Theory* (Roothaan 1960), in which he developed the LCAO method of Lennard-Jones in the form of matrix equations as an application of the self-consistent method of Hartree–Fock. Roothaan's LCAO-MO method used either a Gaussian or a Slater basis to represent the orbitals.

That same year, the same self-consistent method was independently developed by the Irishman George G. Hall, a pupil of Lennard-Jones at Cambridge, who became professor of mathematics at Nottingham University in 1962. For long time, the Roothaan–Hall method was the most used method in quantum chemistry.

Erich Hückel, one of the products of the mathematically sophisticated tradition of the Göttingen School of theoretical physics, used the algorithm at the basis of the LCAO-MO method to develop one of the simplest methods for computing the structure of complex molecules, which was easily utilizable by chemists.

After his graduation, Erich Hückel (1896–1980) traveled to Zurich to work with Peter Debye. There, he proved his talent by developing the Debye–Hückel theory discussed in Chap. 3. After spending the period 1928–1929 first in England and then in Bohr's laboratory in Denmark, Hückel was enrolled at the Technische Hochschule in Stuttgart and later moved to Marburg University. Although he was a physicist by training, Erich Hückel developed a great interest in organic molecules, may be due to the influence of his elder brother Walter Hückel (1895–1973), professor of organic chemistry at Tubingen and author of the organic chemistry treatise *Theoretische Grundlagen der Organischen Chemie*.

In 1930, following on the basic idea of a previous 1929 paper by Lennard-Jones on the fundamental triplet state of the oxygen molecule, Hückel developed an extremely simplified and yet very practical MO method to treat unsaturated organic molecules, suggesting the separation of σ electrons from π electrons to explain the hindered rotation around the double C=C bonds in alkenes (Hückel 1930). In Hückel's treatment, the C=C double bond is formed by superposition of an axially symmetric σ bond with a π bond created by the overlap of the two 2p orbitals of the carbon atoms. In the period 1931–1933, Hückel applied his treatment to benzene and, in 1937, he extended it to other cyclic hydrocarbons (Hückel 1931).

The simplicity of Hückel's method is bound to the postulate of complete separability between σ and π electrons, i.e., to the fact that only 2p orbital electrons are taken into account in the calculation, whereas σ electrons, whose task is that of forming the molecular backbone, are ignored.

The method is a simple form of linear combination of atomic orbitals, which is implemented by standard application of group theory to exploit molecular symmetry. The method's advantage is found in its ability to predict the molecular structure, the electric dipole moment, the number of energy levels, their degeneracy, the corresponding electronic transitions, the electric charge density on the atoms, and even the bond order, by condensing the whole MO formalism into two energy terms, α representing the electron's energy in a 2p orbital and β representing the interaction energy between two π electrons.

Hückel also developed a rule that allows computing the delocalization energy of π electrons in a polygon with n sides, in terms of the energy of a π bond in ethylene and of a quantum number. This rule, which in its simplest form was known as the $4n + 2$ rule, establishes that the properties of aromaticity appear when the number of π electrons in a ring obeys this condition. The rule shows that for rings with 6, 10, and 14p electrons ($n = 1, 2,$ and 3) the aromatic structures of benzene, naphthalene, and anthracene are obtained, whereas in the case of annulenes (rings with 4, 8, and 12π electrons),the aromatic character cannot exist and structures with alternating single and double bonds are thus formed. The Hückel method, which undoubtedly represents a milestone in theoretical organic chemistry, remained almost unknown for a long time to the chemical community not only because it was published in a physics journal not read by chemists but also due to the low communicative ability and the reluctant and hypochondriac personality of Hückel.

An important variant of Hückel's method, known as the extended Hückel method (EHMO), was developed in 1963 by Roald Hoffmann (1937–) (Hoffmann 1963), who was awarded the Nobel Prize in chemistry in 1981. Hoffmann was a Polish Jew, born in Złoczów, Galitzia, a small town with an important Jewish cultural tradition, whose inhabitants were exterminated by the Nazi in 1944. Roald and his mother were the only members of their family who managed to escape the Holocaust and later immigrate to the United States in 1949.

After studying at Columbia University in New York, he received his Ph.D. in chemistry at Harvard in 1960 under the supervision of William N. Lipscomb Jr., who was awarded the 1976 Nobel Prize in chemistry. At Harvard, he had the good fortune to work with Robert Burns Woodward from 1962 to 1965, one of the most famous organic chemists of the time, who was awarded the 1965 Nobel Prize in chemistry for his contributions to organic synthesis. Robert Burns Woodward was his true master, guiding him to the study of molecular structure and encouraging his inclination toward quantum chemistry.

Roald Hoffmann, who has been professor of chemistry at Cornell University since 1965 and, more recently, also professor of humanities, is a man with several cultural interests, from poetry to theater to philosophy. He is very keen to cultivate the relationships between arts and science. His poems have been published in several literary magazines. Two collections of his poems, entitled The Metamict State (1987) and Gaps and Verges (1990), were published by the University of Florida Press. He also wrote a comedy entitled Oxygen, in collaboration with Carl Djerassi, in which Lavoisier, Priestley, Scheele, and their wives meet and examine the methods of scientific discovery, while discussing the priority of the discovery of oxygen. More recently, in 2006, he published a new comedy entitled *Should've*.

The extended Hückel's method of Hoffmann is a variant of the standard Hückel's method where, instead of considering only π orbitals, the valence σ orbitals are also included in the treatment. The method is not very accurate for the calculation of molecular structures but is very convenient for constructing relative energy scales of molecular levels in different configurations. In several cases, the extended Hückel's method was utilized as a preliminary stage for more sophisticated calculations to determine molecular orbitals, for instance, the CNDO/2 or ab initio methods (Simonetta 1968). As discussed in Chap. 8, Hoffmann applied his method to the study of stereochemistry in organic reactions in collaboration with Robert Woodward.

In 1950, with the development of electronic computers, the Hartree–Fock method was extended to molecules. In the original HF method, the wavefunction was approximated by a single determinant, an approximation that would be valid if all electrons were independent from each other. However, this approximation does not correctly account for the fact that electrons try to be as far as possible away from each other to minimize their electrostatic repulsion. Unfortunately, this also means that their motion is correlated. As a matter of fact, in the original HF method, the correlation of the motion of electrons with parallel spins, called Fermi correlation, is partially accounted for since the antisymmetry principle equates the wavefunction to zero when two electrons with parallel spins occupy the same point in space. The method, however, still does not account for the correlation of electrons with antiparallel spins, called Coulomb's correlation. As a consequence, the limiting value of the energy computed by the standard HF method is always greater than its exact value. The difference between the Hartree–Fock limit and the exact energy value, which confers important contributions to several molecular properties, was called correlation energy by Per-Olov Löwdin (1955a, b). The influence of the electron correlation on molecular properties can be significant and the problem of calculating correlated wavefunctions is still a major obstacle to the accurate analysis of molecular properties.

One of the first computational methods introducing the electron correlation in the HF method was developed by Christian Møller (1904–1980) and Milton Spinoza Plesset (1908–1991) in 1934. It consisted in an application of the Rayleigh–Schrödinger perturbation technique, in which the effect of the electron correlation was introduced as a perturbation of the HF Hamiltonian (Møller 1934). In the Møller–Plesset method, the first-order correction to the energy is zero whereas the second- and fourth-order corrections may be significant, depending on the specific system being considered. The electron correlation was introduced in the HF method by several other authors, using a linear combination of determinants instead of a single determinant, thus changing the wavefunctions rather than the Hamiltonian. This method, called the *configuration interaction method* (CI), uses variational functions written as linear combinations of determinants derived from the basic HF theory to promote one or more electrons into excited orbitals (configurations).The number of determinants that can be obtained from N orbitals and M electrons is proportional to the binomial factor (N/M) that can easily reach extremely high values, up to billions of possible configurations. This method gives very accurate results and represents the benchmarks for the different techniques used in the calculation of molecular properties. In the second half of the twentieth century, several authors contributed to the development of the configuration interaction method and of its variants, including Per-Olov Löwdin (1959), Oktay Sinanoğlu (1961), Roy McWeeny (1967), Ernest R. Davidson (1974), Charlotte Froese Fischer (1977), and C.N. Handy (1980), J. Olsen (1990).

Several additional methods were recently developed to compute correlated functions, including the so-called Cluster Expansion method that yields very accurate and compact functions compared to the CI ones, at the cost, however, of introducing nonlinear variational parameters. The Czech Jiří Čížek (1966), professor of applied mathematics and chemistry at the University of Waterloo, Ontario, Canada, made a significant contribution to the derivation of the equations of coupled cluster theory. Rodney J. Bartlett, professor of chemistry and physics at the University of Florida, Gainesville, United States, was another pioneer in the development of rigorous many-body methods for electron correlation. In particular, he suggested combining the many-body perturbation and the coupled cluster methods, which have become key tools for the prediction of accurate electronic structures (Bartlett 1989). In 1982, Bartlett and his coworkers, in particular the

Dutchman Marcel Nooijen (1963–), were the first to formulate and to implement the coupled cluster theory with single and double excitation operators (Nooijen and Bartlett 1995). He also developed a version of Feynman diagrams that greatly helped visualize the physics of electron correlation, and that promoted the concept of size extensivity for many-body theory. This approach scales correctly with the number of particles and is now viewed as an essential element of sound quantum chemistry approximations.

7.7 The English School of Quantum Chemistry

The MO method rapidly spread to England, where it had its most significant developments thanks to John Edward Lennard-Jones (1894–1954), first professor of mathematics at the University of Bristol and later of theoretical chemistry at Cambridge. Similar to several English mathematicians, Lennard-Jones had a deep interest in applying mathematics to concrete problems, in particular to the study of atoms and molecules.

In a famous paper of 1929, wherein he developed the approximation of the linear combination of atomic orbitals (Lennard-Jones 1929) (LCAO method), Lennard-Jones expanded an idea from an earlier paper by Hund that the most internal electrons in a molecule are localized in atomic orbitals and that only valence electrons are delocalized in molecular orbitals that extend over all nuclei.

Similar to Bohr's Aufbau principle, where electrons had been added one at a time to the hydrogen atom to obtain the structure of complex atoms, he suggested adding electrons to the orbitals of the H_2^+ molecular ion, one at a time, to achieve the structure of diatomic molecules with many electrons. Since for the H_2^+ molecule an approximate solution had been obtained using linear combinations, i.e., sums or differences of atomic orbitals, a linear combination of properly normalized atomic orbitals of the hydrogen atom type may provide an acceptable approximation for other molecules as well. Of course, each time an electron was added it was necessary to adjust the shielding constants with a perturbative technique to account for the presence of other electrons.

According to the LCAO method, the overlap of two atomic orbitals gives rise to two molecular orbitals, one of lower energy called bonding orbital, which corresponds to the sum $\psi_a(1) + \psi_b(1)$ of the wavefunctions of the two atomic orbitals and one of higher energy, called antibonding orbital, which corresponds to the difference $\psi_a(1) - \psi_b(1)$ of the wavefunctions of the two orbitals.

The Lennard-Jones method gave excellent results for diatomic molecules with few electrons, confirming the results obtained by Hund and Mulliken. For larger molecules Lennard-Jones used the inverse procedure, starting from the idea that two rare gas atoms cannot form a bond since their external shell is complete. By subtraction of electrons, one at a time from a pair of rare gas atoms, he actually obtained the structure of more complex diatomic molecules. For instance, starting from a pair of neon atoms he reproduced the structure of the fluorine molecule F_2, showing how to form a bond by elimination of two electrons from the antibonding orbital. The name of Lennard-Jones is also associated with a semiempirical potential that he developed to describe the interaction between rare gas atoms and in general between neutral atoms, often utilized to account for the weak van der Waals interactions binding atoms and molecules in the condensed states.

Lennard-Jones established important schools of theoretical chemistry, first at Bristol and then at Cambridge, which counted among their members some highly regarded scientists. Among these, Charles Coulson and John Pople represented the front rank of theoretical chemistry in the United Kingdom in the first half of the twentieth century.

When Lennard-Jones became professor of theoretical chemistry at Cambridge, one of his first Ph.D. students was Charles Alfred Coulson (1910–1974), who under his supervision worked on the first ab initio calculation of a polyatomic molecule, the H_3^+ ion. Coulson was a very religious man, strongly engaged in the English Methodist Church, to the point that he became vice president of the 1959–1960 Methodist meeting and a member of the central committee of the World Council of Churches between 1962 and 1968. At a time when the valence bond theory was very popular among chemists, Coulson realized that the Hund–Mulliken MO method offered a more mathematically consistent picture for polyatomic molecules, even more so than the Pauling VB method. Thus, against mainstream precepts, he decided to develop further the MO theory. He was also the first to use the Hartree–Fock method, which he applied to the calculation of wavefunctions and energies of the hydrogen molecule up to complete convergence (Hartree–Fock limit). At that time, electronic computers were not yet available and calculations were made with infinite patience using mechanical hand-operated computers such as the Friden or the Brunsviga machines.

From 1936 to 1946, he and his coworkers made a series of significant contributions to the semiempirical MO methods. In a 1937 MO calculation of the structure of methane (Coulson 1937a, b), Coulson proved that localized bonding orbitals can be obtained through a unitary transformation and, starting from Slater-type orbitals, he developed the so-called Z matrix method for the calculation of multicenter integrals (Barnett and Coulson 1951a, b). That same year, he also defined the concepts of charge and bond order, investigated the performance of the density matrix method, and performed a systematic application of population analysis. Later, Coulson gave a further boost to the applications of the molecular orbital theory by working on graphite, on conjugated molecules, and on benzene (Coulson 1960a, b) by studying the metallic bonds and by defining the concept of partial valence to explain the nature of bonds in electron-deficient molecules such as diborane. To his large scientific body of work he added two highly successful books: *Valence* (1953) and *Waves: A Mathematical Account of the Common Types of Wave Motion* (1943). Coulson also had a number of pupils who contributed to the development of quantum chemistry. Among them, Christopher Longuet-Higgins, William Moffitt, and Roy McWeeny were undoubtedly the best representatives of the post-war generation of English theoretical chemists branching out from the Lennard-Jones School.

Hugh Christopher Longuet-Higgins (1923–2004) collaborated strongly with Coulson on several applications of the MO method, concerning the use of Green's functions in the calculation of π electron energies in aromatic molecules (Coulson and Longuet-Higgins 1947). Later he became professor of theoretical chemistry, first at London King's College and then at Cambridge. He subsequently moved to Edinburgh where he created a department for the study of artificial intelligence.

William Moffitt (1925–1958), who died prematurely when he was only 33 years old, contributed with Coulson to the study of the excited electronic states of aromatic hydrocarbons (Moffitt and Coulson 1948). In 1951, Moffitt published a famous paper on the subject called *Atoms in Molecules* (Moffitt 1951). He then moved to Harvard University in the United States where he developed the quantum theory of optical rotatory power (Moffitt 1961) in collaboration with Al Moskowitz (who also died prematurely), who, after graduating from Harvard in 1967 under Moffitt's supervision, became professor at the University of Minnesota.

Roy McWeeny (1924–) started a brilliant career in the United Kingdom as professor of theoretical chemistry, first at the University of Keele and then at the University of Sheffield, before moving to the University of Pisa, Italy, in 1982. There, he joined the theoretical chemistry center founded by Eolo Scrocco, one of the most important in Europe (Salvetti, Moccia, Tomasi, etc.). McWeeny is the author of two excellent books entitled *Symmetry: An Introduction to Group Theory and Its Applications* and *Quantum Mechanics: Principles and Formalism* (1972), widely used by scholars.

MO-LCAO semiempirical methods also found applications in molecular biology and biochemistry. In France, in particular, many biological and medical applications were developed by Raymond Daudel (1920–2006), assistant to Irène Joliot-Curie at the Institut du Radium. Daudel was one of the pioneers of the applications of semiempirical potentials to the calculations of physical properties and reactivity of biologically relevant molecules. Thanks to his excellent relationships with the political powers and with ministerial bureaucracy, he succeeded in obtaining large amounts of governmental funds to finance the activity of a CNRS center for theoretical chemistry in Paris that he directed for several years, attracting several scientists from abroad. In 1967, he founded his own Academy at Menton, the International Academy of Quantum Molecular Science, of which he became president. In his scientific career, he pursued several unorthodox ideas. For instance, as a young researcher he was convinced that the half-life of radioactive substances depended on the type of molecule in which they were situated and tried to apply quantum mechanical methods to the study of cancer, without achieving significant results.

Another contributor to the foundation of the International Academy of Quantum Molecular Science was Bernard Pullman (1919–1996) who was born in Wloclawek, Poland, and was later French naturalized. Pullman, who studied at the Sorbonne, was a member of the CNRS for some time, until in 1954 he became full professor and in 1959 was appointed director of the Departement de Biochimie Quantique of the Paris Institut de Biologie Physico-Chimique, an institute that he directed from 1963. Together with his lifelong coworker, his wife Alberte Pullman, he published several semiempirical MO calculations and was among the first to study the application of quantum mechanics to the prediction of the carcinogenic properties of aromatic molecules, defining a so-called K region that, according to their calculations, had selective interactions with proteins and amino acids.

Another contributor to the diffusion of the semiempirical calculations was the Swede Per-Olov Löwdin (1916–2000), who was more focused on the study of ab initio methods and on theoretical developments. Löwdin graduated in physics at the University of Uppsala, under the supervision of Ivar Waller, with a Ph.D. thesis on the ab initio calculation of the elastic constants of alkali halogenides. He became professor of physics in 1960 at the University of Uppsala and was jointly appointed to Florida University, in the United States, until 1993. In 1950, he developed the technique of symmetrical orthogonalization, a method that was the basis for the development of the Zero Differential Overlap, a widely used approximation in quantum chemical calculations. His name is often associated with a coupling scheme, largely applied in Hartree–Fock calculations, that he suggested to George G. Hall, who later expressed it in a correct mathematical form. Per Olov Löwdin also contributed to the development of perturbative techniques in quantum mechanics and even to quantum calculations of the DNA structure.

7.8 The Density Functional Theory

Another pupil of Lennard-Jones, who played a fundamental role in the history of quantum chemistry, was John Anthony Pople (1925–2004), Nobel laureate together with Walter Kohn in 1998. He too was a mathematician persuaded by his master to study theoretical chemistry. After graduating in 1944 at the University of Cambridge, he first worked on the applications of the MO method in statistical mechanics and in nuclear magnetic resonance. However, he soon became interested in semiempirical methods of quantum chemistry (Pople, 1953). He developed a series of approximate MO methods, widely used in chemistry, such as the Pariser–Parr–Pople method in 1953 (Pariser et al. 1953), the CNDO (Complete Neglect of Differential Overlap) method in 1965 (Pople and Segal 1966), and the INDO (Incomplete Neglect of Differential Overlap) method in 1967. He also succeeded in producing a series of high-performance computer programs, easily utilizable also by nonexperts, especially by organic chemists interested in understanding the stability and reactivity of their compounds.

After years of work on semiempirical models, Pople realized that the new powerful computation algorithms that were available, coupled with the disposability of high-speed computers with large memories, offered the possibility of going back to accurate ab initio calculations, taking advantage of the sufficiently high computing speeds to access the numerical evaluation of all integrals necessary to obtain significant results. He developed the series of orbital bases called "STO-3G," acronym for Slater-Type-Orbitals, simulated using three Gaussian functions. In the 1970s, there followed in a relatively short time a more extended series of orbitals labeled with the acronyms 6-31G and 6-31G*. Together with his student Warren Hehre he developed a complex computer program for big computers, GAUSSIAN 70, which radically changed the use of computers in quantum chemistry. Pople and his group continued over the following years to produce sophisticated computation techniques that eventually converged in the development of the density functional method.

At the beginning of the 1960s, the density functional method gained ground and very rapidly became the most used method for the study of molecules and for the prediction of their physical properties. Density Functional Theory (DFT) is a quantum theory for many electrons systems (atoms, molecules, solids, etc.), which is completely different from standard quantum chemistry methods. In contrast to the classical solutions of Schrödinger's equations, based on the use of wavefunctions, density functional theory uses the electron charge density.

Density Functional Theory originated from a statistical model developed in 1927 by Enrico Fermi (1901–1954) (Fermi 1928) and by the Anglo-American physicist Llewellyn Hilleth Thomas (1903–1992) (Thomas 1927). It is a relatively simple model devised to compute the energy of many-electron systems, which approximates the distribution of the electrons in an atom by assuming that they are uniformly distributed in the phase space. In this model, the kinetic and potential energies are expressed in terms of functionals of the electron density, taking into account both nucleus–electron and electron–electron interactions. The Fermi–Thomas model represented an important step in the calculation of the electron density of atoms, even though in its original form it was not sufficiently accurate as it neglected the exchange energy contribution. This limitation was eliminated in 1928 by Paul Dirac, who improved the method by adding an exchange functional.

Modern Density Functional Theory was developed on the basis of two important theorems, both published on 9 November 1964, by Pierre Hohenberg, a Yale professor, and by Walter Kohn (1923–), a University of California, Santa Barbara, professor. The first theorem (Hohenberg and Kohn 1964) proves that all properties of the fundamental state of a nondegenerate multi-electron system are correctly described by functionals of the mono-electronic density. The second theorem (Kohn and Sham 1965) proves that the electron density of the fundamental state minimizes the total electronic energy of the system. In this way, Hohenberg and Kohn demonstrated that the electron energy of the nondegenerate fundamental state of an electron system allows the determination of all physical properties that can be expressed as a density functional. Exact mathematical definition of the density functionals are presented by several authors, in particular in the book by Robert Parr and Weitao Yang (1989). We recall here that, to a good approximation, a functional is a mathematical function that uses functions as argument, i.e., is a function whose domain is an ensemble of functions. The standard methods of quantum mechanics, in particular the Hartree–Fock method, are all based on the use of complex many-electrons wavefunctions.

The aim of the Density Functional Theory, for a molecule with N electrons, is to substitute wavefunctions, which depend on $4 \times N$ variables (three space coordinates and the spin), with the electron density, which depends only on three variables.

Kohn and Sham showed that while electron density obviously depends on the system under consideration, there are density functionals such as the kinetic energy and part of the potential energy that are universal, i.e., valid for any kind of problem.

The most used method for the evaluation of density functionals is that of Kohn–Sham, which reduces the problem of a system of interacting electrons, in an external static potential, to that of a system of noninteracting electrons in an effective potential, which includes the external potential, as well as the exchange and correlation interactions between electrons (Kohn and Sham 1965). The main difficulty of this method consists in finding an appropriate form for these interactions. The simplest approach is the so-called local density approximation, based on the use of the exact exchange energy of a uniform electron gas obtained from the Fermi–Thomas model. This approximation had been widely used for decades in solid-state physics for the calculation of the band structure of electronic states in crystals. Unfortunately it is not too useful in the case of molecules for which the energy differences between electronic states are very significant and of the same order of magnitude of the errors inherent to the Fermi–Thomas model.

At the beginning of the 1980s, several variations of the Kohn and Sham Density Functional method were proposed, each one introducing key improvements in the calculation.

An improvement that was particularly convenient for computer applications was the inclusion of the Density Functional technique in the famous GAUSSIAN program developed by Pople. This program has provided a large number of nonexperts with the opportunity to develop quantum mechanical calculations for a large variety of chemical problems, from vibrational spectroscopy to chemical reactivity, to the prediction of the activity of pharmaceutical products with specific activity, to meteorological projects, and even to astrophysics.

7.9 Molecular Dynamics Simulation

The expression *Molecular Dynamics* (MD) defines computer simulations of the coordinated motion of interacting atoms or molecules. This computational technique, which was originally developed in the 1950s, has become the most widely used computational method to predict the dynamical evolution of molecular systems. In its most widely adopted version, particle trajectories are simply determined by numerically solving Newton's equations of motion. The idea of using a computer for simulating molecular motion was originally conceived in 1955, in the context of a theoretical physics project by Enrico Fermi, John Pasta, and Stanislaw Ulam at Los Alamos Scientific Laboratory (Fermi et al. 1965), using one of the first available high-performance computers, the so-called MANIAC 1 (*M*athematical *A*nalyzer, *N*umerator, *I*ntegrator, *a*nd *C*omputer). This computer was used for the calculation of the normal modes of a simple chain of particles coupled by quasi-harmonic springs, i.e., by a potential that included strong harmonic and weak cubic anharmonic terms. Fermi and his colleagues expected that due to nonlinear interactions, the energy accumulated into the lowest frequency mode $k = 1$ would slowly drift to the other modes, eventually reaching the normal equipartition of energy. Although the initial phase of the calculation indeed suggested that this was the correct evolution of the system, they observed the very strange result that when the simulation was conducted over longer times, the chain was no longer preserving the expected energy repartition. As a matter of fact, one day they let the program run longer by accident and when they came back they found that after remaining in a near equipartition state for a while, the system had eventually departed from it. To their surprise, after 157 periods of the mode $k = 1$, almost all the energy was back to this mode. Further calculations performed later with faster computers showed that the same phenomenon repeated many times, and that the initial state was cyclically recovered in time.

A more complex molecular dynamics (MD) methodology was soon introduced by Berni Julian Alder (1925–) and by Thomas Everett Wainwright (1927–2007) (Alder and Wainwright 1957, 1959), who had produced a highly efficient MD algorithm while working at Lawrence Livermore National Laboratory. Still in the 1950s, another group from Los Alamos Laboratory, directed by Nicholas Metropolis, proposed what is now known as the Metropolis Monte Carlo algorithm, which forms the basis for Monte Carlo statistical mechanics simulations of atomic and molecular systems (Metropolis et al. 1953). Actually, Metropolis himself played no role in the development of the method other than providing computer time on the more advanced MANIAC II machine. Indeed, it was Marshall N. Rosenbluth, who would later become one of the greatest plasma physicists of the twentieth century, who did most of the work.

With the spread of high-performance computers, molecular dynamics simulation started to be popular among physical chemists, who were eager to consolidate their knowledge of empirical force fields calculations with the necessary mastering of computer programs.

In the late 1960s, Bruce Berne (Berne et al. 1968), another pioneer in the development of molecular dynamics, organized at the Department of Chemistry of the Columbia University an important group of computer simulation where the first true molecular dynamics simulation for particles larger than an atom was realized (Bishop et al. 1974). The Columbia group has performed important quantum Monte Carlo calculations over the years, has contributed to the development of mixed quantum-classical methods, has simulated the dynamics of molecules in condensed states of matter, and has played a significant role in the study of vibrational relaxation in condensed media (Berne et al. 1967)

By the early 1970s, a large number of semiempirical force fields, including atom–atom interactions, electrostatic multipoles, and dispersion forces were utilized all over the world for molecular dynamics simulation. Among them, popular and widely applied force fields were developed by Harold A. Scheraga and his colleagues (Burgess 1975), by Norman L. Allinger (1977), by Arieh

Warshel who gave concrete form to the ideas of his mentor Shneior Lifson (Lifson and Warshel 1968), and by Jerry H. Schachtschneider and Robert G. Snyder (1963), which led to the production of additional force fields of broad application (Zerbi and Gussoni 1966; Califano 1969, Shimanouchi 1972).

At that time it became obvious that these methods were ultimately limited by the memory and computational requirements necessary to handle large numbers of particles and depth of interaction. Yet, the development of a new generation of ultrafast computers allowed the extension of molecular dynamics simulations to very large systems, including polypeptides and proteins. The first molecular dynamics simulation of a realistic complex system was performed by Rahman and Stillinger in their simulation of liquid water in 1974 (Stillinger and Rahman 1974) and later extended to larger systems like polypeptides and proteins. A central role in the development of molecular dynamics approaches for the study of proteins was played by the American–Viennese Martin Karplus (1930–), professor at Harvard University, who in 1977, in collaboration with his students Andrew McCammon and Bruce Gelin, performed the first true simulation of the dynamics of a protein (McCammon et al. 1977).

Martin Karplus's group at Harvard developed the famous CHARMM (*C*hemistry at *HAR*vard *M*acromolecular *M*echanics) project for protein dynamics, in collaboration with several groups, including Harold Scheraga's group at Cornell University, Michael Levitt at Stanford, and in particular Schneior Lifson's group at the Weizmann Institute of Israel, including his pupil and colleague Arieh Warshel, who later moved to Harvard, bringing his substantial force field program with him.

At the end of the 1970s, Martin Karplus and his graduate student Bruce Gelin started to assemble a computer program (Gelin and Karplus 1979) that could calculate the energy of a protein system as a function of the atomic positions, starting from a given amino acid sequence and coordinates. In the 1980s, a paper in which the original Gelin's program had been considerably restructured finally appeared as the CHARMM compute protocol (Brooks et al. 1983). CHARMM has continued to grow and the latest release of the executable program was made available in August 2009 as CHARMM35b3. Within the framework of the Born–Oppenheimer approximation, the potentials used in the classical molecular dynamics research discussed above are normally represented as semiempirical force fields. In excited states, however, the potential needs to be calculated from first principles using quantum mechanical methods, for instance, the Density Functional Theory. The computational cost of this kind of simulation is much higher than classical molecular dynamics. Ab initio quantum-mechanical methods may be needed to calculate the correct potential energy of an evolving molecular system, as required, for instance, in the analysis of conformational trajectories. Ab initio quantum-mechanical calculations produce a large body of information that is not available from empirical methods, such as the density of electronic states and other electron properties. A significant advantage of using such ab initio methods is the ability to study reactions that involve breaking or formation of covalent bonds, which correspond to multiple electronic states and cannot be effectively handled by semiempirical force fields.

A popular piece of software for ab initio quantum-mechanical molecular dynamics is the Car–Parrinello Molecular Dynamics (CPMD) package, based on density functional theory and proposed by Roberto Car and Michele Parrinello in 1985. This program was the fruit of their collaboration when they were both working in Trieste, Italy (Car and Parrinello 1985), in the stimulating atmosphere of the new SISSA (Scuola Internazionale Superiore di Studi Avanzati) laboratory that at the time was almost completely overlapping with the well-established International Centre of Theoretical Physics (ICTP). Their collaboration profited from Car's expertise in density functional theory and from Parrinello's knowledge of statistical mechanics to originate one of the most seminal papers in molecular dynamics simulation. In contrast to the Born–Oppenheimer molecular dynamics approach, used in classical simulations, their method explicitly introduced the electronic degrees of freedom as fictitious dynamical variables, leading to a system of coupled equations of motion for both ions and electrons. A fictitious mass is assigned to the electrons to describe their dynamics in time, chosen to be small enough to avoid significant energy transfer from the ionic to the electronic degrees of freedom. This allowed the elimination of a minimization of the electronic configuration at each step of the process, thus keeping them on the electronic ground state corresponding to each new ionic configuration visited along the dynamics evolution of the system.

The method was further extended by Michele Parrinello and Naseem Rahman (1981) to treat systems at constant temperature and pressure and by Shuichi Nosé, William Graham Hoover, and Michael L. Klein to examine related problems (Nosé 1984, 1991; Hoover 1985; Martyna et al. 1992). Comprehensive reviews of recent computer simulations were published by Michael P. Allen and Dominic Tildesley (1987, 1993).

Quantum-mechanical methods of the Car–Parrinello type are very powerful but also very expensive computationally, compared to the faster classical molecular dynamics simulation methods. For this reason, a new family of simulation has emerged that combines the accuracy of the quantum-mechanical methods with the speed of the classical ones. These methods are known as mixed or hybrid quantum-mechanical and molecular mechanics methods. These hybrid methodologies were first introduced by Warshel and colleagues and more recently by several other groups, including those directed by Weitao Yang at Duke University, by Sharon Hammes-Schiffer at Pennsylvania State University, and by Donald Truhlar and Jiali Gao at the University of Minnesota.

References

Alder BJ, Wainwright TE (1957) Phase transition for a hard sphere system. J Chem Phys 27:1208

Alder BJ, Wainwright TE (1959) Studies in molecular dynamics. I. General method. J Chem Phys 31:459

Allen MP, Tildesley DJ (1987) Computer simulation of liquids. Clarendon Press, Oxford

Allen MP, Tildesley DJ (1993) Computer simulation in chemical physics. Kluwer Academic Publishers, Dordrecht

Allinger NL (1977) Conformational analysis. 130. MM2. A hydrocarbon force field utilizing V1 and V2 torsional terms. J Am Chem Soc 99:8127–8134

Andrews Th (1869) The Bakerian lecture: on the continuity of the gaseous and liquid states of matter. Philos Trans R Soc Lond 159:575–590

Barnett MP, Coulson CA (1951a) Evaluation of integrals occurring in the theory of molecular structure, Part I: basic Functions. Philos Trans R Soc Lond A 243:221–233

Barnett MP, Coulson CA (1951b) Evaluation of integrals occurring in the theory of molecular structure, Part II: overlap, resonance, Coulomb, hybrid and other two-centre integrals. Philos Trans R Soc Lond A 243:234–249

Bartlett N (1962) Xenon hexafluoroplatinate $Xe^+[PtF_6]^-$. Proc Chem Soc Lond 6:218

Bartlett RJ (1989) Coupled-cluster approach to molecular structure and spectra: a step toward predictive quantum chemistry. J Phys Chem (Feature Article) 93:1697–1708

Berne BJ, Jortner J, Gordon R (1967) Vibrational relaxation of diatomic molecules in gases and liquids. J Chem Phys 47:1600

Berne BJ, Pechukas P, Harp GD (1968) Molecular reorientation in liquids and gases. J Chem Phys 49:3125

Birge RT (1926) The structure of molecules. Nature 117:300

Bishop M, Berne BJ (1974) Molecular dynamics of one-dimensional hard rods. J Chem Phys 60:893

Boltzmann L (1884) Ableitung des Stefan'schen Gesetzes, betreffend die Abhängigkeit der Wärmestrahlung von der Temperatur aus der electromagnetischen Lichttheorie. Ann Phys Chem 22:291–294

Boltzmann L (1897) Zu Hrn Zermelos Abhandlung Über die mechanische Erklärung irreversibler Vorgänge. Ann Phys 60:392–398

Bohr N, Kramers HA, Slater JC (1924) Über die quantentheorie der strahlung. Z Phys 24:69–87

Born M, Jordan P (1925) Zur Quantenmechanik. Z Phys 34:858

Born M, Heisenberg W, Jordan P (1926) Zür Quantenmechanik II. Z Phys 35:557–615

Born M (1926a) Quantenmechanik der Stoßvorgänge. Z Phys 37:863

Born M (1926b) Quantenmechanik der Stoßvorgänge. Z Phys 38:803

Born M (1926c) Zur Wellenmechanik der Stossvorgänge. Göttinger Nachr Math Phys Kl 146–160

Born M, Oppenheimer R (1927) Zur Quantentheorie der Molekeln. Ann Phys 84:457–484

Brooks BR, Bruccoleri RE, Olafson BD, States DJ, Swaminathan S, Karplus M (1983) CHARMM: a program for macromolecular energy, minimization, and dynamics calculations. J Comput Chem 4:187–217

Burrau O (1927) Berechnung des Energieeigenwertes des Wasserstoffmolekel-Ions H_2^+ im Normalzustand. Det Kongelige Daske Videnskabernes Selskab, Mathematisk-fysiske Meddeelelser, Kobenhavn, bind **7**, 14, 1–18

Buckingham AD, Utting BD (1970) Intermolecular forces. Annu Rev Phys Chem 21:287–316

Buckingham AD, Fowler PW, Stone AJ (1986) Electrostatic predictions of shapes and properties of van der Waals molecules. Int Rev Phys Chem 5(2/3):107–114

Burgess AW, Shipman LL, Scheraga HA (1975) A new approach to empirical intermolecular and conformational potential energy functions. II. Applications to crystal packing, rotational barriers and conformational analysis. Proc Natl Acad Sci USA 72:854–858

Califano S (1969) Force fields of large molecules. Pure Appl Chem 18:353, Butterworths

Car R, Parrinello M (1985) Unified approach for molecular dynamics and density-functional theory. Phys Rev Lett 55:2471–2474

Casimir HBG, Polder D (1948) The influence of retardation on the London–van der Waals forces. Phys Rev 73:360

Cizek J (1966) On the correlation problem in atomic and molecular systems. J Chem Phys 45:4256

Compton AH (1923) A quantum theory of the scattering of X rays by light elements. Phys Rev 21(5):483

Coulson CA (1937a) The electronic structure of methane. Trans Faraday Soc 33:388–398

Coulson CA (1937b) The evaluation of certain integrals occurring in studies of molecular structure. Math Proc Camb Philos Soc 33:104–110
Coulson CA, Longuet-Higgins HC (1947) The electronic structure of conjugated systems. I. General theory. Proc R Soc A 191:39
Coulson CA (1960) Present state of molecular structure calculations. Rev Mod Phys 32:170–177
Davidson ER (1974) Configuration-interaction description of electron correlation. In: Daudel R, Pullman B (eds) The world of quantum chemistry. Reidel, Dordrecht/Holland, pp 17–30
Davisson C, Germer LH (1927) Reflection of electrons by a crystal of nickel. Nature 119:558–560
De Broglie L (1924) Ondes et quanta. CR Acad Sci 177:507
Debye P (1920) Die van der Waalsschen Kohäsionkräfte. Phys Z 21:178
Debye P (1921) Molekularkräfte und ihre Elektrische Deutung. Phys Z 22:302
Debye P, Hückel E (1923) Zur Theorie der Elektrolyte. I. Gefrierpunktserniedrigung und verwandte erscheinungen. Phys Z 24:185–206
Dirac PAM (1925) The fundamental equations of quantum mechanics. Proc R Soc Lond A 109:642
Dirac PAM (1930) The principles of quantum mechanics. Clarendon Press, Oxford
Eckart C (1935) Some studies concerning rotating axes and polyatomic molecules. Phys Rev 46:383–387
Einstein A (1906a) Zur theorie der Lichterzeugung und Lichtabsorption. Ann Phys 20:199
Einstein A (1906b) Die Plancksche Theorie der Strahlung und die Theorie der spezifischen Wärme. Ann Phys 22:180–190
Einstein A (1907) Berichtigung zu meiner Arbeit: Die Plancksche Theorie der Strahlung und die Theorie der spezifischen Wärme. Ann Phys 22:800
Eisenschitz R, London F (1930) Über das Verhältnis der van der Waalsschen kräfte zu den homöopolaren bindungskräften. Z Phys 60:491
Fermi E (1928) Eine statistische Begründung zur Bestimmung einiger Eigenschaften des Atoms und ihre Anwendungen auf die Theorie des periodischen Systems der Elemente. Z Phys 48:73–79
Fermi E, Pasta J, Ulam S (1965) Studies of nonlinear problems. I. Los Alamos report LA-1940, published later. In: Segré E (ed) Collected papers of Enrico Fermi. University of Chicago Press, Chicago
Fock V (1930) Näherungsmethode zür Lösung des Quantenmechanischen Mehrkörperproblems. Z Phys 61:126
Fock V (1932) Konfigurationsraum und zweite quantelung. Z Phys 75:622–647
Fischer CF (1977) The Hartree–Fock method for atoms: a numerical approach. Wiley, New York
Fues E (1922) Die Berechnung wasserstoffunähnlicher Spektren aus Zentralbewegungen der Elektronen. I. Z Phys 11:364–378
Gelin BR, Karplus M (1979) Side-chain torsional potentials: effect of dipeptide, protein, and solvent environment. Biochemistry 18:1256
Gaunt JA (1928) Theory of Hartree's atomic fields. Proc Camb Philos Soc 24:328
Gaunt JA (1929) Triplets of helium. Philos Trans R Soc Lond 228:151–196
Hartree DR (1923) On some approximate numerical applications of Bohr's theory of spectra. Proc Camb Philos Soc 21:625
Hartree DR (1928a) The wave mechanics of an atom with a non-Coulomb central field. Part I-theory and methods. Proc Camb Philos Soc 24:89–110
Hartree DR (1928b) The wave mechanics of an atom with a non-Coulomb central field. Part II. Some results and discussion. Proc Camb Philos Soc 24:111
Hartree DR, Hartree W (1935) Self-consistent field, with exchange, for beryllium. Proc R Soc Lond A150:9–33
Heisenberg W (1925) Über quantentheoretische Umdeutung kinematischer und mechanischer Beziehungen. Z Phys 33:879
Heisenberg W (1926) Quantenmechanik. Naturwissenschaften 14:989–994

Heisenberg W (1927) Über den anschaulichen Inhalt der quantentheoretischen Kinematik und Mechanik. Z Phys 43:172

Heitler W, London F (1927) Wechselwirkung neutraler Atome und homöpolare Bindung nach der Quantenmechanik. Z Phys 44:455–472

Herzberg G (1929) The heat of dissociation of oxygen. Z Phys Chem 4B:223

Hirschfelder JO, Curtiss CF, Bird RB (1954) Molecular theory of gases and liquids. Wiley, New York

Hoffmann R (1963) An extended Hückel theory. I. Hydrocarbons. J Chem Phys 39:1397

Hohenberg P, Kohn W (1964) Inhomogeneous electron gas. Phys Rev B 136:864

Hoover WG (1985) Canonical dynamics: equilibrium phase-space distributions. Phys Rev A 31:1695–1697

Hoppe R (1964) Valence compounds of the inert gases. Angew Chem Int Ed Engl 3:538

Hückel E (1930) Quantentheorie der Doppelbindung. Z Phys 60:423

Hückel E (1931) Quantentheoretische Beiträge zum Benzolproblem. Z Phys 70:204–286

Hund F (1926) Zur Deutung einiger Erscheinungen in den Molekelspektren. Z Phys 36:657

Hund F (1927a) Zur Deutung der Molekelspektren. Z Phys I 40:742–764, II, 42:93–120, III, 43:805–826

Hund F (1927b) Symmetriecharaktere von Termen bei Systemen mit gleichen Partikeln in der Quantenmechanik. Z Phys 43:788

James HM, Coolidge AS (1933) The ground state of the hydrogen molecule. J Chem Phys 1:825

Jeans JH (1905) On the partition of energy between matter and aether. Philos Mag 10:91

Keesom WH (1920) Quadrupole moments of the oxygen and nitrogen and nitrogen molecules. Proc Koninkl Nederland Akad Wetenschap 23:939

Keesom WH (1921a) Van der Waals attractive force. Phys Z 22:129–141

Keesom WH (1921b) Van der Waals attractive force. Phys Z 22:643–644

Kirchhoff G (1860) Über das Verhältnis zwischen dem Emissionsvermöogen und dem Absorptionsvermögen der Körper für Wärme und Licht. Ann Phys Chem 109:275–301

Kohn W, Sham LJ (1965) Self-consistent equations including exchange and correlation effects. Phys Rev A1 140:133

Lennard-Jones JE (1924a) On the determination of molecular fields. I. from the variation of the viscosity of a gas with temperature. Proc R Soc Lond A106:441–462

Lennard-Jones JE (1924b) On the determination of molecular fields. II. From the equation of state of a gas. Proc R Soc Lond A106:463–477

Lennard-Jones JE (1929) The electronic structure of some diatomic molecules. Trans Faraday Soc 25:668–686

Lifson S, Warshel A (1968) A consistent force field for calculation on conformations, vibrational spectra and enthalpies of cycloalkanes and n-alkane molecules. J Phys Chem 49:5116

London F (1930a) Theorie und systematik der molekularkrafte. Z Phys 63:245–279

London F (1930b) Über einige Eigenschaften und Anwendungen der Molekularkräften. Z Phys Chem B11:222–251

London F (1930c) Properties and applications of molecular forces. Z Phys Chem B 11:222–223

London F (1937) The general theory of molecular forces. Trans Faraday Soc 33:8–26

Löwdin P-O III (1955a) Quantum theory of many-particle systems. I. Physical interpretations by means of density matrices, natural spin-orbitals, and convergence problems in the method of configurational interaction. Phys Rev 97:1474–1508

Löwdin P-O (1955b) Extension of the Hartree–Fock scheme to include degenerate systems and correlation effects. Phys Rev 97:1509–1520

Löwdin P-O (1959) Correlation problem in many-electronic quantum mechanics. I. Review of different approaches and discussion of some current ideas. Adv Chem Phys 2:207–322

Martyna GJ, Tuckerman ME, Klein ML (1992) Nose–Hoover chains: the canonical ensemble via continuous dynamics. J Chem Phys 97:2635

McCammon JA, Gelin BR, Karplus M (1977) Dynamics of folded proteins. Nature 267:585–590

McWeeny R (1967) Nature of electron correlation in molecules. Int J Quant Chem Symp 1:351–359

Metropolis N, Rosenbluth AW, Rosenbluth MN, Teller AH, Teller E (1953) Equation of state calculations by fast computing machines. J Chem Phys 21:1087–1092

Mie G (1903) Zur kinetischen Theorie der einatomigen Körper. Ann Phys 316:657–697

Moffitt WE, Coulson CA (1948) The electronic structure and bond lengths of coronene and pyrene. Proc Phys Soc 60:309

Moffitt W (1951) Atoms in molecules. Proc R Soc Lond A210:245

Moffitt W, Moscowitz A, Klyne W, DJerassi C (1961) Structure and the optical rotatory dispersion of saturated ketones. J Am Chem Soc 83:4013–4018

Møller C, Plesset MS (1934) Note on an approximation treatment for many-electron systems. Phys Rev 46:618–622

Mulliken RS (1926) Systematic relations between electronic structure and band-spectrum structure in diatomic molecules, III: molecule formation and molecular structure. Proc Natl Acad Sci 12:144–151, 338

Mulliken RS (1928) The assignment of quantum numbers for electrons in molecules. I. Phys Rev 32:186–222

Mulliken RS (1930) The interpretation of band spectra. Parts I, IIa, IIb. Rev Mod Phys 2:60–115

Mulliken RS (1932) Electronic structures of polyatomic molecules and valence. Phys Rev 40:55–71

Mulliken RS (1937) Solved and unsolved problems in the spectra of diatomic molecules. J Chem Phys 41:5–45

Nooijen M, Bartlett RJ (1995) Equation of motion coupled cluster method for electron attachment. J Chem Phys 102:3629–3647

Nosé S (1984) A unified formulation of the constant temperature molecular dynamics methods. J Chem Phys 81:511–519

Nosé S (1991) Constant temperature molecular dynamics methods, progress of theoretical physics supplement 103 (Molecular Dynamics Simulations, Nosé S (ed)), pp 1–46

Oddershede J, Jørgensen P, Yeager DL (1984) Polarization propagator methods in atomic and molecular calculations. Comput Phys Rep 2:33

Pais A (1991) Niels Bohr's times: in physics, philosophy, and polity. Oxford University Press, Oxford

Pariser R, Parr R (1953) Semi-empirical theory of the electronic spectra and electronic structure of complex unsaturated molecules. I. J Chem Phys 21:466, 767

Parr RG, Yang W (1989) Density functional theory of atoms and molecules. Oxford University Press, New York

Parrinello M, Rahman A (1981) Polymorphic transitions in single crystals: a new molecular dynamics method. J Appl Phys 52:7182

Pauling L (1928) The application of the quantum mechanics to the structure of the hydrogen molecule and the hydrogen molecule-ion and to related problems. Chem Rev 5:173–213

Pauling L (1931a) Quantum mechanics and the chemical bond. Phys Rev 37:1185–1186

Pauling L (1931b) The nature of the chemical bond. J Am Chem Soc 53:1367–1400

Pauling L (1931c) The nature of the chemical bond. II. J Am Chem Soc 53:3225–3237

Pauling L (1932a) The nature of the chemical bond. III. J Am Chem Soc 54:988–1003

Pauling L (1932b) The nature of the chemical bond. IV. J Am Chem Soc 54:3570–3582

Pauling L, Wheland GW (1933) The nature of the chemical bond. V. J Chem Phys 1:362–374

Pauling L, Wilson EB (1935) Introduction to quantum mechanics with applications to chemistry. Dover, New York

Pauling L (1939) The nature of the chemical bond and the structure of molecules and crystals: an introduction to modern structural chemistry. Cornell University Press, Ithaca

Planck M (1900) Entropie und temperatur strahlender Wärme. Ann Phys 1:719

Planck M (1901) Über das Gesetz der Energieverteilung im Normalspectrum. Ann Phys 4:553

Planck M, Wood RW (1931) Archives of the centre for the history and philosophy of physics of the American Institute of Physics in New York City, 7 Oct 1931
Poincaré H (1889) Sur les tentatives d'explication mecanique des principes de la thermodynamique. CR Acad Sci Paris 108:550–553
Pople JA (1953) Electron interaction in unsaturated hydrocarbons. Trans Faraday Soc 49:1375
Pople JA, Segal GA (1966) Approximate self-consistent molecular orbital theory III. CNDO results for AB2 and AB3 systems. J Chem Phys 44:3289
Rayleigh JWS (1900) The law of partition of kinetic energy. Philos Mag 49:98–118
Rayleigh JWS (1905) The dynamical theory of gases and of radiation. Nature 72:54
Roothaan CCJ (1951) New developments in molecular orbital theory. Rev Mod Phys 23:69–89
Roothaan CCJ (1960) Self-consistent field theory for open shells of electronic systems. Rev Mod Phys 32:179
Sayvetz A (1939) The kinetic energy of polyatomic molecules. J Chem Phys 7:383–389
Schachtschneider JH, Snyder RG (1963) Vibrational analysis of the n-paraffins – II: normal coordinate calculations. Spectrochim Acta 19:117–168
Schrödinger E (1926a) Quantizierung als Eigenwertproblem (Erste Mitteilung). Ann Phys 79:361
Schrödinger E (1926b) Quantisierung als Eigenwertproblem (Zweite Mitteilung). Ann Phys 79:489
Schrödinger E (1926c) Quantisierung als Eigenwertproblem (Dritte Mitteilung). Ann Phys 80:437
Schrödinger E (1927) Quantisierung als Eigenwertproblem (Vierte Mitteilung). Ann Phys 81:109
Shimanouchi T (1972) Tables of molecular vibrational frequencies. National Bureau of Standards, Washington
Simonetta M (1968) Forty years of valence bond theory. In: Rich A, Davidson N (eds) Structural chemistry and molecular biology. Freeman, San Francisco
Sinanoglu O (1961) Theory of electron correlation in atoms and molecules. Proc R Soc Lond A 260:379
Slater JC (1928) The self-consistent field and the structure of atoms. Phys Rev 32:339–348
Slater JC (1929) The theory of complex spectra. Phys Rev 34:1293
Slater JC (1931) Directed valence in polyatomic molecules. Phys Rev 37:481–489
Slater JC, Kirkwood JG (1931) The van der Waals forces in gases. Phys Rev 37:682–697
Stefan J (1879) Über die Beziehung zwischen der Wärmestrahlung und der Temperatur. In: Sitzungsberichte der mathematisch-naturwissenschaftlichen Classe der kaiserlichen Akademie der Wissenschaften, vol 79. Wien, pp 391–428
Stefan J (1881) Über die Beziehung zwischen der Wärmestrahlung und der Temperatur. J Phys Theor Appl 10:317–319
Stillinger FH, Rahman A (1974) Improved simulation of liquid water by molecular dynamics. J Chem Phys 60:1545–1557
Swirles B (1935) The relativistic self-consistent field. Proc R Soc Lond A 152(877):625–649
Swirles B (1936) The relativistic interaction of two electrons in the self-consistent field method. Proc R Soc A 157(892):680–696
Swirles B, Hartree DR, Hartree W (1939) Self-consistent field, including exchange and superposition of configurations, with some results for oxygen. Philos Trans R Soc A 238:229
Thomas LH (1927) The calculation of atomic fields. Proc Camb Philos Soc 23:542
Thomson GP, Reid A (1927) Diffraction of cathode rays by a thin film. Nature 119:890
Wang SC (1927a) The diamagnetic susceptibility of hydrogen molecule and of helium in the new quantum mechanics. Proc Natl Acad Sci 13(12):798–800
Wang SC (1927b) The mutual energy of two hydrogen atoms. Phys Z 28:663
Wichelhaus H (1869) Die Hypothesen über die Constitution des Benzols. Berichte 2:197–199
Wien W (1893) Eine neue Beziehung der Strahlung schwarzer Körper zum zweiten Hauptsatz der Wärmetheorie. Sitzungsberichte der preussischer Akademie 55–62, Sitzungsberichte, 9. Februar1893
Wien W (1896) Über die Energievertheilung im Emissionsspectrum eines schwarzen Körpers. Ann Phys Chem 294:662–669

Zerbi G, Gussoni IM (1966) Transferability of valence force constants from "Overlay" calculations of syndiotactic 1,2poly(butadiene). Spectrochim Acta 22:211
Zermelo E (1896a) Über enien Satz der Dynamik und die mechanische Wärmetheorie. Ann Phys 57:485–494
Zermelo E (1896b) Über mechanische Erklärungen irreversibler Vorgänge. Ann Phys 59:4793–4801

Chapter 8
The Mechanisms of Chemical Reactions

Not knowing the mechanism is like seeing the first and the last scene of Hamlet.

(Arne Fredga)

8.1 Reaction Mechanisms and Residual Affinities

In the paradigmatic vision of chemistry presented by the development of Lewis electronic theory, the interpretation of the mechanisms of chemical reactions found its grounds in a few fundamental concepts sufficient to build the basic skeleton of the new theoretical organic chemistry. The carbon atom tetravalence, the benzene ring hexagonal structure, the radicals mobility, the electronegativity, the bond directionality, and the existence of single, double, and triple bonds became the basic ingredients for the development of a mechanistic theory of chemical reactions.

These concepts contributed to the growth of theoretical organic chemistry (often called "physical organic chemistry") allowing the organic chemists to identify classes of molecules occurring all with the same basic mechanisms and opening in this way the route to a systematic interpretation of chemical reactions. The exact knowledge of the route to follow in the synthesis from common ancestors (Jacobs 1997) thus made possible the theoretical prediction of an unlimited number of new molecules.

Many of the figures that animated the scenario of theoretical organic chemistry in England at the end of the nineteenth and the beginning of the twentieth century were educated in Germany and were directly or indirectly connected with Henry Edward Armstrong (1848–1937), professor of chemistry at the University of London who was assistant to Frankland at the Royal College of Chemistry and studied in Kolbe's laboratory in Leipzig from 1867 to 1870. Armstrong did not make any particularly significant contributions to organic chemistry, but helped greatly in improving the teaching of chemistry and in stimulating in the English scientific community an interest in the theoretical aspects of organic chemistry.

S. Califano, *Pathways to Modern Chemical Physics*,
DOI 10.1007/978-3-642-28180-8_8, © Springer-Verlag Berlin Heidelberg 2012

His personality was actually not too open to the new ideas arising within the framework of the electronic theory of valence. Kolbe, who introduced Armstrong to the chemistry of aromatic molecules (Armstrong 1890), had planted in his cultural background the seeds of a virulent dislike against the innovations introduced into chemistry by physical theories. Armstrong thus considered particularly negative the acceptance of the atomic theory and Lewis's electronic doublet that he sarcastically called *Californian thermodynamiter*. Lewis, according to him, indulged *in premature speculation upon electrons as cause of valency*. In the same way he refused to accept the idea of the hydrogen bond as well as the ionic theory that he replaced with a model of association between solute and solvent molecules held together by residual affinities.

The concept of residual affinities was developed in 1899 by the German Friedrich Karl Johannes Thiele (1865–1927), who first studied mathematics at Breslau and then moved to chemistry obtaining a Ph.D. at the Halle University in 1890. From 1893 to 1902, he taught at Munich before becoming professor at Strasburg. Thiele developed the idea of residual affinity, studying the addition reactions of hydrogen or halogens to molecules with double bonds. He realized that the addition to a chain with conjugated double bonds led, in addition to the elimination of one double bond, to a shift of the position of the other double bonds, and argued that this was due to the presence of residual valences on the atoms. These residual valences, which he represented by dotted hyphens, were auto-saturated in the molecule but were easily opened through the attack by a reagent as shown in the following reaction:

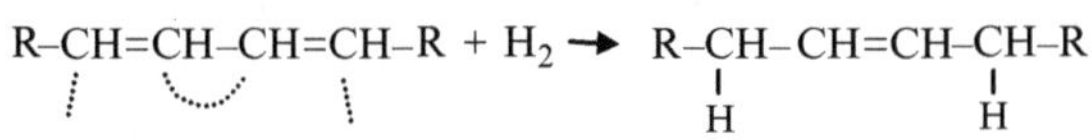

Thiele extended the idea of residual valences to aromatic compounds and in particular to benzene. Kekulé's structure of benzene was in fact not too convincing for several chemists of his generation since it included three double bonds, known to be very reactive, whereas benzene was very stable and gave rise only to substitution reactions. Thiele suggested that when, in a cyclic molecule, single and double bonds alternate, the properties of each one were mixed with those of the others so that the single bonds neutralized the reactivity of the double bonds (Thiele 1899). The same could not occur in an open chain since in this case the end of the chain was available for addition reactions.

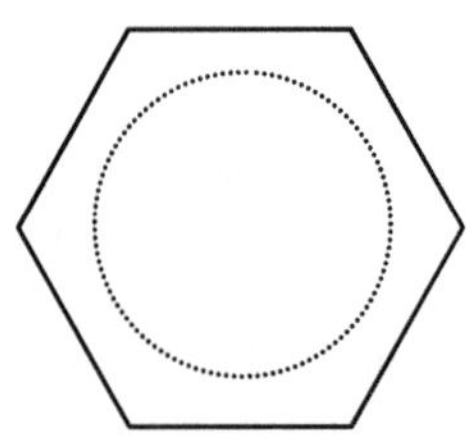

In order to explain the mechanisms of the substitutions at the benzene ring, Thiele proposed a structure represented by a hexagon with a dotted circle inside to indicate the presence of the residual valences, thus anticipating the idea of resonance which would be developed several years later (Thiele 1899). Thiele instilled his ideas on residual valences in his pupil Bernard Flürscheim (1874–1955) who in turn transmitted them to Arthur Lapworth; both played a crucial role in England for the development of the reaction mechanism theory. Incidentally, Thiele also became known for his studies on the condensation of aldehydes and ketones with cyclopentadiene, a reaction that opened the route to the chemistry of fulvenes (Thiele 1900).

John Norman Collie (1859–1942) was another English chemist who made significant contributions to theoretical organic chemistry. He was a peculiar character, on the one hand a famous explorer and mountain climber and on the other an excellent chemistry professor at University College, London. As a chemist, he studied the inert gases, invented the first neon lamp, proposed a dynamic structure for benzene (Collie 1897), discovered the first oxonium salt of dimethylpyrone, and suggested that the $-CH_2-CO-$ group could play a significant role in the biosynthesis of several natural products. His presence in London was, however, overshadowed by that of the two important pupils of Armstrong, Lowry and Lapworth, who were both involved, even if on different sides, in the diatribe that opposed polarizability to residual affinities supporters.

In 1869, the Russian Vladimir Markovnikov (1838–1904), pupil of Butlerov and Kolbe, formulated the empirical rule (Markownikoff 1870) stating that in the attack of the double bond of unsaturated compounds by halogen acids (HCl, HBr, HI), the hydrogen atom connects to the carbon atom bound to the larger number of hydrogen atoms, whereas the halogen binds to the carbon atom with less hydrogen atoms as in the example shown below of the reaction of attack by HCl of the propylene double bond:

$$H^+Cl^- + CH_3 - CH = CH_2 \longrightarrow CH_3 - CHCl - CH_3$$

Markovnikov's rule led to the thought that there were free electric charges in the molecules, orienting addition reactions to the double bond. This problem was reconsidered by the American Arthur Michael (1853–1942), another eccentric character who, even having studied with high level chemists, Hofmann at Berlin, Bunsen at Heidelberg, Wurtz at Paris, and Mendeleev at St. Petersburg, never bothered to obtain a Ph.D. Michael became well known in the organic chemistry community not only for the reaction that bears his name (addition of an enolate such as an acetoacetate to the double bond of a compound of the type R—CH=CH—C_6H_5) but also for his work, often written in German, on theoretical organic chemistry. Michael shared Lowry's idea of the presence of separated charges in unsaturated chains and proposed the *positive–negative* rule, according to which the CH_3 group made the vicinal CH group more positive, giving rise to a charge separation that led to the addition of HCl to the double bond as foreseen by Markovnikov's rule:

$$H^+Cl^- + CH_3 - \underset{+}{CH} - \underset{-}{CH_2} \rightarrow CH_3 - CHCl - CH_3$$

The distribution of single and double bonds in the molecules was obviously a difficult problem to face and was particularly intriguing for benzene. Armstrong, for instance, did not accept the structure of the Kekulé benzene ring with alternate single and double bonds, although recognizing that it was superior to all the other available proposals up to that time. In particular, he agreed with Kekulé's pupils, Adolf Baeyer, Wilhelm Körner, and Albert Ladenburg (1842–1911), in assuming that Kekulé's benzene structure, destroying the hexagonal symmetry, led to the thought that two ortho and two meta isomers could exist, while it was well known that in both cases only a single structure existed. For this reason he proposed a center-symmetric structure with crossed bonds between diagonally opposed atoms that he later transformed into a second structure in which the bonds crossed at the center (Armstrong 1890):

> the (six) centric affinities act within a cycle ... benzene may be represented by a double ring ... and when an additive compound is formed, the inner cycle of affinity suffers disruption, the contiguous carbon-atoms to which nothing has been attached of necessity acquire the ethylenic condition

This second structure is intended to represent the idea of the existence of unsaturated affinities directed toward the center of the ring. Armstrong also extended these ideas to naphthalene which he represented with double affinities located on the two central atoms and directed in two opposite directions, as well as by single affinities directed toward the center for all other atoms.

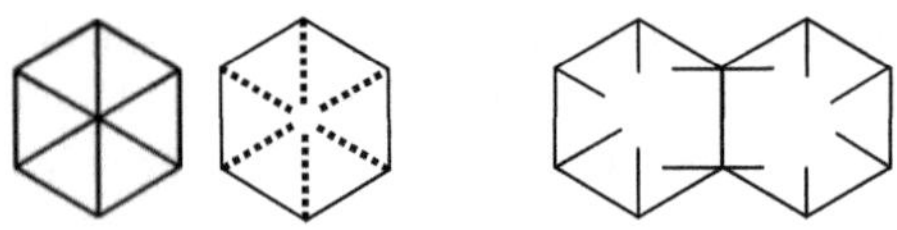

benzene naphthalene

benzene and naphthalene according to Armstrong

Armstrong was also interested in tautomerism, i.e., in the existence of two structural isomers of a molecule in equilibrium, differing in the transfer of one atom from one position to another in the molecule as in the case of the well-known keto–enol tautomerism:

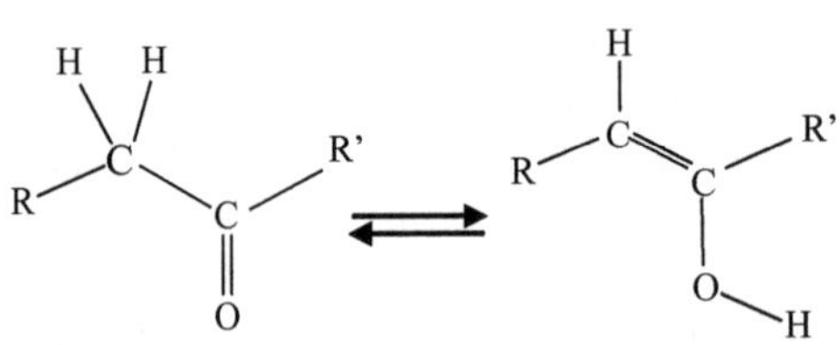

Tautomerism had already been highlighted in 1863 by Butlerov in a paper on diisobutylenes (Butlerov 1863) and later interpreted by Conrad Laar (1853–1929), assistant to Kekulé at Bonn, who in 1886 coined the term "tautomerism" for the equilibrium between two molecular forms (Laar 1885,1886):

$$X{=}Y{-}Z{-}A \leftrightarrow A{-}X{-}Y{=}Z$$

The tautomerism that Armstrong imported from Germany was taken up again by Arthur Lapworth (1872–1941) who in 1893 joined the City and Guilds of London Institute, created in 1878 and transformed in 1910 into Imperial College London. The college's most representative member was Armstrong, but Lapworth was assigned to the care of Frederick Stanley Kipping (1863–1949), another chemist among the many who studied in von Baeyer's laboratory in Munich, Germany. Lapworth obtained his Ph.D. under the supervision of Kipping in 1895 and continued to work with him on the chemistry of camphor and on the mechanism of aromatic substitution until 1897 when he joined Armstrong's group. He started his academic career in 1895 as demonstrator in chemistry at the school of pharmacy in Bloomsbury and after some mixed activities became senior lecturer in 1909 at the University of Manchester, professor of organic chemistry in 1913, and in 1922 director of the department of chemistry at the same university.

In 1898, he decided to study the tautomeric equilibrium reactions between isomers, a problem that Armstrong was acutely aware of. He suggested that the transformation between the two forms was due to the mobility of an atom or of a *labile* monatomic radical able to move within the molecule from one position to another (Lapworth 1898). Lapworth's theory explained well, in the case of the keto–enol tautomerism, why the hydrogen atom bound to the central carbon atom would migrate on the oxygen atom and the double bond would shift between two carbon atoms according to the scheme shown below:

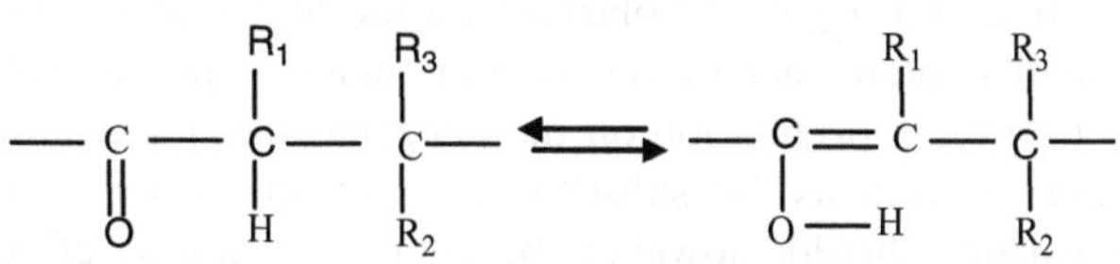

This paper represented the first attempt to interpret chemical reactions in terms of molecular structure. In following papers, Lapworth used the expression *key group* to define the labile group. Later he changed his mind and even if Armstrong was against the concept of ion, he reached the conclusion that the shift of the key group somehow involved dissociation processes (Lapworth 1901):

> It is to electrolytic dissociation, often doubtless in extremely minute amount, that the majority of changes in organic compounds may most probably be assigned.

The idea of an ionic nature of the key group led him to assert that ionized groups were important not only in tautomeric equilibria but also in normal addition reactions. By studying the formation of cyanohydrins obtained by addition of the CN to the carbonyl group of benzaldehydes or ketones in general (Lapworth 1903, 1904a, b, c, 1906a, b) and of unsaturated αβ-ketones in particular, he developed from 1903 to 1912 a theory of addition reactions in terms of alternate and latent polarities (Lapworth 1904d) activated by the presence of the key group exerting an action at a distance by polarizing the atoms. On this basis, he interpreted the formation of the cyanohydrins as the sequence of two steps, the first one in which the double bond opens and the second in which a hydrogen atom adds to the molecule according to the mechanism

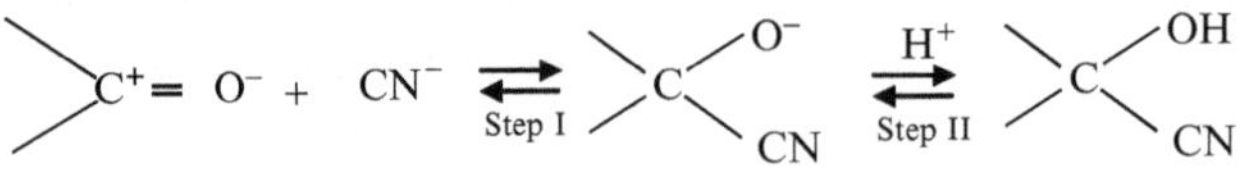

The German Daniel Vorländer (1867–1941), professor at the University of Halle and father of the chemistry of liquid crystals, was convinced that the course of a chemical reaction was determined by the polarity of the bonds and by the presence of charges on the atoms. For example, he maintained that in chlorobenzene, the chlorine atom, being strongly electronegative, attracted electrons, giving rise to an alternative distribution of charges to that imagined by Lapworth. A similar effect was produced by an NO_2 group, but in this case the alternation of positive and negative charges was inverted. As a consequence, a positive group like H^+ attacked chlorobenzene at the ortho and para positions and nitrobenzene at the meta ones while a negative group like Cl^- would do exactly the inverse (Vorländer 1925).

In 1920, Lapworth published a paper on the theory of alternate polarities in an unknown local journal, the *Memoirs of the Manchester Literary and Philosophical Society*, that led to, at that time, great interest on one side and severe criticism on the other (Lapworth 1920). According to the principle of alternate polarities, a heteroatom connected to a chain of conjugated carbon atoms would give rise to alternating positive and negative polarities on the atoms along the chain. This mechanism could explain, according to him, many experimental results, for instance, the substituent orientation on the benzene ring, the addition to double and triple bonds, as well as the substituent effect on the CH group's acidity. Lapworth's polarities should, however, be considered not as effective electric charges localized on the atoms but rather as instantaneous polarities during the reaction, as Robinson pointed out (Robinson 1947):

> It must be emphasized, however, that in attaching the + and – signs to the oxygen and carbon atoms no hypothesis is invoked, nor is it necessary, or even desirable, to assume that electric charges are developed on these two atoms (except perhaps at the actual instant of chemical change). The signs are applied, in the first instance, merely as expressing the relative polar character which the atoms seem to display at the instant of the chemical change in question.

In 1922, Lapworth reformulated his theory of organic chemical reactions in terms of virtual valences, an idea derived from Thiele (Lapworth 1922). It involved polar atoms and alternating positive and negative centers at which reactions would occur. In a relatively short time, however, this vision of the reaction mechanisms became obsolete due to the birth of the electronic theory of valence.

8.2 The Robinson–Ingold Controversy

The most significant contributions to the understanding of reaction mechanisms arose from the passionate polemical confrontation between two great English schools of chemistry, the first directed by Robert Robinson, chemistry Nobel laureate in 1947 for his research on anthocyanins and alkaloids, and the second by Christopher Ingold, the chemist who succeeded in organizing in a single theoretical scheme the whole of organic chemistry and in proving in more than 50 years of scientific activity the importance of chemical physics for the study of reaction mechanisms.

Robert Robinson (1886–1975) was an outgoing and anxious character, a skilled and charming speaker, a lover of music and photography, an avid and passionate chess player who had even spent 3 years as president of the British Chess Federation, and even an excellent mountaineer who had climbed the Alps, Pyrenees, and major mountains in New Zealand and Norway. He started his academic career at the University of Manchester as coworker of Thorpe and Perkin Jr. (1860–1929), the oldest son of W.H. Perkin, the discoverer of synthetic aniline. In 1909, he entered Perkin's laboratory and, influenced by his father who insisted that he should concentrate on practical problems, oriented his research to the study of dyes. In Perkin's group he worked on a dye extracted from Brazilian wood used to make ink and became friends with Lapworth who introduced him to the field of reaction mechanisms, explaining his theory of alternative polarities. In 1912, Robinson agreed to move to the University of Sydney in Australia, where he stayed for 3 years until, in 1915, he went back to England as professor of organic chemistry at the University of Liverpool. In 1920, he accepted the position of research director of the British Dyestuffs Corporation, but after a year went back to the academic world first at St Andrews, then at Manchester in 1922, at London in 1928, and finally at Oxford in 1930.

The ideas developed by Lapworth from 1902 to 1912 of the importance of polarities in the tautomeric equilibrium and in the reactions of addition to the carbonyl group had a profound influence on Robert Robinson in the period 1909–1912 in which both worked together in Perkin's laboratory at London. Robinson combined together the alternative polarities of Lapworth with the residual valences of Thiele, proposing reaction mechanisms in which the double bonds were separated into a normal single bond and an additional one treated in parts in terms of charges on the atoms and in parts as fractional valences.

These theories were undoubtedly useful to practical organic chemists since they suggested how to plan the synthesis of new compounds, but they were based on ad hoc hypotheses without connection to the electron distribution in the molecule and served more to suggest empirically possible reactivity schemes than to interpret reaction paths. Only with the development by Lewis and Langmuir of an electronic theory of valence did the reaction mechanisms finally become understandable in terms of electron pairs.

One of the principal opponents to the Lewis theory was the German Bernhard Jacques Flürscheim (1874–1955) who immigrated in 1905 to England where he opened his own private chemistry laboratory. Flürscheim, born in South Germany into the wealthy family of the economist Michael Flürscheim, received his chemical education first in Switzerland at the University of Geneva and of Zurich where he worked in the laboratory of Adolph Werner, then in Germany at the University of Heidelberg under the supervision of Emil Knoevenagel (1865–1921) and at Strasbourg in the Institute of Thiele. He became rich patenting in 1910 the synthesis of 2,3,4,6-tetranitroaniline, a powerful explosive that he discovered in 1904 and that was largely used by the German army in the First and Second World Wars. Following Thiele's teaching, Flürscheim was also bound to the idea of residual valences that he even considered responsible for the strengths of acids and bases (Flürscheim 1909). The skeleton of Flürscheim's preelectronic theory of valence was represented by the "request of affinity" that an atom exerted on others. If two atoms (or groups of atoms) in a molecule had a high affinity, the bond between them would be strong. Any other group would however be weakly bound to them, since the two atoms had already lost most of their available affinity.

As a consequence, once a strong bond was formed between two atoms in a linear chain, a sequence of strong (thick lines) and weak (thin lines) bonds would alternate as the characteristic structure of systems with conjugated double bonds:

Flürscheim considered phenol as an excellent example of his theory. In the phenol molecule, the oxygen atom had, according to him, a strong tendency to attract affinities from the next carbon atom C_1 and therefore the O–C_1 bond would be strong. Having lost part of its affinities, the C_1 atom would then form weak bonds with C_2 and C_6 and these in turn would form strong bonds with C_3 and C_5, respectively. The atoms C_2, C_4, and C_6, not having used all their affinities, would still dispose of some residual affinities and would be available to accept a substituent in the ortho and para positions.

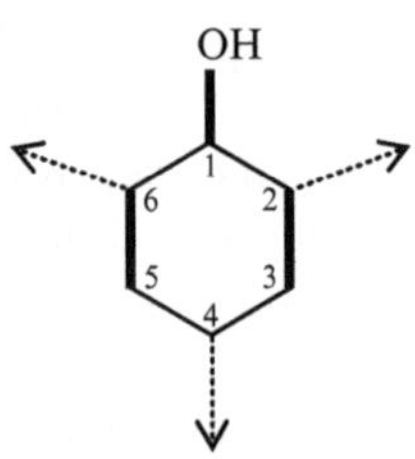

Crucial to the official diffusion of Lewis' electron doublet in English organic chemistry was the meeting organized at Edinburgh in 1921 on September 4–14 by the *British Association for the Advancement of Science* to which Irving Langmuir was invited to present the theory of the covalent bond. Robinson attended the meeting and, thanks to the clarity of Langmuir's exposition, realized that Levis' theory supplied an important tool to interpret concepts such as those of partial valence and polarity on atoms which, together with Lapworth, he had used until then. Lapworth reached the same conclusion following the suggestions of Julius B. Cohen (1859–1935), professor at Leeds and author of an up-to-date book on organic chemistry (Cohen 1907). Both Robinson and Lapworth realized that it was easy to translate their concepts into the electronic theory language; the saturated valences were nothing more than an electron pair shared between two atoms. Thiele's latent valences were free electron doublets whereas virtual valences corresponded to incomplete octets. They decided therefore to reframe their ideas in terms of the electronic theory in two separate papers that appeared one after the other in 1922. Lapworth published his paper alone (Lapworth 1922) while Robinson published his in collaboration with William Ogilvy Kermack (1898–1970), a coworker of his who was blinded in a laboratory accident in 1924, and who shifted later to biochemistry, becoming professor at the University of Aberdeen in 1948 (Kermack and Robinson 1922).

The transformation from the classical formulas with conjugated double bond to those with partial valences was formally realized in the formulation of Lapworth's and Robinson's new theories by substituting one of the hyphens of the double bond with a dotted hyphen and by specifying in addition the presence of alternative polarities on the atoms. In the new representation it was sufficient to substitute a full hyphen with two dots and a dotted hyphen with a single dot as shown in the below scheme:

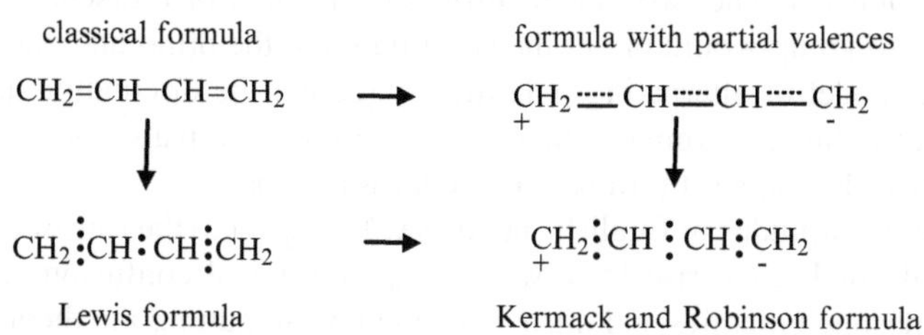

Kermack and Robinson faced the problem of the displacement of electrons in unsaturated systems, representing it with curved arrows, a method invented by Lapworth about 20 years before and which, since then, has became a standard symbol of organic chemistry.

The 1922 Lapworth and Kermack–Robinson papers stimulated an avalanche of contributions in English journals of chemistry. At the *Faraday Discussion* meeting in July 1923 at Cambridge there were even more violent discussions than those at Edinburgh in 1921. The characters were roughly the same, each standing firm on his previous position and unwilling to accept the opinions of the others. There were, however, two new achievements. The first was the conversion of Lapworth and Robinson to the electronic theory and the second the presence of Christopher Ingold at the meeting. As a consequence, although with different nuances and differences, a formation directed by Lowry, Lapworth, and Robinson arose on one side in favor of the electronic theory, whereas on the other side a different group was built up by members of Armstrong's school, in particular by his direct and indirect pupils

Thorpe and Ingold, as well as by Flürscheim, stubbornly attached to his theory of alternating polarities.

The Londoner Christopher Kelk Ingold (1893–1970), having graduated in 1913, worked for his Ph.D. in chemistry at Imperial College London under the direction of Thorpe, who steered him toward the study of tautomerism, an ongoing problem in Armstrong's department. Following Thorpe's suggestions, he left Imperial College in 1918 for 2 years to train at the *Cassel Cyanide Company of Glasgow*, coming back to Imperial College in 1920. During this period he was extremely productive, publishing 12 papers on organic reactions in 1921 and as many as 17 in 1922. This impressive production of scientific papers made him well known to the English academic community and ensured him the Ph.D. in chemistry in 1921, the nomination to membership of the Royal Society, as well as a chair at the University of Leeds in 1924. In 1922, he started a research project on benzene and its derivatives within the framework of the issues concerning the relationship between Kekulè's and Dewar's formulas (Ingold 1922a, b) that led him to disagree with the ideas of Lapworth and Robinson and to assume a negative position with respect to the theory of alternating polarities. Once he arrived in Leeds in 1924, Ingold decided to study the electrophilic reactions and in a series of papers tackled the study of the directive influence of the nitroso group on substitution at the benzene ring (Ingold 1925). According to him, this test should solve definitively the comparison between the Flürscheim model supported by him and that of Lapworth–Robinson. The two models differed since that of Flürscheim predicted that the nitroso group would orient the substitution at the ortho and para positions whereas the model of alternating polarities suggested the orientation at the meta position. Since his experiments favored substitution at para positions, Ingold concluded that the Lapworth–Robinson model was wrong.

The different attitudes of English chemists with respect to the role of electrons in the chemical bonds gave rise for a while to quite a bit of confusion until Ingold succeeded, despite polemics and misunderstandings, in putting the theory in order. The different positions emerged in 1923, when Thomas Lowry, first professor of chemical physics at the University of Cambridge, organized there a meeting of the Faraday Society entitled *The Electronic Theory of Valence*. The opening lecture was given by Gilbert Lewis. The general theme of the meeting was Bohr's atomic theory, but the discussion unavoidably ended with consideration of the concepts of greatest importance for that time, the role of polarities in the structure of organic molecules, and their relationship to the existence of electron doublets in chemical bonds.

At the Cambridge meeting Flürscheim claimed that the atoms were not bound by electrons but by a distribution of affinities in space. Flürscheim's position was shared by Jocelyn Thorpe (1872–1940), at that time director of the laboratories of Imperial College London where the young Ingold worked. Ingold, however, did not participate in the meeting since he was on honeymoon. Thomas Lowry (1874–1936), who in the same year had developed independently from Brønsted the acid/base theory, stood up in favor of the ideas of Lewis, insisting on defending the importance of electrons in chemistry. Lowry was convinced that double bonds were the sum of a covalent and an ionic bond. The double bond nature was still not clear at the beginning of the twentieth century as it was difficult to understand how double and triple bonds could break much easier than single bonds, while it seemed logical that the opposite would apply, i.e., that they would be stronger.

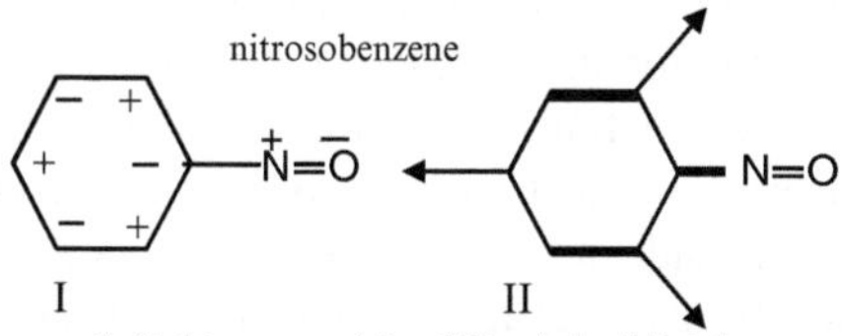

I Lapworth-Robinson model

II Flürscheim Model

In the meantime, however, Lapworth and Robinson had espoused the ideas of Lewis' electronic theory and Robinson started to apply his curved arrows symbolism to describe the transfer of electrons from the free doublet on the nitrogen atom to the benzene ring in order to prove that, shifting from one bond to the next as shown in the figure below, they favored para substitution.

The year 1925 was not lucky for Ingold in his debate with Robinson. Studying the addition reaction of nitrosobenzene to ethylene, Ingold made the mistake of claiming (Ingold and Weaver 1925) that his experimental data proved that the reaction mechanism was in conflict with the predictions of the Lapworth–Robinson theory. In a few months Lapworth and Robinson responded (Lapworth et al. 1925) showing that the reaction products found by Ingold and Weaver had been wrongly identified and therefore that their criticisms was nonsense.

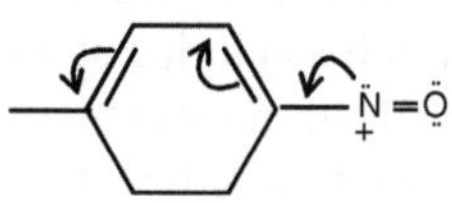

Stubbornly persisting in his dispute, Ingold in collaboration with his wife Edith Hilda Usherwood decided to select an aromatic system in which the oxygen and nitrogen atoms were not directly bound to the benzene ring in order to reduce the effect of electron transfer from their free doublets (Ingold and Ingold 1925). They were actually convinced that in this case the Lapworth–Robinson theory would predict the substitution in the ortho/para positions while that of Flürscheim predicted the meta substitution in agreement with experimental data. Once more, however, Lapworth and Robinson succeeded in showing that their electronic theory predicted the meta substitution. Continuing his fight in defense of the Flürscheim theory, Ingold and his student Eric Leighton Holmes affirmed to have experimentally proved that Flürscheim's predictions that nitration occurred at the meta position in *N,N*-diacetyl-benzylamine and at the ortho-para positions in the corresponding salt were perfectly verified (Holmes and Ingold 1925). Robinson for some time felt the pinch but, still convinced of the veracity of his ideas, decided to repeat Ingold's measurements. In January 1926, he announced that the Holmes and Ingold experiments were not only wrong but even distorted the truth: *N,N*-diacetyl-benzylamine is nitrated at the ortho/para positions and the salt at the meta!

As always happens, neither of the two opponents emerged unaffected from that story. The debate between the two schools of thought helped both, however, to clarify their respective positions. In 1925, Robinson presented to the Royal Society a complete reformulation of his theory in terms of electronic structure, proposing two different mechanisms for the attack of reagents on aromatic and conjugated systems. The first involved the transfer of a free doublet while the second was due to pure effects of electrostatic induction not too different from those of Ingold and Flürscheim. Ingold on his side, accepting Robinson's ideas in parts, introduced the concept of fractional charges that he symbolically represented with δ^- or δ^+ to show an excess of negative or positive charges on the atoms, respectively, as consequence of the presence of a group able to polarize the molecule.

The work of Ingold and his school gave rise to a general theory of the reaction mechanisms that he elaborated with his wife in the period from 1926 to 1933. Ingold also developed a nomenclature that later became the official one in organic chemistry. In 1926, he invented the expression *tautomeric effect* to indicate Flürscheim's electron transfer that Robinson called *conjugation effect* and represented with his curved arrows. The tautomeric effect corresponded to a permanent polarization induced by electron displacements. Ingold characterized two additional polarization effects, a transient and a temporary one that he called *electrometric and inductive effects,* respectively. The electrometric effect was a transient polarization due to the activation of the molecule. The inductive effect was instead the ability of an atom or of a functional group to stabilize with its electronegativity the electron distribution within a molecule. A typical inductive effect, for example, is that in which a chlorine atom attracts electrons, thus making trichloroacetic acid, CCl_3–COOH, stronger than acetic acid, CH_3–COOH.

These effects added or subtracted in different systems, favoring or not favoring substitutions in conjugated molecules. Ingold and his coworkers made great use in their research of physical-chemical techniques such as dipole moments, pH, heat of formation, and polarizability measurements, as well as vibrational, UV, and visible spectroscopy, etc., to emphasize the importance of these effects and to evaluate their specific contribution.

According to Ingold, the tautomeric effect was a convenient tool to overcome the difficulty of representing some types of molecules with a single structural formula. In 1933, however, he changed the name from tautomeric to *mesomeric* to account for the limitations of the classical formalism in representing molecular structures (Ingold 1934a). Ingold's *mesomerism* was nothing else than the resonance of Linus Pauling. In addition, in Ingold's nomenclature, the Lapworth's key groups, classified as *anionic* or *cationic* depending on their ability to borrow or to give electrons, took the name of *nucleophilic* if they supplied or of *electrophilic* if they withdrew a couple of electrons. A nucleophilic substituent which released electrons, like the alkyl group, reduced the positive polarization of the benzene ring and encouraged electrophilic substituents to attack at a para or ortho position. In contrast, an electrophilic substituent which attracted electrons made ortho or para substitution difficult and that at the meta position easier. The theory of nucleophilic and electrophilic substitutions was the "inizio modulo."

This modulo was masterfully summarized by Ingold in a famous review article in 1934 (Ingold 1934b).

In the fall of 1930, the Welshman Edward David Hughes (1906–1963) went to work with Ingold who had left Leeds to become professor at University College London. A profound friendship began between them, leading to a collaboration that lasted for 33 years and was interrupted only by the premature death of Hughes. During his Ph.D. work, Hughes became an expert on kinetics and his competence in this field played a fundamental role in the collaboration with Ingold. Their collaboration started with the study of the decomposition of quaternary ammonium salts in a series of papers in 1933. In this research, Ingold and Hughes distinguished three reaction mechanisms, the E2, S_N1, and S_N2 mechanisms where E meant elimination, S substitution, and the numbers 1 and 2 indicated the number of molecules involved in the rate-determining step of the reaction (Hughes and Ingold 1933). The S_N2 symbol (*S*ubstitution *N*ucleophilic *B*imolecular) indicated, in Ingold formalism, a bimolecular substitution reaction in which a nucleophilic group (nucl) colliding with an electropositive atom produced the separation of an outgoing group X. The result of this type of reaction is therefore the expulsion of a chemical group from a molecule and at the same time the introduction of an external group at a different position. Ingold and Hughes realized that in the reaction of a nucleophilic group with a tetrahedral carbon atom, the nucleophilic groups attack the molecule always from the side opposite to where the outgoing group is located. The reaction thus always leads to an inversion of configuration, going through an intermediate transition state (‡) with a triangular bipyramidal structure

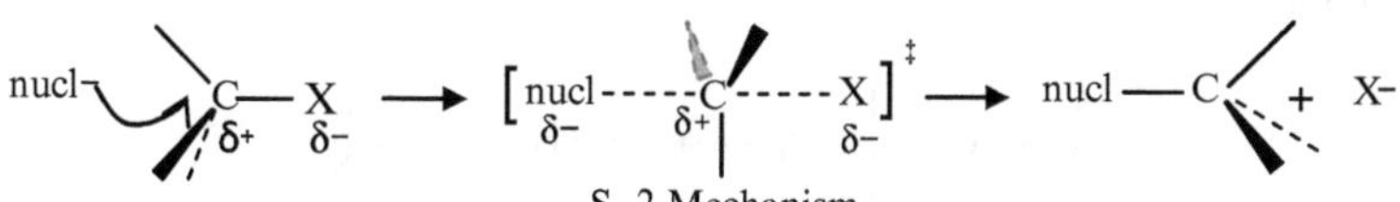

S_N2 Mechanism

thus interpreting in this way the previously mysterious Walden inversion reaction in a chiral molecule from one enantiomeric form to the other (Hughes 1938; Ingold 1938, 1953). The study of the effect of the solvent polarity on the reaction rates led Ingold and Hughes to define the general conditions by which the S_N2 mechanisms give rise to an inversion process in optical isomerism reactions, while the S_N1 mechanisms lead to racemization:

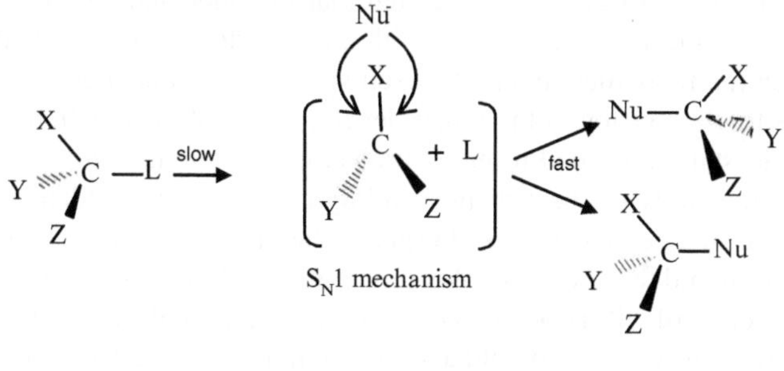

S_N1 mechanism

The S_N2 mechanism characterizes a reaction in which the molecules of an active solute randomly move among the solvent molecules until a high-energy collision occurs, leading to the formation of a transient species and then to the following reaction. The S_N1 mechanism instead involves a nucleophilic monomolecular substitution process in which the rate-determining step depends on the scission of a single molecule around which the solvent molecules arranged themselves to facilitate its heterolysis into a solvated and an unstable ion. The solvated ion is excluded by the successive reaction whereas the unstable one reacts with the solvent molecules (solvolysis) or with another solute molecule. Hughes and Ingold suggested that in this last case the reaction path took place in two steps, a slow step (I) consisting of the scission of the outgoing group X from the molecule with formation of a transient carbocation:

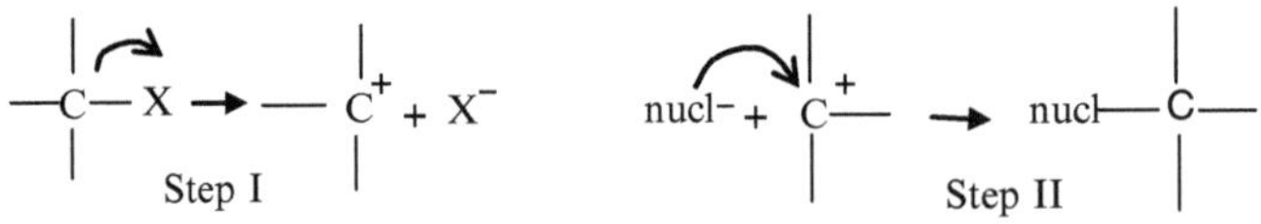

and a second fast step (II) in which a nucleophilic substituent attacks the carbocation with the formation of a new bond.

Finally they classified as E2 the bimolecular elimination reactions in which an atom or ion X^- substitutes an atom or an atomic group in a molecule. In this case the overall mechanism is a concerted one since at the same time a different atom Y is released by the molecule which rearranges its structure with the formation of a double bond:

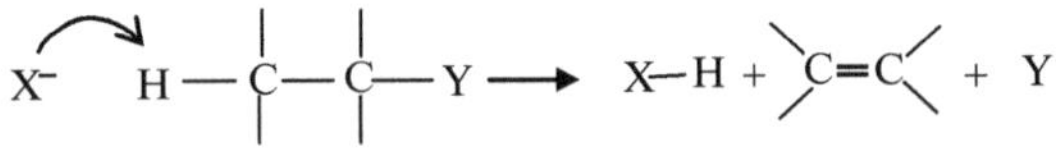

In the 1930–1940 period, Hughes, Ingold, and their coworkers completely restructured the theoretical organic chemistry, building ab initio the full interpretative scheme of large classes of organic reactions.

To this impressive body of activities, Ingold associated a new research subject, born from his interests in chemical physics and in particular from his attempts to connect his reaction mechanisms theory to the newly rising quantum chemistry. As a matter of fact, in just these few years, Eyring's theory of the activated complex developed and several new facts suggested that the understanding of the reaction mechanisms was intimately connected with the knowledge of the structural difference between the fundamental and the excited states of molecules.

With this fearless vision of the future developments of the reaction mechanisms theory, Ingold started a broad research project of vibrational and electronic spectroscopy for the study of the structure and dynamics of the fundamental state of benzene, one of the key molecules of organic chemistry. In a series of papers from 1934 to 1948, Ingold faced the complex problem of assigning to precise frequencies all normal modes of vibration of the benzene ring (Angus et al. 1936). In this impressive body of research, Ingold and his group made great use of the technique of isotopic substitution, consisting in selectively replacing the atoms of hydrogen with deuterium in the molecule. The interest of this technique is that all molecular modes of vibrations involving a displacement of the hydrogen atoms are more or less shifted to lower frequencies depending upon the relative importance of the hydrogen motion in each specific normal mode. This work thus required the synthesis of several isotopically substituted benzene derivatives such as hexadeuterobenzene (Bailey et al. 1936), different trideuterobenzenes (Bailey et al. 1946), and several bi- and monodeuterobenzenes, as well as the interpretation of their infrared and Raman spectra. This work that represents a milestone of vibrational spectroscopy was implemented with the interpretation of several spectra of the benzene excited electronic states.

For classical organic chemists the intrusion of chemical physics concepts into their territory obviously had the taste of invasion. Those more reactionary even questioned the experiments of Ingold and his school and ridiculed their attempts to generalize chemical reactions which to the majority of them appeared as completely disconnected from one another. The less aggressive critics limited their comments to maintaining that it was simply a matter of confusing language that had nothing to add to the understanding of chemical reactions.

For the members of the new generation of chemical physicists, used to the mathematical formalism, this orgy of curved arrows, of electrons represented by dots, of dreamed reaction mechanisms, and of ad hoc hypothesis to justify experimental facts, often even contradictory, looked like childish play to be at best considered with indulgent smugness. This situation was by no means simplified by the pugnacious and nonconformist temperament of a typical Welshman like Hughes who saw everywhere plots machinated by Robinson and his acolytes and who continued to treat the English with his traditionally fierce ironic behavior. With patience and courage, however, Ingold and Hughes won the scientific battle and by 1950 the acceptance of theoretical organic chemistry had been realized. Ingold's book *Structure and Mechanism in Organic Chemistry* (1953) (Ingold 1953) became in a short time one of the fundamental textbooks of the new organic chemistry.

8.3 Short Living Molecules: Carbanions and Radicals

An important merit of Ingold and his school was that of suggesting the existence of unstable molecular species, with very short lifetimes, which occurred in the intermediate steps of several reactions. Their existence could not be proved experimentally and was only guaranteed by the logic of interpretation of the reaction paths. In reality, the idea of the existence of unstable molecules appearing and disappearing in the course of the reaction had already been proposed by the American Julius Stieglitz (1867–1937) who studied in Germany in Berlin where he obtained the Ph.D. in chemistry in 1889. In 1899, Stieglitz suggested the existence of carbocations, positive ions of carbon compounds (Stieglitz 1899). Even before, in 1891

Georg Merling (1856–1939), professor at the Maximilian Universität of Munich, realized that by addition of bromine to cycloheptatriene and successive heating, one obtained a crystalline material of formula C_7H_7Br, for which he did not propose a structure (Merling 1891). William von Eggers Doering (1917–2011) and Lawrence H. Knox proved 60 years later that this was the aromatic *tropilium* ion (Doering and Knox 1954). In 1901, James Flack Norris (1871–1940), professor of chemistry at the MIT (Norris 1901), and independently F. Kehrmann and F. Wentzel in Germany (Kehrmann and Wentzel 1901), discovered that the colorless triphenylmethanol $(C_6H_5)_3C{-}OH$ gave rise to strongly yellow-colored solutions in concentrated sulfuric acid, and in the same year Adolf von Baeyer showed that the compound obtained in the reaction

$$(C_6H_5)_3C - OH + H_2SO_4 \rightarrow (C_6H_5)_3C^+HSO_4^- + H_2O$$

was the salt of the intermediate carbocationic species $(C_6H_5)_3C^+$ and invented the term *halochromie* to indicate the relationship between the formation of the salt and the emergence of the color.

One of the guiding concepts of classical organic chemistry was based on Laurent's principle of minimal structural change in the stabilization of unstable molecular structures. In 1860, this principle was put into discussion by the discovery of the *pinacol rearrangement*, a type of reaction that revolutionized several basic concepts of theoretical organic chemistry and involved several leading chemists of the nineteenth and twentieth century. Among these the Russian Egor Egorovich Vagner was the first who clearly showed the nature of the process and it was Hans Meerwein who elucidated the complex rearrangement mechanism.

Egor Egorovich Vagner (1849–1903) is better known in the chemical literature as Georg J. Wagner since Egor is the Russian pronunciation for George and Wagner is the German transcription for Vagner, a name that he used when publishing papers in Europe. In 1874, Vagner graduated at the famous Siberian University of Kazan where he collaborated in 1875 with A.M. Butlerov and from 1876 with N.A. Menshutkin before becoming assistant to Aleksandr Mikhailovich Zaitsev (1841–1910), one of the greatest Russian organic chemists of that time. In 1886, Vagner became professor at the polytechnic school of the Russian Imperial University of Warsaw where in 1888 he discovered a general oxidation method with permanganate of the ethylenic double bond and, where working on terpenes, identified the structure of α-pinene (1895–1896), a compound unstable in air that represents the major component of turpentine essence (Wagner 1899).

The idea of the existence of carbocationic intermediate species was later developed in the period 1914–1922 by the German Hans Meerwein (1879–1965) who started as laboratory technician at the Fresenius company and later became professor of chemistry at Bonn in 1914. He studied the transformation of camphene hydrochloride into isobornyl chloride, finding that the reaction rate increased with the increase of dielectric constant of the solvent and that Lewis acids, such as the

$SbCl_5$, $SnCl_4$, $FeCl_3$, $AlCl_3$, and $SbCl_3$ chlorides, speeded up the process (Meerwein 1914). He concluded that the isomerization reaction did not proceed by migration of a chlorine atom but by rearrangement of the cation as shown below:

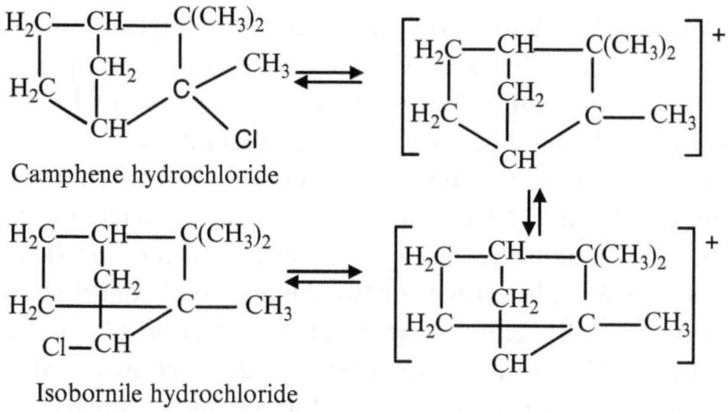

This reaction is known in the literature as the Wagner–Meerwein rearrangement. Later Meerwein moved to Königsberg where he studied the transformation of carbocations in oxonium ions (Meerwein and van Emster 1922).

The existence of an intermediate carbo-ion had already been shown by Moses Gomberg (1866–1947), who was born in central Ukraine but immigrated with his family to Chicago in 1884 to escape the pogroms following the assassination of Czar Alexander II. Gomberg obtained his Ph.D. in chemistry in 1894 at the University of Michigan where he worked for the rest of his academic career. In the period 1896–1897, he went to Germany to postdoctoral work first with von Baeyer and Thiele at Munich and then with Victor Meyer at Heidelberg where he succeeded in synthesizing tetraphenylmethane $(C_6H_5)_4C$. Back at the University of Michigan, he improved tetraphenylmethane synthesis and, in the attempt to prepare the hexaphenylethane, $(C_6H_5)_3C—C(C_6H_5)_3$, the next completely phenylated hydrocarbon, he discovered (Gomberg 1900, 1902) the triphenylmethyl radical $(C_6H_5)_3C^-$. In reality, Gomberg did not isolate the carbo-ion but its peroxide, owing to its extremely high reactivity with oxygen:

$$(C_6H_5)_3CCl \xrightarrow{Zn} (C_6H_5)_3C^- \xrightarrow{O_2} (C_6H_5)_3C-O-O-C(C_6H_5)_3$$

The first stable carboanion, the tris(4-biphenylyl)methyl radical $(C_6H_5–C_6H_5)_3C^-$, was instead isolated in 1914 by Wilhelm Johann Schlenk (1879–1943), professor at Berlin, in collaboration with E. Marcus (Schlenk and Marcus 1914). Schlenk also proposed the structural formula of the benzophenon radical ion $(C_6H_5)_2\dot{C}–O^-$, first obtained by E. Beckmann and T. Paul in 1891 by reacting benzophenone with metallic sodium. For his research on radicals, Schlenk was proposed as a serious candidate for the Nobel Prize in 1918, although without success.

In parallel to the English one, another important research school of reaction mechanisms developed in the 1930s on the other side of the Atlantic thanks to Louis

Plack Hammett (1894–1987), professor at the Columbia University, New York. Louis Hammett originated from a typical Yankee family of Maine in New England, living in the myth of the Harvard and MIT educational system. Hammett was therefore educated, as his father, at Harvard, where he graduated in chemistry in 1916 before going to Zurich to work in Hermann Staudinger's laboratory. Back in America in 1917, after military service and a short period spent in industrial research, in 1920 he joined the Columbia University where in 1923 he obtain his Ph.D. in chemistry and where he spent all his academic life.

Louis Hammett is considered the true founder of "organic chemical physics," a term that he invented and that has since been used in the scientific literature since the publication of his famous treaty *Physical Organic Chemistry* (Hammett 1940), written when he was already known for two fundamental contributions to chemistry: the idea of superacidity and the theory of the effect of substituents in reaction mechanisms. As Ingold was an organic chemist interested in chemical physics, so was Hammett a chemical physicist interested in organic chemistry.

The first important contribution of Louis Hammett to theoretical organic chemistry was the development of the concept of superacid and the definition of the acidity function bearing his name. In 1928, he suggested that hydrochloric acid is stronger in benzene where it cannot ionize than in water where it is completely ionized (Hammett 1928). The term superacid was coined in 1927 by his friend James Bryant Conant (1893–1978), 1 year older than him, who he met when he was a student at Harvard, and who also worked on superacids.

A superacid is a substance with an acidity greater than that of pure sulfuric acid, able to add protons to molecules, or as said in laboratory jargon, able to protonate almost all organic compounds. Known superacids are trifluoromethanesulfonic acid (CF_3SO_3H), also called triflic acid, and fluorosulfonic (FSO_3H) acid, both about a thousand times more acidic than sulfuric acid. A superacid is often not a pure compound but rather a mixture of several strong acids mixed together to obtain higher acidity.

Hammett, studying concentrated solutions of sulfuric acid, realized that they have an acidity superior to that indicated by their concentration and that their chemical behavior is completely different from that of the diluted solutions. Together with his student Alden J. Deyrup, using *p*-nitroaniline as indicator, he defined an acidity scale in which the 100% pure sulfuric acid has an acidity 10 billion times (10^{10}) greater than the 10% solution and defined an acidity function (Hammett and Deyrup 1932)

$$H_0 = pK + \log(\mathrm{B})/(\mathrm{BH}^+)$$

where (B) and (BH^+) are the concentrations of the indicator (*p*-nitroaniline) and of its conjugated acid, respectively, quantities easily measurable by simple colorimetric methods even in pure sulfuric acid (Hammett and Paul 1934). As we shall see, George Olah obtained the 1995 Nobel Prize in chemistry for having shown the catalytic effect of superacids and their ability in stabilizing carbocations.

The second important contribution of Hammett to organic chemical physics concerns the application of the mass action law and of the kinetic theory equations to organic reactions. In particular, Hammett studied the effect of meta and para substituents on the rate and on the equilibrium constants in substitution reactions at the benzene ring, convinced that the free energy was the ideal quantity to quantize the influence of the molecular structure on reactivity (Hammett 1935).

As a start, Hammett selected the series of the meta- and para-substituted benzoic acids and defined a constant σ (Hammett's constant) specific for each substituent, according to the equation $\sigma = \log K_s/K_o$, where K_s is the ionization constant of the specific s substituted benzoic acid and K_o is the corresponding ionization constant of pure benzoic acid.

The most spectacular result of Hammett's work was that this equation was utilizable for both the equilibrium constants K of substitution reaction and for the corresponding rate constants k in a series of meta- and para-substituted aromatic compounds simply by multiplication by a constant ρ, unique for each type of reaction and independent from the type of substituent.

For the equilibrium constants, the Hammett equation therefore has the form

$$\log \frac{K}{K_0} = \sigma\rho$$

where K is the equilibrium constant of a given reaction involving a substituent R on the benzene ring and K_0 is the equilibrium constant for the corresponding reaction of benzoic acid in which obviously the substituent R is a hydrogen atom.

In the case of the rate constants of a series of reactions involving meta- and para-substituted benzene derivatives, the equation is identical:

$$\log \frac{k}{k_0} = \sigma\rho$$

where now k_0 is the reaction rate constant for the nonsubstituted benzene derivative and k the reaction rate constant relative to the substituted compound. By plotting $\log(K/K_0)$ of a given equilibrium reaction as a function of $\log(k/k_0)$ for the corresponding reaction rate for several substituent, one obtains a straight line.

The general validity of Hammett's equation is simply the consequence of the fact that for two reactions with two different reactants, the activation free energy is proportional to the variation of the Gibbs' free energy (Hammett 1937).

The golden period of the development of organic chemical physics, associated with the names of Lapworth, Hantzsch, Ingold, and Hammett, found a rightful heir in Paul Doughty Bartlett (1907–1997), who at Harvard created one of the most important schools of organic chemical physics, a noteworthy rival of that of Ingold and undoubtedly superior in number of pupils as well as in scientific production and synthetic practice. Bartlett graduated in chemistry at Harvard in 1931 under the supervision of James Bryant Conant with whom he published in 1932 a study on the formation of semicarbazones ($R_2C = N(NH)C(=O)NH_2$) in water solutions and on the evaluation of their rate and equilibrium constants (Bartlett and Conant 1936). Harvard's policy in that period was strictly adherent to the rule that a teacher who graduated in the same university could not be hired before having proved his value with some relevant publications. The basic idea was to avoid an important personality filling the university with his pupils. Therefore, Bartlett, even if considered the official successor to Conant, was forced to work first for 1 year as *postdoctor* at the Rockefeller and then for 2 years as instructor at the University of Minnesota before being accepted in the group of Harvard's faculty members.

Bartlett's contributions to the study of reaction mechanisms cover a large variety of applications. Bartlett discovered that the electrophilic addition to the C=C double bond occurs in two steps, pointing out the importance of the configuration inversion in the Wagner–Meerwein molecular rearrangement and of the coplanarity in the stabilization of carbocation structures. He also proved the transfer of H^+ ions in alkylation reactions and the occurrence of a sequence of chain transfer processes in polymerization, reaction mechanisms today normally taught to students. The school that he created at Harvard hosted more than 250 students and produced more than 50 chemists that later became professors in the most important American universities.

In 1939, Bartlett and his coworker Lawrence Knox (1906–1966), one of the first colored men with a degree in chemistry, published a famous paper on bridgehead halogens, synthesizing 1-bromonorbornane, a molecule in which the bromine atom is unable to react with alkali or with silver nitrate since the back side of the carbon bearing the halogen is sterically protected by the bicyclic structure which makes impossible an attack from the back by an OH^- ion followed by a Walden inversion process (Bartlett and Knox 1939):

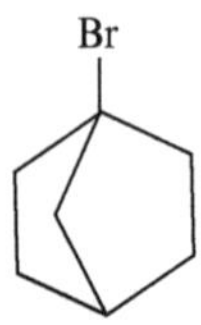

Bromonorbornane

In this molecule, the bromine atom could not be detached from the carbon atom by ionization through an S_N1 mechanism, leaving a positively charged norborane$^+$ carbocation with a pyramidal structure. Bartlett thus made the hypothesis that carbocations with only six electrons in their external shell cannot be stable unless they can assume a planar or semiplanar structure. After a few years, he synthesized, in collaboration with Saul Cohen (1916–2010) 1-bromotripticene (Bartlett et al. 1950), another molecule with a bridgehead bromine that confirmed his ideas on the structure of the carbocations.

Bromine was a favored factor in his research. In 1944, he discovered in collaboration with his students Francis E. Condon and Abraham Schneider the hydrogen-halogen fast exchange reaction in which, in the presence of the Lewis acid aluminum tribromide as catalyst, *tert*-butyl chloride reacts with isopentane to yield *tert*-amyl bromide and isobutane.

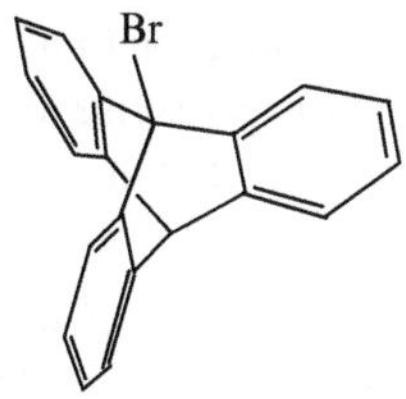

Bromotripticene

Substantial contributions to the study of carbocations were also due to Saul Winstein (1912–1969), pupil of Bartlett, a Canadian Jew who immigrated to the United States when he was very young and whose academic career took place almost completely at the University of Southern California, Los Angeles (UCLA) where he became professor in 1947.

Winstein, in addition to being a very good organic chemist, had a firm tendency to coin new terms to add to the chemical vocabulary, a passion shared in the past by several high-level chemists from Lavoisier to Laurent to Liebig to Ingold. To Winstein we owe the invention of a series of terms such as *neighboring group, solvent participation*, *internal return*, *anchimeric assistance* (when part of a molecule helps another part to react), *intimate ion pair*, *ion-pair return, bridged ions, nonclassical ions, and homoaromaticity* (Winstein 1969) which entered the modern lexicon of organic chemistry together with the carbocation classification.

Winstein is essentially known for his theory of the effect of the *intimate ion pair* and of the *vicinal group* (Winstein et al. 1956). The concept of intimate ion pair described the interaction between a cation, an anion, and the solvent molecules around them. According to Winstein, while in aqueous solutions ions are completely solvated and separated from the counterions of opposite charge, in nonpolar solvents, two counterions can be in close contact since there are no solvent molecules between them. When solvation increases, the ionic bond among them decreases and a separated ionic pair is formed. Winstein utilized the concept of *intimate ion pair* to explain their tendency to favor the production of configuration inversion processes in S_N1 reactions. He suggested that the separation of the outgoing group in the formation of a carbocation was assisted by the solvent molecules and by association with the cationic intermediate. According to his theory, the association of the outgoing group with the solvent molecules or with the ion blocks the side of the rising carboanion, favoring attack from the back by a nucleophilic group and thus giving rise to a small excess with inverted stereochemistry in an S_N1 reaction that should have instead led to a racemic product. In 1938, Winstein developed the idea of the *vicinal group* just after having obtained his Ph.D. from Caltech. When a molecule subject to a nucleophilic attack already contains a nucleophilic group, one observes a profound influence on the stereochemistry of the reaction. Winstein selected the term *vicinal group* to indicate a nucleophilic group already preexisting in the molecule which influences the reaction (Winstein et al. 1948a). Winstein and coworkers identified a classical example of vicinal group effect in the case of *trans*-2-acetoxycyclohexyl *p*-toluenesulfonate, showing that the speed of solvolysis differed from the *cis* to the *trans* isomer by a factor 670, with the *trans* isomer more reactive than the *cis* (Winstein et al. 1948b). Winstein also resumed the problem of addition of a halogen molecule to the ethylenic double bond to explain why the reaction gives rise to the *trans* instead of the *cis* isomer as expected if the two halogen atoms reacted in synchrony:

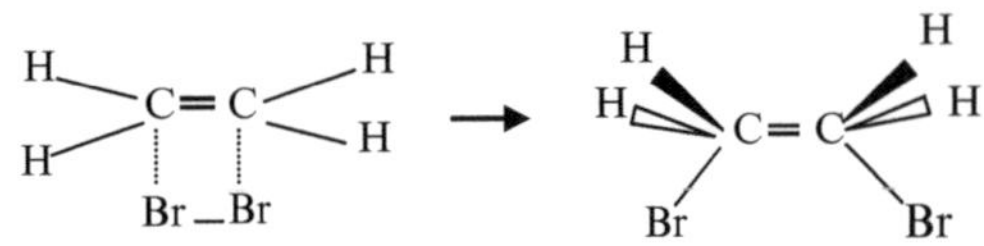

Winstein explained the formation of the *trans* isomer assuming that the reaction takes place in two steps with the formation of a bridged bromonium ion between the two carbon atoms as shown below:

1° step: formation of an intermediate with a bridged bromonium ion

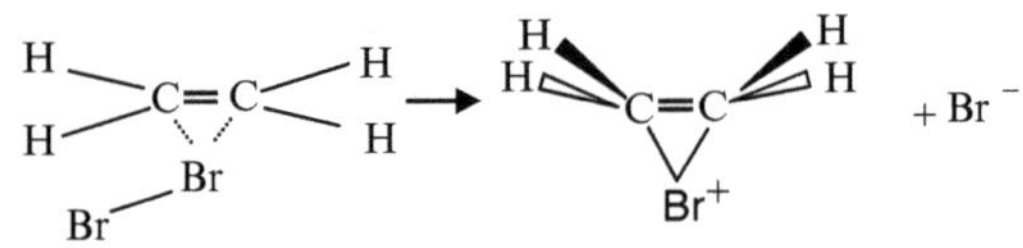

2° step: the Br^- attack from the side opposite to the bromonium ion breaks the threefold ring and produces the trans form.

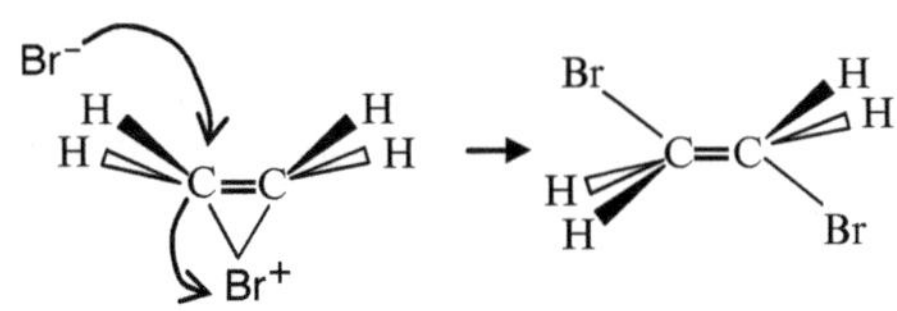

The nonclassical ions postulated by Winstein were at the origin of a large number of polemical discussions essentially with Herbert Brown, 1979 Nobel laureate in chemistry for his research on the hydroboranes. In his Nobel Prize lecture, George Olah recalls the period in which the debate on the carbocations was growing strongly and emphasizes Winstein's contribution to the definition of nonclassical ions (Olah 1994):

> The scheduled "main event" of the meeting was the continuing debate between Saul Winstein and Herbert C. Brown on the classical or non classical nature of some carbocations (or carbonium ions as they were still called at the time).... Sure he (Brown) won the Nobel Prize for his work with hydroborations, but his contributions to physical-organic chemistry were just as important as those to synthesis. His epic battles with Saul Winstein over the nature of carbocations (classical vs. nonclassical) forced chemists at the time to think critically about how to disprove a mechanism and the existence of a particular reactive intermediate.

Herbert Charles Brown (1912–2004), born as Herbert Brovarnik into a Ukrainian Jewish family, immigrated to London as a young boy. In 1914, the family moved to the United States and in 1936 he became an American citizen with the last name of Brown. In 1938, he obtained his Ph.D. in chemistry at the University of

Chicago under the supervision of Hermann Irving Schlesinger (1882–1960). At that time Schlesinger was involved with his coworker Anton B. Burg (1904–2003) in the study of exotic molecules such as the diborane B_2H_6 and the boron hydrides which at that time represented a real problem for the Lewis theory of the free electronic doublet, owing to the reduced number of available electrons (Schlesinger and Burg 1942). In his memoires, Brown remembers that he went to work with Schlesinger after having read the book on the boron hydrides that his girlfriend had selected as a gift for him since it was the most economical book of chemistry available!

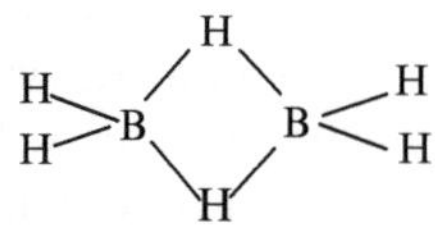

Diborane

During the Second World War, Brown, working with Schlesinger, discovered how to prepare sodium borohydride $NaBH_4$, an extremely reactive substance that can be used to produce boranes, compounds of hydrogen and boron. The discovery of the diboranes opened up a new field in organic chemistry because of the facility with which they react with unsaturated compounds to produce a new family of molecules, that of the organoboranes (Brown 1982). The organoboranes can be synthesized with different functional groups and often give rise to stereospecific compounds. In these compounds, the boron atom can be easily substituted with a variety of functional groups, thereby producing a large number of new organic molecules. The study of these reactions showed that the organoboranes can transfer alkyl groups to the majority of molecules of organic or biological interest, still preserving their stereospecificity (Brown 1988). For this research, Brown obtained in 1979 the Nobel Prize in chemistry together with Georg Wittig (1897–1987).

The development of stereospecific syntheses stimulated the interest in reaction mechanisms and bridged molecules in Brown. In this respect, it is certainly strange that, being used to dealing with boranes possessing nonconventional bridged hydrogen atoms, he took such a firm stand against Winstein's nonclassical ions with bridged halogen atoms (Brown 1976).

The study of carbocations was taken up at the beginning of the 1960s by the Hungarian George Andrew Olah who proved that the very short living carbocations can be stabilized by superacids. This very important discovery allowed lengthening of carbocations' lifetime, thereby making possible their structural analysis and their applications to complex reactions.

George Andrew Olah (1927–) graduated in Budapest in 1949 and became assistant of Prof. Geza Zemplen (1883–1956), a pupil of Emil Fischer. Following the research activity of his master, Zempel continued to be interested in carbohydrates, essentially glycosides, while Olah wanted to concentrate on fluorine chemistry, fascinated by the works of Fritz Seel (1915–1987), professor at the University of Saarland and one of the true great experts of fluorine chemistry.

In this period, following in Seel's footsteps, he started to be interested first in acylation reactions with acyl fluorides R-COF and then in alkylation with alkyl fluorides, using BF_3 as a catalyst. This research stimulated his interest in reaction mechanisms, in particular of the complexes R-COF and R-F with Lewis acids:

$$\mathrm{ArH + RCOF \xrightarrow{BF_3} ArCOR + HF}$$

$$\mathrm{ArH + RF \xrightarrow{BF_3} ArR + HF}$$

and from these studies originated his deep involvement with carbocation chemistry. In 1954, Olah joined the Central Chemical Research Institute of the Hungarian Academy of Sciences, where together with his wife Judith he organized a small group of organic chemists. After the tragic events in Hungary of October 1956 which gave rise to the popular uprising, soon put down by the Soviet tanks, Olah abandoned his Fatherland with the family and immigrated first to London and then in 1957 to Canada. In 1957, he was appointed by Dow Chemical, an important American chemical company, which opened new research laboratories at Sarnia in Ontario and after a short while was promoted to *scientist*, a position that made it possible for him to do research without administrative involvement. In 1964, he moved to the United States to the *Eastern Research Laboratories* of the Dow Chemical in Massachusetts. The following year, he returned to academic activity, joining the Western Reserve University of Cleveland, Ohio, as teacher and department director. In 1977, he finally moved to the University of California Los Angeles, where he still works today. In 1991, he assumed the office of director of the Loker Hydrocarbon Research Institute, an important organization in the field of hydrocarbons.

In Hungary, Olah started to synthesize fluorine compounds as the superacid FSO_3H and boron trifluoride BF_3. After arriving in Canada at the Dow Chemical research laboratories he finally had the chance to resume his previous work on fluorine compounds and to organize a systematic study of the carbocations.

In the 1950s, there were several observations indicating that the carbocations were intermediate products in several chemical reactions, but that they had extremely short lifetimes, less than a billionth of a second (nanosecond), owing to their high reactivity. Their existence had been suggested several times although nobody had succeeded in proving their presence even with the most sophisticated spectroscopic techniques available at the time. Without experimental evidence of their existence, it was impossible to establish whether they were the product of the imagination of researchers or a true physical reality. In the 1960–1970 period, a great number of more or less fanciful interpretations flourished, until George Olah with a series of brilliant experiments definitively solved the problem, developing the methods to prepare stable carbocations with a relatively long lifetime, sufficient to study their structure and their physical properties with spectroscopic techniques. Olah realized that by using superacids as solvents, with an acidity greater than that of sulfuric acid by up to 18 orders of magnitude, able to protonate any organic molecule, the carbocations remained stable in solution. Under these conditions, in fact, they could not react with the solvent molecules supplying protons in rearrangement reactions of the type

$$(CH_3)_3C^+ \rightleftarrows (CH_3)_2C{=}CH_2 + H^+$$

to a milieu that resolutely refused to accept them.

Olah found that two types of carbocations exist, those that are trivalent named *carbenium ions* in which the positive carbon atom is connected to three atoms and those that are penta-coordinated called *carbonium ions* with five substituents. On the basis of research that definitely proved the existence of the carbonium ions, Olah disproved the belief that in organic compounds the carbon atoms could be at maximum quadri-coordinated.

Looking for Lewis acids, Olah found that (Olah et al. 1962) antimony pentafluoride SbF_5 was an extremely powerful Lewis acid, able to ionize the alkylfluorides in solution giving rise to stable alkyl cations with sufficiently long lifetime, according to the equations

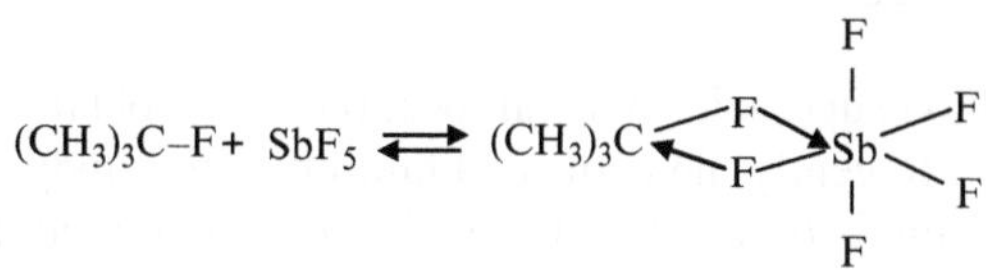

or

$$(CH_3)_3C{-}F + SbF_5 \rightleftarrows (CH_3)_3C^+ + SbF_6^-$$

Olah succeeded in finding the first evidence of the occurrence in solution of a stable carbocation using NMR spectroscopy. He then studied several important superacids such as FSO_3H (fluorosulfuric acid, a typical Brønsted–Lowry acid), CF_3SO_3H (triflic acid), as well as the very powerful fluoroantimonic acid HF-SbF_5 and the magic acid FSO_3HSbF_5, a mixture of antimony pentafluoride (Lewis acid) and fluorosulfonic acid, that his group largely utilized to ionize several precursors. Also, superacids based on fluorides such as AsF_5, TaF_5, NbF_5, and other powerful Lewis acids such as $B(O_3SCF_3)$ were successfully utilized. The name *magic acid* for the FSO_3H–SbF_5 mixture was conceived by Olah's coworker Joachim Lukas, a German researcher that worked with him at Cleveland in the 1960s, who, after a Christmas party in the laboratory, slipped a piece of a candle in the acid. The candle dissolved completely quite rapidly, giving rise to a solution that when examined with ^{1}H-NMR was found to contain the *tert*-butyl cation (Olah et al. 1967).

One of the most significant results of the carbocation stabilization with superacids was the preparation of protonated methane (Olah and Schlosberg 1968) using the magic acid:

$$CH_4 + H^+ \rightarrow CH_5{}^+$$

Olah and his school also proved that penta-coordinated methane can be used to prepare long hydrocarbon chains. Their research on penta-coordinated carbonium ions showed that these can produce new reactions, a result that has greatly contributed to the development of petrol chemistry. It allows in particular the transformation of linear chains of saturated hydrocarbons at low octane number into branched chains at high octane number. Natural gasoline of low octane value derived from natural gas is easily upgraded to higher octane value by means of catalysts consisting essentially of trifluoromethanesulfonic acid at mild temperature conditions. In addition, the branched chain hydrocarbons are important materials for several industrial syntheses. Furthermore, the use of superacids as catalysts makes possible the cracking of heavy oils and their transformation into liquid fuels in mild conditions.

Olah's group also disclosed an original method for producing branched aliphatic ketones in hydrocarbon mixtures from isoalkanes by a superacid catalyzed formylation-rearrangement reaction. The method can be used to isomerize and formylate hydrocarbons simultaneously in complex hydrocarbon mixtures such as refinery streams, alkylate mixtures, and natural gas liquids. Liquefied natural gases are upgraded and oxygenated by addition or by direct production of branched aliphatic ketones.

For his research on superacids and carbocations, Olah obtained the 1994 Nobel Prize for chemistry. Recently, he dedicated himself to the study of methanol as an ideal fuel for the future (Olah 2005), convinced that this could lead to a new economy, more sustainable and less polluting.

8.4 Theoretical Organic Chemistry

Research on the reaction mechanisms experienced an important evolution at the international level when it became possible to transform the classical treatment, essentially phenomenological and based on inductive procedures connecting a huge amount of empirical data, into quantum calculations starting from the electron distribution in the molecular orbitals. This new direction led to the research of Michael J.S. Dewar, probably the first English organic chemist who really mastered the quantum mechanics methods, in particular those of the molecular orbital theory, to apply them correctly to the almost infinite series of organic reactions.

Michael James Steuart Dewar (1918–1997), born in India to Scottish parents, introduced into theoretical organic chemistry several of the concepts that today constitute its main structure and that in more than 40 years of unceasing activity contributed to the rise of semiempirical models of quantum calculation to the present level of sophistication and diffusion in chemical culture.

After graduating in chemistry at Oxford, he went to work in the last years of the Second World War with Robert Robinson at the *Dyson Perrins Laboratory* of the same university, on a project of synthesis of penicillin. Dewar rapidly became a favorite coworker of Robinson, above all for his critical position with respect to the resonance theory and thus to Ingold's mesomerism. This dislike for the resonance theory is clearly illustrated in the preface to his book *The Electronic Theory of Organic Chemistry* (Dewar 1949):

> This approach is most unsuitable from the organic chemist's point of view since it involves a new symbolism and a novel and uncongenial outlook (A cursory examination of the literature shows how difficult it is for chemists to distinguish intuitively between resonance and tautomerism). For these reasons the molecular orbital theory has much to offer. It provides a picture of molecular structure closely akin to that of classical organic chemistry; in both, charge migration, residual affinity and the like, appear in similar forms.

During his Ph.D. work, Dewar discovered the existence of stipitatic acid, a metabolic product of the mold of the *Penicillium* species and of the aromatic ring from which it is derived, a seven-membered aromatic ring that he named *tropolone* (Dewar 1945b). From the discovery in 1945 of the tropolone structure originated the field of non-benzenoid aromatic molecules and of new π electron systems which he proposed as an intermediate in the benzidine rearrangement reaction (Dewar 1945a).

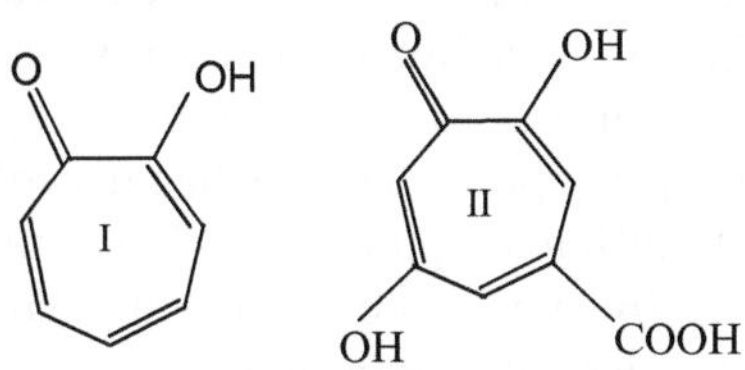

Tropolone (I) and Stipitatic acid (II)

In 1945, Dewar entered the research laboratories of the *Courtaulds Ltd.* company which produced rayon and acetate fibers, still preserving tight research relationships with his master Robinson. In 1946, they programmed together the drawing up of a book that in principle was supposed to translate Robinson's electronic theory into the molecular orbital formalism. Robinson, however, did not find the time for serious writing, and at the end of 1949 Dewar published it alone (Dewar 1949). Dewar also presented his ideas on a new formulation of organic chemistry at an international meeting held at Montpellier in 1950 and further developed them in final form in 1952 (Dewar 1952). Dewar's presentation that reformulated the molecular orbital theory in a semiquantitative form was undoubtedly superior to the resonance theory. However, neither this revolutionary book nor the Montpellier conference succeeded in converting the organic chemists to the new belief, owing essentially to the extremely condensed and difficult reading used by Dewar.

In 1844, the Russian chemist Nikolai Nikolaevich Zinin (1812–1880), founder of a great school of organic chemistry at Kazan in Siberia, which had as students Butlerov and Borodin, discovered a reaction known as the benzidine rearrangement in which benzidine is formed by the reaction of 1,2-diphenylhydrazine with acids (Zinin 1845).

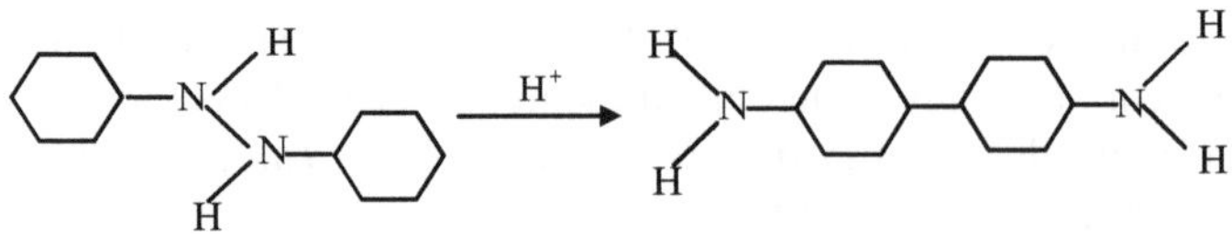

The mechanism of this reaction had been studied first by Hofmann (1862–1863) and later by Ingold (Ingold and Kidd 1933) and Robinson (1941) who all proposed complex and unrealistic molecular folding models. In 1946, Dewar suggested a very simple explanation of Zinin's reaction (Dewar 1945a), based on the formation of π complexes between the benzene rings. According to Dewar, the two benzene rings form dative bonds around which the rings rotate until they break the N–N bonds and form new C–C bonds that bind them together. Dewar's idea of the formation of π complexes supplied the first correct explanation for the electronic structure of transition metal complexes with olefins (Dewar 1951) known today as the *Dewar–Chatt–Duncanson model* (Chatt and Duncanson 1953).

In 1951, Dewar was appointed professor at the Queen Mary College of London University where he organized an important research group of theoretical organic chemists. In this period he wrote a famous series of six articles on the molecular orbitals theory in organic chemistry which represent an important generalization of Hückel's method. The first paper presented the general theory (Dewar 1952) and the following ones dealt with different applications. Using the selection rules developed by Woodward and Hoffmann, he then started the study of pericyclic reactions in terms of aromatic and antiaromatic transitions.

In 1959, he decided to move to America to the University of Chicago where he was involved in the development of a series of semiempirical molecular orbitals calculations of organic molecules. The offer of a considerable number of computing hours on large computers convinced him to go in 1963 to the University of Texas at Austin where in few years, with the help of his coworker Rowland Pettit (1927–1981), he made the University of Texas at Austin an internationally known center for the theoretical study of reaction mechanisms. Dewar dedicated himself to the development of more and more sophisticated and accurate semiempirical electronic computing methods. In addition, he also developed a series of experimental problems in different fields, from carbenium ions to semiconductors and to liquid crystals.

Today, electronic computer technology allows one to approach problems that could not even be imagined at the time of Dewar's semiempirical computing methods. Nevertheless, his *Modified Intermediate Neglect of Differential Overlap MINDO* (Baird et al. 1969), *Modified Neglect of Differential Overlap MNDO* (Dewar et al. 1977), *DEWAR-PI* (Dewar et al. 1989), *Austin model 1 AM-1* (Dewar et al. 1985), and *Semiempirical ab initio Model 1 SAM-1* (Dewar et al. 1993) computing methods are still used in organic chemistry laboratories for fast calculations and for very large molecular systems.

In the same years an important step forward was taken for the transformation of the concepts of the reaction mechanisms theory into quantum-mechanical formalism with the use of electronic density. In this new approach, electron density became the basic quantity for the interpretation of attractive and repulsive electrostatic interactions determining all processes connected to electron displacements inside the molecules. George Wheland and Linus Pauling had already proposed in 1935 that an electrophilic group would attack a molecule in regions of high electronic density whereas a nucleophilic group would instead be inserted at low electronic density sites, explaining in this way the orientation of substitution reactions at the benzene ring (Wheland and Pauling 1935). Not everything, however, could be so easily accounted for. For example, in benzene without substituents, both electrophilic and nucleophilic substitution reactions take place and therefore the same charge density can lead to opposite results. Also, in the case of naphthalene, it was difficult to explain in this way why an electrophilic substitution reaction such as nitration occurs always at the α position.

The idea that the electron density could explain every chemical process had already started to be doubted in the 1940s. This led several researchers to implement the idea of the distribution of the electron density within the molecules with additional effects contributing to thicken it in some specific positions during electrophilic attacks. A brilliant attempt was made by Coulson and Longuet-Higgins, introducing a transient polarization induced by the substituent approach (Coulson and Longuet-Higgins 1947) and by Wheland, computing the energy necessary to localize electrons at the reaction site (Wheland 1942). In 1952, the Japanese Kenichi Fukui (1918–1998), professor of chemical physics at the University of Kyoto, and his coworkers T. Yonezawa and H. Shingu obtained a brilliant result introducing the approximation of the frontier orbitals (Fukui et al. 1952). In the calculation of the chemical reactivity, their method neglects all molecular orbitals except two, the occupied one of higher energy (HOMO) and the vacant one of lowest energy (LUMO). Here the acronyms HOMO and LUMO mean *Highest Occupied Molecular Orbital* and *Lowest Unoccupied Molecular Orbital*, respectively. According to Fukui, these two orbitals behave in a molecule as the valence orbitals in an atom. In particular, the HOMO containing electrons with the highest energy possesses the character of electron donor whereas the LUMO that of electron acceptor.

Since a chemical reaction is nothing other than an exchange of electrons among the reactants, the knowledge of the frontier orbitals is equivalent to the knowledge of the reactivity of the molecule. Fukui's theory is very simple but extremely powerful. When molecule A approaches molecule B, their external molecular orbitals start to overlap and to interact. According to perturbation theory, the larger the overlap and the lower the difference in energy between the two orbitals, the greater will be the stabilization of the reacting system. If then an electron is transferred from the HOMO of molecule A to the LUMO of molecule B, the bond between the overlap region, called the reaction center, weakens while that between the reaction center and molecule B becomes stronger. As a consequence, the HOMO of molecule A is destabilized with respect to the LUMO of molecule B.

The interaction between the two orbitals therefore always gives rise to an orbital of lower energy that will host the two electrons and an orbital of higher energy as shown in the schematic representation of the picture below.

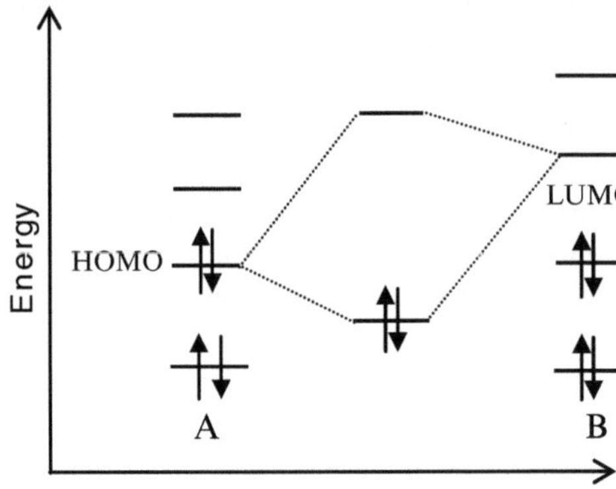

Correlation between the HOMO and LUMO orbitals of two molecules A and B.

Fukui's theory explains very clearly the sequence of events for a nucleophilic reaction. At the beginning, the nucleophilic group A attacks molecule B by placing its excess of electrons, normally a free doublet, in the LUMO of molecule B. The free doublet of the nucleophilic group therefore acts as electron donor and the LUMO of molecule B as electron acceptor. Robert Woodward, professor of organic chemistry at Harvard and 1965 Nobel laureate for his contributions to organic chemistry, soon realized the significance of the frontier orbitals theory of Fukui and asked Roald Hoffmann, at that time a young researcher at Harvard, to collaborate with him to verify his ideas about the interpretation of the stereochemistry of electrocyclic organic reactions. Hoffmann was already an expert in approximate quantum mechanical computing methods, having personally developed the extended Hückel method. He performed the calculations using the extended Hückel method and confirmed Woodward's previsions perfectly. Hoffmann's results led to the publication of a famous paper (Woodward and Hoffmann 1965) in which three important rules, called the Woodward–Hoffmann rules, allow one to predict the stereochemistry of pericyclic reactions from the knowledge of the orbital symmetry. Pericyclic reactions take place without the intervention of intermediate reactive species (anions, cations, radicals) but through a concerted mechanism of breaking old and forming new bonds. These rules, which apply to stereospecific ring opening and closing processes by reactions of conjugated polyenes induced by the heat or by photons, establish that:

1. In a non aromatic open chain molecule containing 4n electrons, a bonding interaction between the termini of the chain must involve the overlap between orbital envelopes on opposite faces of the system. This is possible only if both extreme faces of the chain perform a rotation in the same direction.
2. In an aromatic molecular open chain system with 4n + 2 electrons, the bonding interaction between the extreme terms of the chain require instead their rotation

in two opposite directions, one in the clockwise and the other in the anticlockwise direction or vice versa.

3. In a photochemical reaction, one electron of the HOMO of the reactant is promoted to an excited state that produces the inversion of the symmetry relationships and the inversion of the stereo-specificity.

These reactions include electrocyclic, cycloadditions, and sigmatropic reactions. Thanks to this work Hoffmann shared the chemistry Nobel Prize in 1981 with Kenichi Fukui. Unfortunately, Woodward died 2 years before he could receive his second Nobel Prize.

Woodward and Hoffmann developed correlation diagrams connecting the symmetry of the starting orbitals to that of the final ones. These diagrams, easily built using simple symmetry considerations, allow one to predict that if the occupied orbitals do not undergo significant energy variations, the reaction can easily take place. In this way, they introduced the concept of orbital symmetry preservation that, in association with the application of approximate methods of calculation, allowed one to establish the Woodward–Hoffmann rules described above as true orbital selection rules.

The development of the Woodward–Hoffmann rules gave rise to controversy involving three Nobel laureates, Robert Woodward and Roald Hoffmann on one side and Elias James Corey on the other, the latter claiming priority for the explanation of the symmetry of the perturbed (HOMO) molecular orbitals that led later to the definition of these famous rules. In 2004, Corey, in his official speech on the occasion of the Priestley award, said that in May 1964 he had suggested to his colleague Woodward an explanation of the symmetry of the perturbed (HOMO) molecular orbitals for some stereoselective conversions reactions. Woodward on his side always maintained that the explanation of the orbital symmetry had been his own idea and never mentioned a discussion with Corey on this topic. Corey instead insisted that he had also discussed the problem of the orbital's symmetry privately with Hoffmann since 1964. Hoffmann rebutted Corey's statement in an article published in *Angewandte Chemie* (Hoffman 2004) asking Corey why he had waited so long to make the issue public. Corey's answer was that he had omitted to claim for priority earlier to avoid sullying the name of Harvard. The dispute went on even after the death of Woodward in 1979 on the basis of a series of letters between Corey and Hoffmann that did not dissipate Corey's doubts about Woodward's behavior. However, it confirmed that personal disputes originated through scientific competition for international notoriety, even at the level of giants of science and culture, will always leave a bitter taste in the mouth by revealing the weaker sides of our human personality.

Despite this sad history, Corey remains one of the greatest living chemists, who developed important synthetic reagents and methodologies that have advanced the science of organic synthesis considerably. Born in 1928 into a Christian Lebanese family who immigrated to the United States, he was educated at MIT where he obtained his Ph.D. in 1951 before joining the University of Illinois at Urbana-Champaign where he became professor of chemistry in 1956. In 1959, he moved to

Harvard University to become professor of chemistry and colleague of Woodward and Hoffmann at the chemistry department. During his research activity, Corey developed several new reagents such as, for instance, the pyridinium chlorochromate used for the oxidation of alcohols to aldehydes, the alcohol protecting groups *tert*-butyldimethylsilyl ether and triisopropylsilyl ether, as well as a series of boron-based heterocycles for the asymmetric catalysis of the Diels–Alder reaction and reduction of ketones. In addition, he and his coworkers have invented a number of new types of reactions widely used in modern synthetic organic chemistry. Among these are the Corey–Winter olefin synthesis for converting 1,2-diols into olefins (Corey et al. 1963), the special case of the Wittig reaction known as the Corey–Fuchs reaction (Corey et al. 1972), and the Corey–Bakshi–Shibata reduction (Corey et al. 1987) in which a ketone is enantio-selectively reduced to the corresponding chiral alcohol.

8.5 Organic Photochemistry

At the beginning of the twentieth century, the study of chemical processes produced by light absorption in the visible and near ultraviolet regions started to develop within the framework of the reaction mechanisms theory, thus giving rise to the branch of chemical physics called photochemistry. Reactions produced by sunlight have been known for a long time, but the nature of the chemical processes involved in them was practically unknown. In 1817, Theodor Grotthuss already understood that a reaction could be produced by light if this was absorbed by the molecules (Grotthuss 1820). Grotthuss' idea remained unknown until it was restarted in 1842 by the English, American naturalized, John W. Draper (1811–1882), professor at the New York University since 1837 and an expert in chemistry and photographic processes. For this reason, Grotthuss' idea is better known today as the Grotthuss–Draper law.

As early as 1834, the chemist and apothecary Johann Bartholomäus Trommsdorff (1770–1837), founder of the first German pharmaceutical institute at Erfurt in 1795, observed a photochemical reaction of solid (−)-α-santonin (Trommsdorff 1834). He discovered that crystals of α-santonin, when exposed to sunlight, turned yellow and burst, attributed later to a large change in crystal volume on dimerization. In a 2007 study, this reaction was described as a succession of three steps taking place within a single crystal (Natarajan et al. 2007).

The first who developed true photochemical concepts, however, was Johannes Stark (1874–1957), professor at the *Rheinisch-Westfälische Technische Hochschule* of Aachen who in 1908 distinguished between primary and secondary processes taking place in a chemical system by light absorption (Stark 1908). Stark defined as a primary process the immediate absorption of a photon by a molecule or an atom and as a secondary process the whole collection of "obscure" reactions started by the primary process.

In a famous 1912 paper, Albert Einstein (Einstein 1912) established the photochemical equivalence law. In the foreword Einstein wrote:

> Im folgenden wird auf wesentlich thermodynamischem Wege gleichzeitig das Wiensche Strahlungsgesetz und das photochemische Äquivalentgesetz abgeleitet. Unter dem letzteren verstehe ich den Satz, das es zur Zersetzung eines Grammäquivalentes durch einen photochemischen Vorgang der absorbierten Strahlungsenergie Nhν bedarf, falls man mit N die Zahl der Moleküle im Gramm-mol, mit h die bekannte Konstante in Plancks Strahlungsformel, mit ν die Frequenz der wirksamen Strahlung bezeichnet. Das Gesetz erscheint im wesentlichen als eine Konsequenz der Voraussetzung, das die Zahl der pro Zeiteinheit zersetzten Moleküle der Dichte der wirksamen Strahlung proportional ist.

This paper clearly shows how Einstein realized that the basis of a photochemical process is built by the connection between Planck's law and the absorption process transforming radiant into molecular internal energy and then into kinetic energy. The chemical-physical processes consequent to the absorption or emission of electromagnetic radiation were connected with the structure of the molecular vibro-electronic levels by the Ukrainian (Polish after 1918, when Poland once again became an independent state after over 120 years of occupation) physicist Alexander Jablonski (1898–1980) who, in 1935, represented schematically the absorption, fluorescence, and phosphorescence processes in a famous diagram (Jablonski 1935), since then included in all spectroscopy and photochemistry books.

In the below diagram, the letters S and T indicate electronic singlet with paired and triplet states with unpaired spins, respectively. A concept of fundamental importance for photochemistry was the introduction of the existence of metastable states in which the energy is trapped for sufficiently long times to avoid the ignition of chemical reactions. Later, Gilbert Lewis and Michel Kasha identified these metastable states as triplet states.

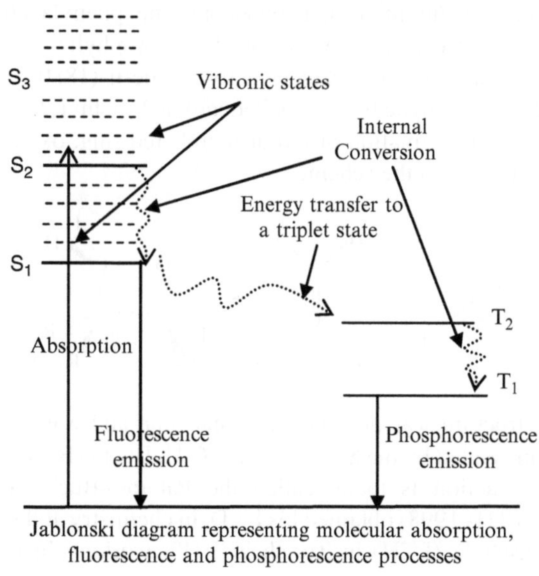

Jablonski diagram representing molecular absorption, fluorescence and phosphorescence processes

In the second half of the nineteenth and at the beginning of the twentieth century, two Italian chemists, Emanuele Paternò and Giacomo Ciamician, both pupils of Cannizzaro at the University of Rome, constructed the true basis of organic photochemistry.

The Palermitan Emanuele Paternò (1847–1935), marquis of Sessa, was a student of Cannizzaro at the University of Palermo. In 1871, when he was only 22 years old, he was appointed as chemistry teacher at the University of Turin where he contributed to the foundation of the Gazzetta Chimica Italiana. Back in Palermo in 1872, he succeeded his master Cannizzaro after the transfer of the latter to the University of Rome. Later, Paternò also moved to the University of Rome in 1892, first as analytical chemistry teacher and then in 1910, upon the death of Cannizzaro, as professor of *general chemistry*.

While Paternò frequented Cannizzaro's laboratory at Palermo, the Viennese Jew Adolf Lieben (1836–1914), who studied at Vienna, Heidelberg, and Paris and held a position of *Privat dozent* at the University of Vienna, went to Palermo to collaborate with Cannizzaro for whom he was assistant from 1863 to 1867. Lieben studied the chloral (trichloroacetaldehyde, CCl_3–CHO) synthesis from alcohol and hydrochloric acid, and the young Paternò, following Lieben's work, learned how to prepare chlorine derivatives of methane and ethane, among them trichloroacetate, dichloroacetaldehyde, and in 1869 pentachloroethane, reacting chloral with phosphorus pentachloride. In this period, he was the first to suggest a tetrahedral conformation for methane and built a series of diagrams to prove the existence of *cis–trans* isomerism in dichloroethylene.

In 1875, Paternò realized that under the effect of sunlight, 3-nitrocuminic acid was converted into a red amorphous product. He then continued the study of the chemical action of sunlight and after a short while proved the photochemical synthesis of propyl butyrate by action of sunlight on butyric acid. In addition, he studied the effect of sunlight on benzophenone and phenylacetic acid and the photochemistry of the green substance of plants, chlorophyll.

In 1909, Paternò and his coworker Generoso Chieffi (1880–1923) discovered that a solution of benzaldehyde (or benzophenone) in 2-methyl-2-butene, if exposed to sunlight, induces the formation of a four-membered ring, oxetane (Paternò and Chieffi 1909), according to the scheme

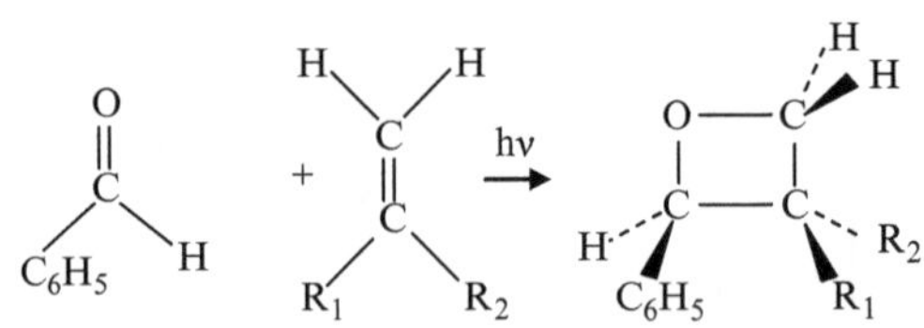

The oxetane structure was not known at that time and was confirmed only in 1954 by the group headed by the Swiss chemist G.H. Büchi (Buchi et al. 1954). For this reason, the reaction is today called the Paternò–Büchi reaction. George Hermann Büchi (1921–1998) obtained his Ph.D. in chemistry at the Eidgenössische Technische Hochschule (ETH) in Zurich in 1947, working in the laboratory of

Leopold Ruzicka, one of the most important centers of organic chemistry in Europe. Having immigrated to the United States in 1951, he joined the MIT where he became professor in 1958. For several years, the Paternò–Büchi reaction was largely ignored by organic chemists until it was disclosed in 1963 by George S. Hammond and Nicholas J. Turro in a review of organic photochemistry (Hammond and Turro 1963) and by J.S. Searles in 1964 in his book on heterocyclic compounds (Searles 1964) who evidenced its utility and clarified its reaction mechanism.

Even more important was the contribution to photochemistry of Giacomo Luigi Ciamician, professor of chemistry at the University of Bologna, who, impressed by the capacity of green plants to exploit sunlight energy, carried out a wide-ranging research project on the chemical effects of light absorption in the first decade of the twentieth century, together with his strict coworker, the German chemist Paul Silber (1851–1932).

Giacomo Ciamician (1857–1922), born into a Triestine family of Armenian origin which immigrated to Italy from Istanbul around 1850, studied first at Vienna and then at Giessen where he received his Ph.D. in 1880 with a thesis on chemical affinity. He then went to work at the University of Rome as assistant to Stanislao Cannizzaro and in 1887 was made professor of general chemistry at the University of Padua. After 2 years, he moved to the University of Bologna where he stayed for the rest of his life.

An important chapter in the scientific activity of Ciamician concerns the chemistry of pyrrole and its derivatives (Ciamician 1887, 1904) and in particular the transformation of pyrrole into pyridine that he realized in collaboration with Maximilian Dennstedt (Ciamician and Dennstedt 1881, 1882). Ciamician is, however, better known for his photochemistry research for which he was proposed for the Nobel Prize nine times (Ciamician and Silber 1909). The Ciamician experiments of photochemistry, mostly in collaboration with his friend and coworker Paul Silber, were made by exposing to sunlight tubes, flasks, glass pipes, and Erlenmeyer bottles on the terrace of the Chemistry Institute of Bologna (Ciamician and Silber 1901). These experiments gave rise to 85 papers, published in the series *Chemisches Lichtwirkungen* of the *Berichte der Deutschen Chemischen Gesellschaf* in the period 1899–1913 and listed in review articles of the same Ciamician (Ciamician 1908). This research led to the discovery of several new reactions (Ciamician and Silber 1896), among them the photoreduction of aldehydes, ketones (Ciamician and Silber 1885), quinines, and nitrocompounds, as well as the photodimerization and cycloaddition of olefins. Ciamician was also a pioneer of ecology, convinced that the exploitation of solar energy would set humanity free from the use of fossil fuels (Nebbia and Kauffman 2007). In a famous 1912 article in *Science* (Ciamician 1912), he wrote:

> civilization is the daughter of coal, for this offers to mankind the solar energy in its most concentrated form; that is, in a form in which it has been accumulated in a long series of centuries. Modern man uses it with increasing eagerness and thoughtless prodigality for the conquest of the world and, like the mythical gold of Rhine, coal is today the greatest source of energy and wealth. The earth still holds enormous quantities of it, but coal is not inexhaustible. The problem of the future begins to interest us ... Is fossil solar energy the only one that may be used in modern life and civilization?

With the outbreak of the First World War, photochemistry research in Italy ceased, both for lack of funds and personnel and for the relative absence of interest of the academic milieu, and migrated first to England and Germany and then to the United States. It was taken up in Italy only after several years in the same city of Bologna where Ciamician had started his photochemical adventure.

With the development of the modern chemical-physics instrumentation, photochemistry recovered its original kinetic vocation and in the 1920s transformed into the study of the velocity of fast organic reactions, a problem strongly connected to the spectroscopic identification of intermediate short-living species, ions, or radicals, whose existence was more and more postulated in the theoretical study of reaction mechanisms.

The study of fast reaction kinetics started in England in the medical community before the chemical-physics one. In 1921, Hamilton Hartridge (1886–1976), professor of physiology at the University of Cambridge and his young coworker Francis John Worsley Roughton (1899–1972) invented the "stopped flow" technique that allows one to mix very rapidly two different solutions in order to study their reaction kinetics (Hartridge and Roughton 1923a). Using an ingenious system invented by Hartridge for the spectroscopic measurement of the absorption variations of solutions they measured the reaction velocity of oxygen and carbon oxide with hemoglobin. Their work, published in 1923 (Hartridge and Roughton 1923b), proved for the first time the possibility of measuring reaction times of the order of milliseconds. A significant jump of quality was realized at Cambridge, only 30 years after Ronald Norrish, professor of chemical physics, and his student George Porter invented in 1949 the "flash" photolysis technique with which they succeeded in measuring, after some improvements of the initial system, times of the order of 10^{-6} s (millionth of second) and in identifying unstable intermediate species with very short lifetimes. Ronald George Wreyford Norrish (1897–1978) started to be interested in photochemistry in 1915 when he was still a young student at Cambridge. His research activity under the supervision of Eric Redeal was interrupted by the First World War in which he was made prisoner of war by the Germans in March 1918 and shut up in a concentration camp, first at Rastatt in Germany and then at Graudenz in Poland. Repatriated in 1919, he returned to Cambridge, and once he obtained his Ph.D. in 1925 was hired at the University first as demonstrator then in 1930 as chemical physics lecturer and finally in 1937 as professor of chemical physics upon the death of Thomas Martin Lowry.

At Cambridge, he started a series of work on organic photochemistry, being interested in chain reactions and in polymerization kinetics. To him we owe the study of a class of chemical reactions, called Norrish reactions, catalyzed by ultraviolet light in which aldehydes and ketones produce a great variety of compounds through the formation of intermediate radical species (Norrish and Bamford 1936, 1938). The Norrish reactions are also important for the study of processes occurring in the upper layers of the atmosphere by action of the sunlight.

At the beginning of the Second World War, Norrish was again forced to interrupt his photochemistry research and could restart it only in 1945, when the young George Porter, who, during the war, had worked at the development of radars in the

Royal Navy, came to work with him at Cambridge. Porter's experience of complex electronic devices for the generation and control of short pulses turned out to be valuable in the study of the kinetics of fast reactions. In 1946 Porter had the idea of using fast light pulses to generate free radicals following their evolution in time by means of spectroscopic techniques. Starting from this idea, Porter and Norrish invented the flash photolysis technique that opened the way to research on fast kinetics at times of millionths of second (microseconds). The flash photolysis technique consists in subjecting a gas or liquid to a light flash of high intensity and very short duration that gives rise to a photochemical dissociation of the molecules. Then a second light flash is applied to the sample with time delays in order to follow the reactions that take place in the system before reaching equilibrium.

The first flash photolysis instrument was operative in 1947 and this started a long-lasting collaboration between Norrish and Porter that led to results for which they obtained the Nobel Prize in 1967.

George Porter (1920–2002) started his research activity with Norrish studying the presence of transient species in chemical reactions, in particular the presence of free radicals and excited molecular states in the gas phase (Norrish and Porter 1949). Later, he extended his research to solutions and to photosynthetic processes in green plants. In 1950, his instrument had already reached the time resolution of a few microseconds (10^{-6} s).

The Norrish and Porter flash photolysis soon had to face competition from a different time-resolved technique, the relaxation method developed in Germany by Manfred Eigen, one of the finest talents of German chemistry after the war. Manfred Eigen (1927–), son of a chamber musician of Bochum, after having served in the antiaircraft army during the Second World War, graduated under the supervision of Arnold Eucken (1884–1950) at Göttingen University in 1951 with a thesis on the specific heat of heavy water and of electrolytic solutions. He then worked for 2 years at the university as assistant to Ewald Wicke (1914–2000) until he joined as researcher at the *Max-Planck Institut für physikalische Chemie*, built at Göttingen in 1949 under the direction of Karl Friedrich Bonhoeffer, in affiliation with the *Kaiser-Wilhelm-Instituts für physikalische Chemie und Elektrochemie* of Berlin.

The outstanding working conditions of the Max-Planck Institut that Bonhoeffer put at his disposal allowed Eigen to prosper in different fields from chemical physics to biophysics. In 1957, he became first researcher of the Max-Planck-Gesellschaft within the *Arbeitsgruppe für Biochemische Kinetik*, in 1964 director, and in 1967 managing director of the institute that later became the *Max-Planck Institut für biophysikalische Chemie*. In the same year, he was made member of the Scientific Council of the German Federal Republic. Eigen started the study of ultrafast reactions in solution, in particular of electrolytes, using a technique experimentally completely different from that of Norrish and Porter, although conceptually similar from a different point of view. Until then, fast reactions in solution were studied using the *stopped flow* method consisting in mixing the reactants and in following their evolution toward equilibrium. Both the Eigen and the Norrish–Porter groups started instead from systems already in equilibrium that they perturbed with appropriate techniques to measure the time required to establish it again. The basic difference between the experiments of Norrish and Porter and those of Eigen concerned the way in which the perturbation was applied and its intensity. Norrish and Porter used high-energy flashes of ultraviolet radiation, producing rather large deviations from equilibrium of the examined systems. Eigen instead used ultrasound waves or electric pulses to increase rapidly, but only weakly, the pressure and temperature of a solution, i.e., to shift the system slightly from the original equilibrium conditions. The propagation of waves in liquids and in solutions had been studied before by Nernst and his coworkers, in particular in a 1910 Ph.D. thesis relative to the propagation of sound waves in nitrogen tetroxide, N_2O_4. In the 1951–1963 period, Eigen dedicated his attention to the study of the relaxation to equilibrium of electrolytic solutions using the technique of ultrasound production developed by his colleagues Konrad Tamm (1913–1986) and Walter Kurtze at the Max-Planck Institute (Eigen et al. 1953). In the following years, he developed a series of new techniques (Eigen 1954) that allowed him to measure relaxation times of the order of nanoseconds (10^{-9} s).

In 1954, the Belgian Leo de Maeyer joined his group, giving rise to a long-standing collaboration involving relaxation measurements associated with thermal conductivity and ultrasound absorption applied to the study of a variety of problems, including thermodynamic properties of water and aqueous solutions (De Maeyer and Kustin 1963), proton transfer, and metallic complexes reactions, as well as multistage chemical processes (Eigen and de Maeyer 1955, 1956). In particular, in collaboration with Leo de Maeyer, Eigen determined the self-dissociation velocity and the anomalous charge conduction of protons in water and ice crystals (Eigen and de Maeyer 1958) and studied metallic ions reactions in relationship to their position in the periodic table.

Around 1960, his research activity was oriented toward organic chemical physics, explaining the individual steps of a series of reaction mechanisms and verifying experimentally the general theory of acid-basic catalysis. For his research on ultrafast reactions he shared the 1967 Nobel Prize with Norrish and Porter.

In the 1960s, his interests shifted toward biological systems that quickly became his principal research activity. He faced the problem of hydrogen bridges in nucleic acids, the dynamics of transfer of biological information, and the study of enzymes and lipid membranes. Overall he concentrated his interests on the storage of biological information in the central nervous system and the control and regulation of biological functions (Eigen 1971).

The name of Manfred Eigen is also bound to the theory of the chemical hypercycle, the association of cyclic reactions that explain the auto-organization of prebiotic systems. He discussed the hypercycle in a book written in collaboration with the Viennese Peter Schuster, professor of theoretical chemistry at the University of Vienna (Eigen and Schuster 1979). The hypercycle is a macromolecular self-reproducing system, in which enzymes and RNA molecules cooperate to ensure the cycle working functions. In the cycle, an RNA molecule **a** codes for the enzyme **A**, which in turn, acting as catalyst, increases the replication rate of an RNA molecule **b** which codes for the enzyme **B**. The latter increases the replication rate of the

RNA molecule **c** and so on until the cycle closes with the **n**th enzyme increasing the production rate of the first RNA molecule **a**. The hypercycle is therefore the envelope of a series of interconnected cycles in which each enzyme created in a cycle catalyzes the reaction in the next connected cycle.

Manfred Eigen and Peter Schuster have also developed the concept of quasi-species to characterize the huge number of mutants in virus populations, especially in the case of RNA viruses (Eigen et al. 1989). A viral quasi-species is a highly structured and interconnected viral population in continuous evolution whose stability is controlled by the percentage error of genomic replica depending on the critical error threshold controlling its limits and regulating the diversity of the viral population. When this percentage overcomes the limiting threshold, one reaches the genomic catastrophe. Manfred Eigen has also written a beautiful book that presents the laws that regulate games (Eigen and Winkler 1978).

Organic photochemistry also had a luxuriant development in the United States, in close symbiosis with the great progress in electronic spectroscopy definitively assessed on the solid bases of the molecular orbitals theory which found in the group theory the ideal classification system of the electronic levels and the most efficient combination of selection rules for allowed or forbidden transitions.

Fundamental for the diffusion of photochemistry in the organic chemistry departments was its ability to produce free radicals, i.e., molecules with unpaired electrons, endowed with extremely high reactivity that were quickly recognized as very important intermediates in several key reactions of synthetic organic chemistry.

At the beginning, photochemistry was confined to the United States in a few specialized centers at the Caltech and the University of California, Los Angeles (UCLA), on the West coast, at the University of Chicago in the center, and at Columbia and Harvard on the East coast. However, it quickly started to spread throughout the country thanks to the large number of high-level graduate students that these universities produced and that occupied key positions in other American universities and in industry. The original aspect of American photochemistry research was that of tightly associating the applicative and industrial research to the academic one. The affirmation of this photochemical research with strong industrial and applicative components was essentially due to George Simms Hammond who in a 30-year period succeeded in assuring the supremacy of American photochemistry over the European one from which it was born by direct affiliation.

George Simms Hammond (1921–2005) graduated in 1943 and joined Harvard University where he obtained his Ph.D. in chemistry in 1947 under the supervision of Paul Bartlett with a thesis on the polymerization processes initiated by free radicals (Bartlett et al. 1947). From that time radicals became an important part of his scientific life both as a young researcher at the University of California, Los Angeles (UCLA), and at the Iowa State College where he went in 1948 and where he stayed for 10 years until in 1958 he was hired by the Caltech, the most important center of applied chemistry on the West coast. At Caltech, Hammond carried on an intense research project on free radicals and on ionic reactions that greatly contributed to the affirmation of photochemistry as an academic research in several American universities. During his time at the Iowa State College, he published a series of papers in the *Journal of the American Chemical Society* (Hammond et al. 1957), essentially on the organic reaction mechanisms of radical types. An important piece of work of this period, known as the "Hammond postulate," concerns the transition state theory and connects reaction rates to the structure and energy of the quantum levels involved in the reaction (Hammond 1955). In particular, the postulate establishes that if two states close in energy follow each other in the reactive process, the molecular structure rearrangement in the interconversion mechanism must necessarily be small.

In the Iowa State College period, Hammond also studied in detail photosensitive reactions, i.e., reactions primed by a photo initiator which absorbs the radiation and transmits to the molecules of another compound the energy excess necessary to start the reaction. Today, these reactions have important industrial applications, for example, in the production of integrated microcircuits on silicon plates (computer chips).

As professor of organic chemistry at Caltech, Hammond continued from 1958 to work on photochemical reactions. The research that he carried out with his students in the period 1959–1962 laid the foundations of modern applicative photochemistry. Among the numerous pieces of research of Hammond, particularly important was the study of the chemistry of the azobisisobutyronitrile (AIBN) (Hammond et al. 1960) which revealed important properties of this photosensitizer and in particular of the cage effect created by the solvent molecules around those of AIBN which blocks their ability to react with other chemical compounds. By acting on this cage effect, Hammond and his coworkers succeeded in controlling the activity of AIBN, transforming it into a powerful radical initiator widely used to photoinitiate free radical polymerizations and other radical reactions, especially for the preparation of vinyl plastics.

In 1978, Hammond left the university to become director of an important chemical industry, the Allied Chemical Corporation. Crucial for the scientific development of Hammond as a modern organic chemist was the friendship and collaboration with Donald Cram, organic chemist at the UCLA and Nobel laureate in chemistry in 1987. Cram and Hammond wrote together in 1959 a treatise on organic chemistry that within a few years became the most used book in all American and in several European universities for its original presentation of organic chemistry in which the reactions were classified on the basis of their mechanisms and not of the nature of reactants and products. Donald Cram was one of the founders of the new branch of chemistry which was named "supramolecular chemistry."

In the last 30 years, photochemistry has gone through a theoretical and experimental development that could not be even conceived a few decades earlier, thanks to the availability of lasers with ultrafast pulses allowing access to time pulses first of picoseconds (10^{-12} s), then of femtoseconds (10^{-15} s), and more recently of attoseconds (quintillionth of a seconds, 10^{-18} s). With the construction of solid-state lasers (titanium–sapphire, neodymium-doped YAG, etc.,) and with the development of modern technologies of parametric amplification, frequency tuning, and pulses time compression, a completely new era started in the study of reaction mechanisms and of molecular dynamics.

The number of physical, chemical, and biological processes investigated with the help of ultrafast lasers is very high and new problems are added every day. Complete classes of ultrafast reactions from isomerization to proton transfer in excited electronic states, from the identification of short-living radicals to ionic intermediates to the characterization of transition states or of the elementary processes involved in photosynthesis, are currently studied in the framework of the reaction mechanism theories.

Fundamental contributions have been made in several advanced laboratories all over the world by an interdisciplinary community of physicists, chemists, biologists, and highly qualified technicians collaborating in pioneering the new discipline of femtochemistry, i.e., the study of chemical reactions across femtoseconds. Using a rapid ultrafast laser technique (consisting of ultrashort laser pulses), the technique allows the description of reactions on very short timescales, short enough to follow the evolution of excited transition states in chemical reactions.

Among this research, crucial was that of the Egyptian-American spectroscopist Ahmed Hassan Zewail (1946–), professor at Caltech and 1999 Nobel laureate in chemistry who first coined the word "femtochemistry" by fusion of the terms femto and chemistry.

Ahmed Zewail, born in Egypt at the City of Disuq about 60 km from Alexandria, graduated in chemistry at the Alexandria University. He then immigrated to the United States following the advice of his professors. There he joined the University of Pennsylvania where he received a Ph.D. in chemistry in 1973. After the Ph.D., he moved to California where he accepted a position at the University of California, Berkeley. This gave him the opportunity to start his career as a laser spectroscopist, working with Charles Harris (1940–) on several projects, including building a picosecond laser. In 1976, he became assistant professor at Caltech where he has remained ever since, becoming in 1990 Linus Pauling professor of Chemical Physics. In 1999, Zewail was awarded the Nobel Prize in chemistry for his research on femtosecond spectroscopy. In his research activity, Ahmed Zewail has produced fundamental contributions to the time evolution of chemical reactions and biological functions and to the structural analysis, at an atomic scale resolution, of three-dimensional structures in systems ranging from small molecules to crystals, and from DNA and proteins to viruses. Latest among his long stream of recognized achievements is the work on the structure of the ribosome protein-synthesis machine. He collected in 1994 his scientific results, produced over several years of spectroscopic research, and published more than 250 research papers in a two-volume book entitled *Femtochemistry: Ultrafast Dynamics of the Chemical Bond* (Zewail 1994). More recently, he focused on the use of coherent, femtosecond single-electron packets or electron pulses to combine time with spatial domain resolution. This has led to the creation of four-dimensional (4D) electron microscopy, which is well suited to studying phase transitions (Grinolds et al. 2006), surface dynamics (Schäfer et al. 2010), evolution of mechanical deformations, time evolution of melting and crystallization processes, and overall protein structural variations in time and space (Zewail and Thomas 2009). Zewail is currently director of the National Science Foundation Laboratory for Molecular Science at Berkeley.

The chemical physics techniques for the study of ultrafast processes have found a broad field of application in the study of the rate of transfer of electrons in oxidation-reduction processes occurring in several reactions of biochemical and photochemical interest. Pioneer in this field of research were Robin M. Hochstrasser, one of the fathers of time resolved spectroscopy, and Rudolph Marcus another staff member of Caltech, a specialist of the study of reaction rate of monomolecular reactions and coauthor of the famous theory (Marcus 1952) RRKM (Rice, Ramsperger, Kassel, Marcus) for which he obtained in 1992 the Nobel Prize for chemistry. Marcus's research activity is discussed in detail in Chap. 1.

References

Angus WR, Ingold CK, Leckie AH (1936) Structure of benzene Part III Raman spectra of liquid benzene and liquid hexadeuterobenzene. J Chem Soc 925. doi: 10.1039/JR9360000925

Armstrong HE (1890) The structure of cycloid hydrocarbons. Proc Chem Soc 6:95–106

Bailey CR, Hale JB, Ingold CK, Thompson JW (1936) Structure of benzene. Part IV. Infra-red absorption spectra of benzene and hexadeuterobenzene both as vapour and as liquid. J Chem Soc 931. doi: 10.1039/JR9360000931

Bailey CR, Gordon RR, Hale JB, Herzfeld N, Ingold CK, Pool HG (1946) Structure of benzene. Part XX. The Raman and infra-red spectra of monodeuterobenzene: description and analysis. J Chem Soc 288–299

Baird NC, Dewar MJS (1969) 1 version of MINDO. J Chem Phys 50:1262

Bartlett PD, Conant JB (1936) A quantitative study of semicarbazone formation. J Am Chem Soc 54:288

Bartlett PD, Knox LH (1939) Bicyclic structures prohibiting the Walden inversion. Replacement reactions in 1-substituted apocamphanes. J Am Chem Soc 61:3184–3192

Bartlett PD, Hammond GS, Kwart H (1947) The inhibition and retardation by certain quinones of the peroxide-induced polymerization of allyl acetate and vinyl acetate. Discuss Faraday Soc 2:342–352

Bartlett PD, Cohen SG, Cotman JD Jr, Kornblum N, Landry JR, Lewis ES (1950) The synthesis of 1-bromotriptycene. J Am Chem Soc 72:1003

Brown HC (1976) Explorations in the non-classical ion area. Tetrahedron 32:179

Brown HC (1982) In: Wilkinson G, Stone FGA, Abel E (eds) Comprehensive organometallic chemistry, vol 7. Pergamon Press, Oxford, UK

Brown HC (1988) Asymmetric synthesis made easy. Chemtracts Org Chem 1:77

Buchi G, Inman CG, Lipinsky ES (1954) Light catalyzed organic reactions. 1. The reaction of carbonyl compounds with 2-methylbutene. J Am Chem Soc 76:4327

Butlerov A (1863) Über die verschiedenen Erklärungsweisen einiger Fälle yon Isomerie. Z Chem 6:500–534

Chatt J, Duncanson LA (1953) Olefin co-ordination compounds. Part III. Infra-red spectra and structure: attempted preparation of acetylene complexes. J Chem Soc 2939. doi: 10.1039/JR9530002939

Ciamician G, Dennstedt M (1881) Synthesis of pyridines from pyrroles. Berichte 14:1153

Ciamician G, Dennstedt G (1882) Trasformazione del pirrolo in piridina. Rend R Acc Lincei XII:1

Ciamician G, Silber P (1885) Über Pyrrilendimethyldiketon. Berichte 18:1466–1468
Ciamician G (1887) Il pirrolo e i suoi derivati. Atti R Acc Lincei Mem 4:274–377
Ciamician G, Silber P (1896) Zur Kenntniss der Tropinsäure. Berichte 29:2975–2976
Ciamician G, Silber P (1901) Chemische Lichtwirkungen. Berichte 34:1530
Ciamician G (1904) Über die Entwicklung der Chemie des Pyrrols in lezten Vierteljahrhundert. Berichte 37:4200–4255
Ciamician G (1908) Sur les Actions Chimiques de la Lumière. Bull Soc Chim France, IV Série, Tome III, pp 1–27
Ciamician G, Silber P (1909) Actions chimiques de la lumière. Ann Chim Phys 16:474–520
Ciamician G (1912) The photochemistry of the future. Science 36:385–394
Cohen JB (1907) Organic chemistry for advanced students I and II, vol 116. Macmillan, New York
Collie JN (1897) A space formula of benzene. Trans Chem Soc 71:1013–1023
Corey EJ, Winter RAE (1963) A new, stereospecific olefin synthesis from 1,2-Diols. J Am Chem Soc 85:2677
Corey EJ, Fuchs PL (1972) A synthesis method for the formyl to ethynyl conversion. Tetrahedron Lett 13:3769–3772
Corey EJ, Bakshi RK, Shibata S (1987) Highly enantioselective borane reduction of ketones catalyzed by chiral oxazaborolidines, mechanism and synthetic implications. J Am Chem Soc 109:5551
Coulson CA, Longuet-Higgins HC (1947) The electronic structure of conjugated systems. I. General theory. Proc R Soc A 191:39
De Maeyer L, Kustin K (1963) Fast reactions in solution. Annu Rev Phys Chem 14:5–34
Dewar MJS (1945a) Mechanism of the benzidine and related rearrangements. Nature 156:784
Dewar MJS (1945b) Structure of stipitatic acid. Nature 155:50–51
Dewar MJS (1949) The electronic theory of organic chemistry. Clarendon Press, Oxford, UK
Dewar MJS (1951) A review of the π-complex theory. M Bull Soc Chim Fr 18:C79
Dewar MJS (1952) A molecular orbital theory of organic chemistry, I. General principles. J Am Chem Soc 74:3341–3345
Dewar MJS, Thiel W (1977) Modified neglect of diatomic overlap. J Am Chem Soc 99:4899–4907
Dewar MJS, Zoebisch EG, Healy EF, Stewart JJP (1985) Austin model 1. Newly parametrized MNDO version. J Am Chem Soc 107:3902
Dewar MJS, Dennington RD II (1989) DEWAR-PI study of electrophilic substitution in selected polycyclic fluoranthene hydrocarbons. J Am Chem Soc 111:3804–3808
Dewar MJS, Jie C, Yu J (1993) SAM1 – the 1st of a new series of general-purpose quantum-mechanical molecular models. Tetrahedron 49:5003–5038
Von Doering WE, Knox LH (1954) The cycloheptatrienylium (tropylium) ion. J Am Chem Soc 76:3203–3206
Eigen M, Kurtze G, Tamm K (1953) Zum Reaktionsmechanismus der Ultraschallabsorption in wässrigen Elektrolytlösungen. Z Elektrochem 57:103
Eigen M (1954) Methods for investigation of ionic reactions in aqueous solutions with half lives as short as 10^{-9} s. Discuss Faraday Soc 17:194–205
Eigen M, de Maeyer L (1955) Die Geschwindigkeit der Neutralisierungsreaktion. Naturwissenschaften 42:413
Eigen M, de Maeyer L (1956) Ein stationäres Feldverfahren zur Untersuchung von Dissoziationsprozessen in Flüssigkeiten und Festkörpern. Z Elektrochem 60:1037
Eigen M, de Maeyer L (1958) Self-dissociation and protonic charge transport in water and ice. Proc R Soc A 247:505–533
Eigen M (1971) Molekulare Selbstorganisation und Evolution. Naturwissenschaften 58:465–523
Eigen M, Winkler R (1975) Das Spiel Naturgesetze steuern den Zufall. Piper Verlag, Munich
Eigen M, Schuster P (1979) The hypercycle: a principle of natural self-organization. Springer, Berlin
Eigen M, McCaskill J, Schuster P (1989) The molecular *Quasi-Species*, vol 75, Advances in Chemical Physics. Wiley, New York

Einstein A (1912) Thermodynamische Begründung des photochemischen Äquivalentgesetz. Ann Phys 37:832
Flürscheim BJ (1909) The strength of acids and bases and the quantitative distribution of affinity in the molecule. J Chem Soc Trans 95:718–734
Fukui K, Yonezawa T, Shingu HJ (1952) A molecular orbital theory of reactivity in aromatic hydrocarbons. Chem Phys 20:722
Gomberg M (1900) Triphenylmethyl, ein Fall von dreiwerthigem Kohlenstoffe. Chem Ber 33:3150
Gomberg M (1902) Über Triphenylmethyl. Ein Beitrag zur Kenntniss der Carboniumsalze. Chem Ber 35:2397
Grinolds MS, Lobastov VA, Weissenrieder J, Zewail AH (2006) Four-dimensional ultrafast electron microscopy of phase transitions. Proc Natl Acad Sci USA 103:18427
Grotthuss T (1820) Physikalische Chemische Forschungen. Ostwald's Klassiker der Exakten Wissenschaften 152
Hammett LP (1928) The theory of acidity. J Am Chem Soc 50:2666
Hammett LP, Deyrup AJ (1932) A series of simple basic indicators. J Am Chem Soc 54:2721
Hammett LP, Paul M (1934) The relation between the rates of some acid catalyzed reactions and the acidity function, Ho. J Am Chem Soc 56:830
Hammett LP (1935) Some relations between reaction rates and equilibrium constants. Chem Rev 17:125–136
Hammett LP (1937) The effect of structure upon the reactions of organic compounds. Benzene derivatives. J Am Chem Soc 59:96–103
Hammett LP (1940) Physical organic chemistry. McGraw-Hill, New York
Hammond GS (1955) A correlation of reaction rates. J Am Chem Soc 77:334–338
Hammond GS, Roberts JD, Cram DJ (1957) Organic reaction mechanisms. Annual reviews of physical chemistry for the year 1956. Annu Rev 8: 299–330
Hammond GS, Wu CS, Trapp OD, Warkentin J, Keys RT (1960) The mechanism of decomposition of azo compounds. II. Cage effects in the decomposition of a, a′-Azoisobutyronitrile and related compounds. J Am Chem Soc 82:5394
Hammond GS, Turro NJ (1963) Organic photochemistry. Science 142:1541
Hartridge H, Roughton FJW (1923a) The velocity with which carbon monoxide displaces oxygen from, combination with hemoglobin. Proc R Soc B 94:336–367
Hartridge H, Roughton FJW (1923b) A method of measuring the velocity of very rapid chemical reactions. Proc R Soc A 104:376–394
Hoffman R (2004) A claim on the development of the frontier orbital explanation electrocyclic reactions. Angew Chem Int Ed 43:6586–6590
Hofmann AW (1862–1863) Notes of researches on the polyammonias, No. XXIII. Hydrazobenzol, a new compound isomeric with benzidine. Proc R Soc Lond 12:576
Holmes EL, Ingold CK (1925) The nature of the alternating effect in carbon chains Part III. A comparative study of the directive efficiencies of oxygen and nitrogen atoms in aromatic substitution. J Chem Soc 127:1800–1821
Hughes ED, Ingold CK (1933) Influence of poles and polar linkings on the course pursued by elimination reactions. Part XIII. Decomposition of some quaternary ammonium salts containing the methyl, benzyl and benzhydryl groups. J Chem Soc 1:69–75
Hughes ED (1938) Aliphatic substitution and the Walden Inversion. Trans Faraday Soc 34:202–221
Ingold CK (1922a) The structure of the benzene nucleus. Part I. Intranuclear tautomerism. Trans Chem Soc 121:1133
Ingold CK (1922b) The structure of the benzene nucleus. Part II. Synthetic formation of the bridged modification of the nucleus. Trans Chem Soc 121:1143
Ingold CK (1925) The nature of the alternating effects in carbon chains. Part I. The directive influence of the nitroso-group. J Chem Soc 127:513

Ingold CK, Ingold EH (1925) The nature of the alternating effects in carbon chains. Part II. The directing influence of the a-methoxyvinyl group in aromatic substitution. J Chem Soc 127:870–875
Ingold CK, Weaver SD (1925) The additive formation of four membered rings. Part VI. The addition of azo-compounds to ethylene and some transformations of the dimethylene-1,2-di-imine ring. J Chem Soc 127:378–387
Ingold CK, Kidd HV (1933) The benzidine rearrangement. J Chem Soc 984
Ingold CK (1934a) Mesomerism and tautomerism. Nature 133:946
Ingold CK (1934b) Principles of an electronic theory of organic reactions. Chem Rev 15:225–274
Ingold CK (1938) General discussion. Trans Faraday Soc 34:221–222
Ingold CK (1953) Structure and mechanism in organic chemistry. Cornell University Press, Ithaca
Jablonski A (1935) Zur Theorie der Polarisation der Photolumineszenz von Farbstofflosungen. Z Phys 96:236–246
Jacobs A (1997) Understanding organic reaction mechanisms. Cambridge University Press, Cambridge, UK
Kehrmann F, Wentzel F (1901) Über die basischen Eigenschaften des Kohlenstoffs und die Constitution des sogenannten Triphenylmethyls. Chem Ber 34:3815–3819
Kermack WO, Robinson R (1922) An explanation of the property of induced polarity of atoms and an interpretation of the theory of partial valences on an electronic basis. J Chem Soc 121:427–440
Laar C (1885) Über die Möglichkeit mehrerer Strukturformeln für dieselbe chemische Verbindung. Berichte Deutsch Chem Gesellschaft 18:648
Laar C (1886) Über die Möglichkeit mehrerer Strukturformeln für dieselbe chemische Verbindung. Berichte Deutsch Chem Gesellschaft 19:730
Lapworth A (1898) A possible basis of generalization of intramolecular changes in organic compounds. J Chem Soc 73:445
Lapworth A (1901) The form of change in organic compounds and the function of the alpha-meta orientating groups. J Cell Sci 79:1265–1284
Lapworth A (1903) Reactions involving the addition of hydrogen cyanide to carbon compounds. Part I. J Chem Soc 83:995
Lapworth A (1904a) Reactions involving the addition of hydrogen cyanide to carbon compounds. Part II. J Chem Soc 85:1206
Lapworth A (1904b) Reactions involving the addition of hydrogen cyanide to carbon compounds. Part III. J Chem Soc 85:1214
Lapworth A (1904c) Reactions involving the addition of hydrogen cyanide to carbon compounds. Part IV. J Chem Soc 85:1355
Lapworth A (1906a) Reactions involving the addition of hydrogen cyanide to carbon compounds. Part V. J Chem Soc 89:945
Lapworth A (1906b) Reactions involving the addition of hydrogen cyanide to carbon compounds. Part VI. J Chem Soc 89:1869
Lapworth A (1904d) Note on the addition of hydrogen cyanide to unsaturated compounds. Proc Chem Soc 20:245
Lapworth A (1920) Latent polarities of atoms and mechanism of reaction, with special reference to carbonyl compounds. Manch Mem 64(3):1
Lapworth A (1922) A theoretical derivation of the principle of induced alternate polarities. J Chem Soc 121:416–427
Lapworth A, Burkhardt GN, Robinson R (1925) The supposed formation of 1:2:4-oxadi-imine rings from nitroso compounds and methylenearylamine. J Chem Soc 127:1742
Marcus RA (1952) *Unimolecular Dissociations and Free Radical Recombination Reactions*, J. Chem. Phys. 20:359
Markownikoff W (1870) Über die Abhängigkeit der verschiedenen Vertretbarkeit des Radicalwasserstoffs in den isomeren Buttersäuren. Ann Pharm 153:228–259
Meerwein H (1914) Über den Reaktionsmechanismus der Umwandlung von Borneol in Camphen; [Dritte Mitteilung über Pinakolinumlagerungen.]. Liebig's Ann Chem 405:129–175

Meerwein H, van Emster K (1922) Gleichgewichtsisomerie zwischen den acyclischen und cyclischen Carbonium-Ionen, die über die Oxoniuni-Ionen ineinander übergehen konnen. Berichte der Deutschen Chemischen Gesellschaft 55:2500–2529

Merling G (1891) Tautomerisierung des ungesattigten Acyl-amidins XVII. Chem Ber 24:3108

Natarajan A, Tsai CK, Khan SI, McCarren P, Houk KN, Garcia-Garibay MA (2007) The photoarrangement of -santonin is a single-crystal-to-single-crystal reaction: a long kept secret in solid-state organic chemistry revealed. J Am Chem Soc 129:9846–9847

Nebbia G, Kauffman GB (2007) Prophet of solar energy: a retrospective view of Giacomo Luigi Ciamician (1857–1922), the founder of green chemistry, on the 150th anniversary of his birth. Chem Educ 12:362–369

Norris JF (1901) Carbocation history. Am Chem J 25:117

Norrish RGW, Bamford CH (1936) Photodecomposition of aldehydes and ketones. Nature 138:1016

Norrish RGW, Bamford CH (1938) Photodecomposition of aldehydes and ketones. J Chem Soc 1521, 1531, 1544

Norrish RGW, Porter G (1949) Chemical reactions produced by very high light intensities. Nature 164:658

Olah GA, Kuhn SJ, Tolgyesi WS, Baker EB (1962) Stable carbonium ions II. J Am Chem Soc 84:2733

Olah GA, Bollinger JM, Cupas CA, Lukas J (1967) Stable carbonium ions. XXXIV. 1-Methylcyclopentyl cation. J Am Chem Soc 89:2692–2694

Olah GA, Schlosberg RH (1968) Method for measuring acid strength on solid superacid catalyst. J Am Chem Soc 90:2726

Olah GA (1994) Nobel lecture

Olah GA (2005) Beyond oil and gas: the methanol economy. Angew Chem Int Ed 44 (18):2636–2639

Paternò E, Chieffi C (1909) Sintesi in chimica organica per mezzo della luce. Nota II. Composti degli idrocarburi non saturi con aldeidi e chetoni. Gazz Chim Ital 39:341

Robinson R (1941) Mechanism of the benzidine transformation and some allied topics. J Chem Soc 220:1941

Robinson R (1947) A. Lapworth 1872–1941. Obit Not Fell R Soc 5(15):554

Schäfer S, Liang W, Zewail AH (2010) Structural dynamics and transient electric-field effects in ultrafast electron diffraction from surfaces. Chem Phys Lett 493:11

Schlenk W, Marcus E (1914) Über Triaryl-methyle. XII. Über Metalladditionen an freie Radikale. Chem Ber 47:1664

Schlesinger HI, Burg AB (1942) Recent developments in the chemistry of the boron hydrides. Chem Rev 31:1–41

Searles S (1964) Oxetanes. In: Heterocyclic compounds with three- and four-membered rings, Part 2. Interscience, New York, p 984

Stark J (1908) Weiter Bemerkungen iiber die thermische und chemische Absorption im Banden Spektrum. Z Phys 9:889–894

Stieglitz J (1899) On the constitution of the salts of imido-ethers and other carbimide derivatives. J Am Chem Soc 21:101

Thiele J (1899) Zur Kenntniss der ungesättigten Verbindungen. 1. Theorie der ungesättigten aromatischen Verbindungen. Justus Liebigs Annalen der Chemie 306:87–142

Thiele J (1900) Über Ketonreactionen bei dem Cyclopentadiën. Berichte 33:666–673

Trommsdorff JB (1834) Über Santonin. Ann Chem Pharm 11:190–207

Vorländer D (1925) Säurer, Saltz, Ester und Addukt. Berichte 58:118–143

Wagner G (1899) Реакция Вагнера-Меервейна (In Russian: Wagner-Meerwein Rearrangement) Phys Chem Soc 31:690

Wheland GW, Pauling L (1935) A quantum mechanical discussion of orientation of substituents in aromatic molecules. J Am Chem Soc 57:2086

Wheland GW (1942) A quantum mechanical investigation of the orientation of substituents in aromatic molecules. J Am Chem Soc 64:900

Winstein S, Grünwald E, Buckles RE, Hanson C (1948a) The role of neighboring groups in replacement reactions. XI. Some reactivities involving neighboring groups. J Am Chem Soc 70:816

Winstein S, Hanson C, Grünwald E (1948b) The role of neighboring groups in replacement reactions. X.[1] Kinetics of solvolysis of *trans*-2-acetoxycyclohexyl *p*-toluenesulfonate. J Am Chem Soc 70:812

Winstein S, Clippinger E, Fainberg AH, Heck R, Robinson GC (1956) Salt effects and ion pairs in solvolysis and related reactions III common ion rate depression and exchange of anions during acetolysis. J Am Chem Soc 78:328–335

Winstein S (1969) Centenary lecture. Nonclassical ions and homoaromaticity. Q Rev Chem Soc 23:141–176

Winstein S, Hoffmann R (1965) Stereochemistry of electrocyclic reactions. J Am Chem Soc 87:395–397

Woodward RB, Hoffmann R (1965) Stereochemistry of electrocyclic reactions. J Am Chem Soc 87:395–397

Zewail A (1994) Femtochemistry: ultrafast dynamics of the chemical bond. World Scientific, Singapore

Zewail A, Thomas JM (2009) 4D electron microscopy: imaging in space and time. Imperial College Press, London

Zinin N (1845) Benzidine Rearrangement. Rearrangement of hydrazobenzene to yield benzidine. J Prakt Chem 36:93–107

Index

A
Abegg, Richard Wilhelm Heinrich, 122, 123, 125, 128
Abegg's rule, 122
Abney, William, 93
Abragam, Anatole, 104
Académie des Sciences, 136, 137, 143
Accademia del Cimento, 93
Accademia Nazionale dei Lincei, 153
Acid/base theory, 232
Actinium,143, 144, 150, 155, 156,
Action of mass law, 65
Activation energy, 33, 34, 37, 48, 50
Adlineation theory, 49
Aepinus Franz Maria Ulrich Theodosius, 113
Affinity, 9–11, 27–32, 41, 63, 64, 70–72, 74, 76, 77, 81, 109, 122, 124, 125, 224, 226, 230, 249, 257
Alder, Berni Julian, 214
Allen, Michael P., 216
Allinger, Norman L., 214
α-rays, 116, 140, 145, 146
Alternate polarities, 228
Alvarez, Luis Walter, 101
Ambarzumian, (Hambardzumyan) Viktor, 157
Anchimeric assistance, 243
Andrews, Thomas, 9, 10, 187, 229
Angewandte, Chemie, 253
Angular momentum quantization, 120
Anharmonicity, 95
Annalen der Naturphilosophie, 64
Annalen der Physik, 5, 15, 170
Antibonding orbital, 203, 208
Antisymmetric, 131, 185, 189, 196
Armstrong, Henry Edward, 66, 67, 223–227, 231, 232
Arrhenius equation, 33, 34, 68
Arrhenius, Svante, 11, 32–34, 36, 37, 62–70, 77, 109, 142
Aston, Francis William, 112, 145, 150, 1151
Atomic orbitals, 180, 192, 195, 196, 203, 205, 208
Aufbau principle, 128–129, 171, 180, 201, 202, 208

B
Badische Anilin und Soda Fabrik (BASF), 43
Baeyer (von), Adolf, 226, 227, 238, 239
Balard, Antoine Jérôme, 10
Balmer, Johann Jacob, 119, 120, 177
Balzac (de), Honoré, 71
Banks, Joseph, 2
Barkla, Charles Glover, 88
Barks, 128
Bartlett, Neil, 197
Bartlett, Paul Doughty, 241–243, 261
Bartlett, Rodney, J., 197, 207, 208
BASF. *See* Badische Anilin und Soda Fabrik (BASF)
Basov, Nikolay Gennadiyevich, 99, 100
Becker, Herbert, 157
Beckmann, Ernst Otto, 75, 239
Becquerel, Alexander Edmond, 134
Becquerel, Antoine Ce´sar, 134
Becquerel, Antoine Henri, 136–141, 144, 146
Becquerel, Jean-Antoine-Edmond, 138
Bellamy, John Lionel, 94
Bell, Ronald Percy, 48
Bemont, Gustave, 140
Berget, Alphonse, 141
Berg, Otto, 154
Berkelium, 162
Bernard, Claude, 41

S. Califano, *Pathways to Modern Chemical Physics*,
DOI 10.1007/978-3-642-28180-8, © Springer-Verlag Berlin Heidelberg 2012

Berne, Bruce, 212
Berthelot, Marcellin Pierre Eugéne, 9–11, 28, 29, 75
Berthollet, Claude-Louis, 27, 29–31, 72, 73
Berzelius, Jöns Jacob, 9, 10, 32, 42, 59, 71, 76, 139
Beta decay, 162
Beta rays β, 116
Bethe, Hans, 21
BET theory, 46
Bijvoet, Johannes Martin, 90
Binnig, German Gerd, 50
Bird, Robert Byron, 189
Birge, Raymond, 200, 201
Bjerrum, Niels, 69, 70, 81, 95
Black body, 116, 167–170
Blagden, Charles, 73, 74
Bleaney, Brebis, 97, 104
Bloch, Felix, 98, 100, 101, 103
Bloembergen, Nicolaas, 100, 102
Blondlot, Rene´, 137, 138
Blume, Richard, 104
Bodenstein, Max Ernst August, 35, 75
Bodländer, Guido, 122
Boerhaave, Herman, 72
Bohr, Niels, 100, 101, 103, 117–123, 128, 129, 131, 154, 155, 160, 171–173, 177–179, 181, 183, 191, 196, 201–203, 205, 208, 232
Bohr–Sommmerfeld theory, 118–120, 128–131, 203
Boiling point, 33, 75
Bolometer, 93
Boltzmann equation, 13
Boltzmann H theorem, 13
Boltzmann, Ludwig, 13, 14, 17, 19, 22, 37, 67, 140, 167, 169, 175
Bonding orbital, 208
Bonino, Giobatta, 94
Boreskov, Georgii K., 48
Born, Max, 173–175, 178, 179, 186, 201
Born–Oppenheimer approximation, 186, 215, 216
Boscovich, Ruder, 187
Bothe, Walther, 157
Bowers, Klaus, 105
Brackett series, 119
Bragg's law, 90
Bragg, William Henry, 88–90
Bragg, William Lawrence, 90
Branching processes, 35
Brauner, Bohuslav, 82, 153
Bridges, Calvin, 198
Bridgman, Percy William, 195
British Association for the Advancement of Science, 5, 8, 230
Brockhouse Bertram Neville, 92
Broglie (de), Louis, 49, 120, 172, 176
Bronsted, Johannes Nicolaus, 48, 81, 232, 247
Brookes, Harriet, 147,
Brown, Charles Herbert, 244, 245
Brown, George Ingham, 2
Brunauer, Stephen, 45
Büchi, George Hermann, 256
Buckingham, David, 189
Buff, Frank P., 78
Buffon (de) Georges-Louis Leclerc, 72
Bulletin de la Sociéte´ Chimique de France, 44
Bunsen-Gesellschaft fu€r angewandte physikalische Chemie, 64
Buontalenti, Bernardo, 73
Burg, Anton B., 245
Burrau, Oyvind, 183, 191
Burrus, Charles A., 97
Bury, Charles R., 128

C

Califano, 95, 215
Californium, 162
Caloric, 1–4, 8
Cannizzaro, Stanislao, 256, 257
Canonical ensemble, 14
Carathéodory, Constantin, 12
Carnot cycle, 4
Carnot, Lazare, 7, 8, 16
Carnot, Nicolas Leonard Sadi, 3, 4, 6, 7
Car–Parrinello method, 216
Car, Roberto, 216
Casimir, Hendrik Brugt Gerhard, 21, 189
Catalysis, 40–52, 66, 254, 260
Catalytic processes, 40, 46, 48
Cathode rays, 110, 111
Chadwick, James, 132, 158
Chain reactions, 34–36, 258
CHARMM. *See* Chemistry at HARvard Macromolecular Mechanics (CHARMM), 215
Chebyshev, Pafnuty Lvovich, 182
Chemical forces, 32, 77, 126
Chemical potential, 14, 15, 20, 78
Chemical primer, 43
Chemistry at HARvard Macromolecular Mechanics (CHARMM), 215
Chieffi, Generoso, 256
Christie, Samuel Hunter, 62

Ciamician, Giacomo Luigi, 256–258
Claës, Balthazar, 71
Clapeyron Benoît Paul Emile, 4, 6, 8
Clausius, Rudolf Julius Emanuel, 4, 6–8, 14, 16, 20, 62, 187
Cleeton, Claud Edwin, 97
Clementi, Enrico, 204
Cleve, Per, 67
Cluster Expansion method, 207
CNDO. *See* Complete Neglect of Differential Overlap (CNDO) 206, 211
Coblentz, William, 93
Cohen Ezechiel Godert David, 13
Cohen, Julius B., 231
Cohen, Saul, 242
Cohen, Victor W., 100
Coherent Raman spectroscopy, 95
Colding, Ludwig August, 6
College de France, 10
Collie, John Norman, 145, 225
Colligative properties, 68, 69, 73–74
Complementarity principle, 179
Complete Neglect of Differential Overlap (CNDO), 206, 211
Comptes, Rendus, 44, 140
Compton, Arthur H., 171
Compton effect, 171, 172, 178
Conant, James Bryant, 240–242
Condon, Francis E., 242
Conductivity, 61, 63–65, 67, 69, 79, 80, 139, 260
Configuration interaction, 182, 207
Conformational analysis, 96, 215
Conjugation effect, 234
Coolidge, Albert Sprague, 186
Coppet (de), Louis, 74, 75
Cordus, Valerius, 40
Corey, Elias James, 253, 254
Coryell, Charles D., 154, 198
Coster, Dirk, 154, 155
Coulomb correlation, 207
Coulomb energy, 36, 76, 77, 183, 205
Coulomb force, 79
Coulomb integral, 185
Coulomb interaction, 78
Coulson, Charles Alfred, 209, 210, 251
Covalent bond, 126, 127, 195–198, 215, 230
Cram, Donald, 262
Crawford, Bryce L. Jr, 95
Cremer, Erika, 49, 51
Crookes, William, 88, 110, 144, 147, 148
Cryoscopy, 74, 75
Crystallography, 90, 91, 194, 198
Curie, Eve, 137
Curie, Ire'ne, 156, 157
Curie, Marie, 138–144, 146, 156
Curie, Paul-Jacques, 138
Curie, Pierre, 138–144
Curium, 162
Curtiss, Charles. F., 189
Cyclotron, 101, 154, 161, 162
Cýzek Jirì, 205

D

Daniell, John Frederic, 60, 61
Danne, Jacques, 141
Darwin Charles, 117
Darwin, Charles Galton, 117, 118
Darwin George Howard, 117
Daudel, Raymond, 210
Daudet, Léon, 142
Davidson, Ernest R., 92, 207
Davisson, Clinton Joseph, 49, 172
Davy, Edmund, 41
Davy, Humphry, 1, 41, 59, 60
Debierne, André, 140, 143, 144, 156
De Broglie, Louis, 49, 120, 172, 176
Debye–Hückel theory, 17, 78–81, 205
Debye, Peter, 17, 76, 80, 90, 96, 175, 188, 203, 205
De Courmelles, Foveau, 141
De Groot, Sybren Ruurds, 21
De Hevesy, George Charles, 154, 155
De Lisle, Armet, 141
De Maeyer, Leo, 260
Dennison, David Mathias, 93
Dennstedt, Maximilian, 257
Density functional theory (DFT), 211–213, 216
Despretz, César Mansuète, 74
Determinant, 182, 195, 196, 206, 207
Deutsche Elektrochemische Gesellschaft, 64
Dewar James, 197, 198, 232
Dewar, Michael James Steuart, 248–250
DEWAR-PI, 250
Dewar structures, 197, 198
Deyrup Alden J., 240
DFT. *See* Density functional theory (DFT)
Dickinson, Roscoe Gilkey, 188
Dickinson, W.C., 102
Diffraction, 91–92
Diffraction patterns, 89, 90
Diluted solutions, 32, 75, 77, 79, 240
Dirac, Paul Adrien Maurice, 95, 175, 177, 180, 182, 196, 212
Dispersion forces, 188, 214

Distribution function, 14, 78
Djerassi, Carl, 206
Döbereiner, Johann Wolfgang, 41, 42
Dobzhanski, Theodosius, 198
Doering von, Eggers William, 238
Donnan, Frederick George, 82
Dorn, Friedrich Ernst, 144
Dorpat University, 9, 63
Dowden, Dennis Albert, 48
Draper, John W., 254
Driving force, 4, 12, 15, 40
Duclaux, Jacques, 44
Duhem, Pierre Maurice Marie, 17
Dulong, Pierre, 41
Dutrochet, René Joachim Henri, 76

E

Ebullioscopy, 75
Eckart, Carl, 17, 187
Eckhardt, Wilfried, 97
Ecole Nationale des Ponts et Chaussées, 136
Ecole Normale Superieure of Paris, 146
Ecole Normale Supérieure of Sèvres, 140
Ecole Superieure de Pharmacie, 10
Eddington, Arthur, 169
Edlung, Eric, 67
EHMO. *See* Extended Hückel method, 206
Ehrenfest, Paul, 95, 130, 131, 168, 181
Eidgenössische, Technische Hochschule, 78, 256
Eigenfunction, 176, 177, 185
Eigen, Manfred, 259–261
Eigenvalue, 176, 177
Einstein, Albert, 88, 98, 125, 129, 143, 171–173, 175, 178, 255
Einsteinium, 162, 163
Eisenschitz, Robert, 189
Eka-actinium, 145
Eka-cesium, 155
Eka-radium, 145
Eka-thorium, 145
Electrochemistry, 11, 17, 59, 62, 64, 67, 68, 74, 78, 81, 82
Electrolytic dissociation, 61, 62, 67–69, 227
Electrometric effect, 234
Electron nuclear double resonance (ENDOR), 105
Electron spin echo envelope modulation (ESEEM), 105
Electron spin resonance spectroscopy, 103, 104
Electrophilic reactions, 232
Electrophoresis, 81–83
Electrophoretic effect, 79, 80
Electrostatic interaction, 79, 188
Eley, Daniel Douglas, 47, 52
Eley-Rideal, mechanism, 47
Elster, Julius, 143
Emanium, 144
E2 mechanism, 235
Emmett, Paul Hugh, 45
ENDOR. *See* Electron nuclear double resonance (ENDOR), 105
Energy levels, 128, 173, 177, 200, 201, 205
English Garten, 2
Enthalpy, 7, 12, 14, 20, 34
Entropy, 6–9, 11–17, 20, 22, 34, 169, 170
Enzyme, 41, 83, 260, 261
Equilibrium constant, 33, 45, 65, 241, 242
Equilibrium reactions, 28, 227
Erlenmeyer, Friedrich, 127, 257
Ertl, German Gerhard, 50–51
ESEEM. *See* Electron spin echo envelope modulation (ESEEM) 105
Esperanto, 66
Espil, Leo, 44
Esson, William, 28, 75
Evans, Meredith Gwynne, 38, 48
Ewald, Peter Paul, 89, 183
Exchange energy, 6, 182, 185, 192–196, 212, 213, 349
Exchange integral, 185
Extended Hückel method (EHMO), 206, 252
Extensive property, 20
Extensive quantities, 18, 20
Eyring, Henry, 36–38, 40, 236

F

Fabrikant, Valentin Aleksandrovich, 98
Fairbrother, Fred, 162
Fajans, Kazimierz, 149
Falk, Kaufman George, 124
Faraday, Michael, 42, 52, 59–61, 109, 110, 138, 203, 231, 232
Faraday Society, 203, 232
Farjas, Henri, 141
Farkas, Adalbert, 46
Farkas, Ladislas, 46
Favre, Pierre Antoine, 10
Femtochemistry, 263
Femtosecond, 95, 262, 263
Fermi–Dirac statistic, 131
Fermi, Enrico, 89, 99, 131, 159–162, 180, 212–214
Fermions, 131

Fermium, 162, 163
Fernandes, Lorenzo, 153
Festing, Edward Robert, 93
First principle of thermodynamics, 5–7
Flame propagation, 35
Flash photolysis, 258, 259
Flerov, Georgij Nikolaevič, 162
Florentium, 153, 154
Fluorescence, 88, 110, 135, 136, 255
Flürscheim, Bernard Jacques, 45, 227, 230, 232–233
Fock operator, 180
Fock space, 180
Fock, Vladimir Aleksandrovich, 180, 182
Force constants, 95, 96
Force fields, 214, 215
Fowler, Ralph, 182
Francium, 156
Frankland, Percy, 150, 223
Free energy, 6–9, 11, 12, 14–16, 20, 33, 48, 125, 241
Friedel, Charles, 140
Friedrich, Walter, 89
Frisch, Otto, 160
Froese, Fischer Charlotte, 207
Fues, Erwin Richard, 181
Fukui, Kenichi, 251–253
Fuoss, Raymond Matthew, 18, 21, 80
Fusinieri, Ambrogio, 42
Fuson, Nelson, 93

G
γ-rays, 89, 146, 149, 157–159, 178
Gao, Jiali, 216
Garner, William E., 203
Gassendi, Pierre, 71
Gaudion, Georges, 46
Gaunt, John Arthur, 181
GAUSSIAN, 204, 211–213
Gazzetta, Chimica Italiana, 256
Gedanken Experimenten, 178
Geiger counter, 116
Geiger Johannes Wilhelm (Hans), 116, 117, 146, 158
Geissler, Heinrich, 110
Geitel, Hans Friedrich, 143
Gelin, Bruce, 215
Genzel, Ludwig, 97
Germain, George (Lord), 2
Germer, Lester Halbert, 49, 172
Ghiorso, Albert, 161, 162
Ghosh, Jnanendra Chandra, 78
Giauque, William Francis, 13, 151
Gibbs, Josiah Willard, 13–15, 34, 66, 69, 78, 241
Giesel, Friedrich Oskar, 140, 144
Glauber salt, 32, 76
Glendenin, Lawrence E., 154
Gmelin Handbuch der anorganischen Chemie, 49
Göhrling, Otto, 149
Goldstein, Eugen, 110, 112
Gomberg, Moses, 239
Gordon, James, 99, 104
Gordy, Walter, 97
Görter, Cornelis Jacobus, 100
Goudsmit, Samuel Abraham, 129–131, 181, 191
Graham, Thomas, 32, 76, 216
Greiling, Walter, 43
Grignard, Victor, 44
Grosse (von), Aristide, 149
Grotthuss conduction mechanism, 61
Grotthuss (von), Theodor, 60, 61, 71, 254
Group theory, 95, 182, 202, 205, 210, 261
Gruppenpest, 202
Guggenheim, Edward Armand, 15, 190, 191, 196, 203
Guldberg, Cato Maximilian, 29, 30, 64
Gun cotton, 42
Gwinn, William Dulaney, 97

H
Haas, Arthur Erich, 120
Habann, Erich, 97
Haber Bosch ammonia synthesis, 43, 44, 46, 47, 50, 51
Haber, Fritz, 43, 46, 160
Hadermur (von), Leopold Pfaundler, 29
Hafnium, 154, 155
Hahn, Otto, 149, 159, 160, 162
Hale, George Hellery, 190
Hall, George G., 204, 211
Halochromie, 238
Hamiltonian operator, 177, 182, 185
Hammett constant, 241
Hammett, Louis Plack, 240, 241
Hammond, George Simms, 257, 261, 262
Hammond postulate, 262
Handy, Nicholas C., 207
Hansen, Hans Marius, 119
Hansen, William Webster, 98, 100
Harcourt, Augustus George Vernon, 28, 41, 75
Harmonic oscillator, 177

Harris, J. Allen, 91, 154
Harteck, Paul Karl Maria, 46
Hartley, Walter Noel, 93
Hartree, Douglas Rayner, 180–182
Hartree–Fock limit, 207
Hartree–Fock method, 182, 206, 209, 212
Hartree method, 181
Hartridge, Hamilton, 258
Hasenoεhrl, Fritz, 175
Hauffe, Karl, 48
Hauksbee, Francis, 187
Hauptmann, Herbert Aaron, 91
Hehre, Warren, 212
Heisenberg principle, 179
Heisenberg, Werner, 95, 100, 101, 130, 173–179, 182, 184, 191, 196, 202, 203
Heitler-London method, 186, 192, 196
Heitler, Walter, 185, 190, 192
Helmholtz Hermann Ludwig Ferdinand, 6, 8, 11, 14, 19, 109, 111, 112, 115
Henry, William, 42
Herschbach, Dudley Robert, 38, 98
Herschel, William, 92, 93
Hertz Gustav, 160
Hertz, Heinrich, 88, 110, 111
Herzberg, Gerhard, 95, 203, 204
Herzfeld, Karl F., 204
Hess, Germain Ivanovich, 9, 10
Hess law, 9, 10
Heterogeneous catalysis, 41, 43–46, 49, 51, 52
Heyrovsky, Jaroslav, 82
Highest Occupied Molecular Orbital (HOMO), 251–253
Hilbert, David, 174, 182
Hinshelwood, Cyril, 34–36, 39, 47
Hirschfelder, Joseph Oakland, 189
Hittorf, Johann Wilhelm, 61, 62, 88, 110
Hoffmann, Roald, 158, 206, 250, 252–254
Hohenberg, Pierre, 212
Hollnagel, Herbert, 96
Holmes, E. L., 233
Holtsmark, Johan Peter, 80
Homogeneous catalysis, 40, 42, 46, 48
HOMO. *See* Highest Occupied Molecular Orbital (HOMO), 251–253
Hönl, Helmut, 183
Hood, John. J., 75
Hoover, William Graham, 216
Hope, Thomas Charles, 74
Hoppe, Rudolf, 197
Hückel, Erich, 78, 205, 206
Hückel method, 205, 252
Hückel, Walter, 205
Hughes, Edward David, 235–237
Hull, Albert, 90, 97
Hund, Friedrich, 196, 199, 201–203, 208
Hund's rules, 201
Huntingdon, A. K., 93
Hybridization sp2, 195
Hybridization sp3, 194–196
Hybridization sp, 195
Hybrid orbitals, 194, 195
Hypercycle, 260, 261
Hyperfine structure in NMR,101, 104

I

IBM research laboratories, 50
Ideal gases, 33, 65, 77
Ido, 66
Illinium, 153, 154
Indeterminacy relationship, 178
Inductive effect, 234
Infrared luminescence, 38
Infrared spectra, 93, 94
Ingold, Christopher Kelk, 229, 231–237, 240, 241, 243, 249, 250
Ingold-Usherwood Edith Hilda, 233
Inter-ionic interactions, 69, 78
Intermediate species, 35, 238
Internal quantum number, 130
Intimate ion pair, 243
Inverted region, 40
Ionescu, Theodor V., 97
Ipatieff, Vladimir Nikolaevich, 47
Irreversible processes of thermodynamics, 16–18, 21, 22
Isomerism, 96, 160, 235, 256
Isomorphism, 32, 76, 144
Isosterism, 127
Isotope, 132, 145, 148–154, 156, 158, 160–162
Isotopic substitution, 235
Iwanenko, Dmitri, 157

J

Jablonski, Alexander, 255
James, Hubert M., 186
Jaumann, Gustav, 17
Jeans, James, 168
Jevons, Wilfred, 200
Joliot, Jean, 156, 157
Jordan, Pascual, 174, 175
Josien, Marie Louise, 94
Joule, James Prescott, 4–6, 8
Joule law, 4

K
Kahlenberg, Louis, 69
Kaiser Wilhelm Institut, 36, 38, 189, 259
Kanellopoulos, Nikolaos, 49
Kant, Immanuel, 8, 155
Karle, Jerome, 91
Karolinska Institutet, 32
Karplus, Martin, 215
Kasha, Michael, 204, 255
Kassel, Louis S., 39, 40, 264
Katchalsky, Aharon Katzir, 21
Kaufman, Bruria, 21
Keesom, Willem Hendrik, 188
Kehrman, F., 236
Kekulè, 30, 122, 197, 224, 226, 232
Kekulè structures, 196, 222
Kelvin (Lord). *See* Thomson, William
Kemble, Edwin Crawford, 95, 200
Kendrew, John, 91
Kennedy, Joseph W., 161
Kermack, William Ogilvy, 231
Kipping, Frederick Stanley, 227
Kirchhoff, Gustav, 167–169
Kirchhoff, Sigismund Konstantin, 40, 41
Kirkwood, John Gamble, 78, 189
Klaproth, Martin Heinrich, 72
Klein, Felix Christian, 174
Klein, Michael L., 216
Klemperer, William, 98
Knipping, Paul, 89
Knoevenagel, Emil, 230
Knox, Lawrence H., 238, 242
Kohlrausch, Friedrich Wilhelm Georg, 62, 65, 67
Kohn, Walter, 189, 211–213
Kolbe, Adolph Wilhelm Hermann, 75, 2233–225
Königliche Gesellschaft der Wissenschaften zu Göttingen, 12
Königlich Preussische Akademie der Wissenschaften, 12
Kopfermann, Hans, 98
Kopp, Herman, 73
Körner, Wilhelm, 226
Kosmium, 153
Kossel, Albrecht, 123
Kossel, Walther, 122–125, 128
Kramers, Hendrik Anthony, 21, 172, 196
Kratzer, Adolf, 200
Krishnan, Kariamanikam Srinivasa, 94
Kronig de, Laer Ralph, 130
Kuhlmann, Charles Frédéric, 43
Kühne, Wilhelm, 41
Kunckel, Johann, 70
Kupelwieser, Karl, 141
Kurtze, Walter, 260

L
Laar, Conrad, 226
Lademburg, Rudolf, 100
Ladenburg, Albert, 226
Landsberg, Grigorii Samuilovich, 94
Langevin, Paul, 142, 143
Langley, Samuel Pierpont, 93
Langmuir–Hinshelwood mechanism, 47
Langmuir, Irvin, 47, 127–131
Lanthanum, 144, 153
Laplace, Pier Simon, 187
Lapworth, Arthur, 225, 227–229, 231–234, 241
Larmor, Joseph, 111
Laser, 38, 94, 95, 98, 102, 262, 263
Laue (von), Max, 89, 90
Lavoisier, 1, 2, 9, 71, 206, 243
Lawrence, Ernest Orlando, 154, 161
Le Bon, Gustave, 137
Le Chatelier, Henri Louis, 14, 34
Lecomte, Jean, 94
Lee, Yuan Tseh, 39
Lemberg, Johann, 63
Lembert, Max, 149
Lénárd (von), Philipp Eduard Anton, 110
Lennard-Jones, John Edward, 188, 203–205, 208, 209, 211
Levitt, Michael, 215
Lewis acids, 238, 242, 246, 247
Lewis, Gilbert Newton, 123–126, 161, 180, 190, 200, 224, 229, 232, 255
Lewis–Langmuir theory, 127
Lide, David R. Jr., 97
Lieben, Adolf, 256
Liebig (von), Justus, 32, 42, 63, 76, 241
Lifson, Shneior, 215
Lindemann, Frederick Alexander, 39
Lindsay, Robert Bruce, 181
Linear combination of atomic orbitals (LCAO), 192, 203, 205, 208
Lippmann, Gabriel, 138
Livingood, J., 162
Loi générale de la congelation, 74
London, Fritz, 36, 183, 187, 192
Longuet-Higgins, Hugh Christopher, 204, 209, 210, 251
Loomis, Francis Wheeler, 200
Loschmidt, Josef, 17
Lotka, Alfred James, 17

Löwdin, Per-Olov, 207, 211
Low energy electron diffraction (LEED), 49, 50
Lowry, Thomas Martin, 81, 197, 225, 231, 232, 258
Lukas, Joachim, 247
LUMO, 251, 252
Lutetium, 153, 155
Lyapunov, Aleksandr Mikhailovich, 182
Lyman series, 119
Lyman, Theodore, 200

M
Mabery, Charles F., 93
MacDonald, William, 146
MacDougall, Duncan P., 13
Machlup, Stefan, 21
Magnetic quantum number, 129
Magnetron, 97, 98
Mailhe, Alfonse, 44
Main-Smith, John David, 128, 129
Malaguti, Faustino, 31
Mandel'shtam, Leonid Isaakovich, 92
MANIAC, 213–214
Many electron systems, 120, 212
Marcus, E., 239
Marcus, Rudolph Arthur, 39, 40, 48, 264
Marinsky, Jacob Akiba, 154
Markovnikov, Vladimir, 225
Marschak, Jacob, 38
Marsden, Ernest, 116, 117
Maser, 98, 99
Mass spectrograph, 112, 151
Masurium, 154, 159
Maxwell, 13, 14, 109, 113, 118
Mayer, Alfred Marshall, 114
Mayer, Hans, 140
Mayer (von), Julius Robert, 5, 6, 42
Mazur, Pavel, 21
McCammon, Andrew, 215
McMillan, Edwin Mattison, 161
McWeeny, Roy, 207, 209, 210
Mechanical equivalent of heat, 5, 6
Mecke, Reinhard, 94, 201
Meerwein, Hans, 238, 239
Meitner, Lise, 149, 159, 160
Meixner, Josef, 21
Melloni, Macedonio, 93
Melting point, 8, 33, 73
Mendeleev, Dmitrij Ivanovic, 67, 68, 115, 123, 145, 153–155, 225
Mendelevium, 162, 163
Menshutkin, Nikolai Aleksandrovich, 238
Mensing, Lucy, 95
Mercury pump, 110
Merling, Georg, 238
Mesomerism, 234, 249
Metropolis, Nicholas, 214
Meyer, Stefan, 140, 141
Meyer, Victor, 239
Michael, Arthur, 225
Microscopic reversibility criterion, 19–21
Microwave spectroscopy, 38, 97
Mie, Gustav Adolf Feodor Wilhelm Ludwig, 188
Miller, William Allen, 61
Millikan, Robert A., 190, 200
Milner, Samuel Roslington, 78
Mims, William B., 104, 105
MINDO, 250
Minkowski, Hermann, 174
Mitscherlich, Eilhard, 31, 32, 76, 77
Mittasch, Alwin, 43
Mizushima, Sanichiro, 94, 96
MNDO, 250
Moccia, Roberto, 210
Modes of vibration, 94, 95, 236, 237
Moffitt, William, 209, 210
Moissan, Henry, 44, 45
Molecular dynamics (MD), 87, 92, 213–216, 263
Møller, Christian, 207
Moller–Plesset method, 207
Monolayer, 45, 47
Monte Carlo algorithm, 214
Morgan, Thomas Hunt, 190, 198
Moseley, Gwyn Jeffreys, 152, 153, 155
Moskowitz Al, 210
Moureu, Charles, 44
Mulliken, Robert Sanderson, 179, 180, 183, 199–204, 208
Mulliken, Samuel Parsons, 199
Murat, Marcel, 44
Myers, Richard J., 97

N
Nagaoka, Hantaro, 115
Natanson, Władysław, 17
Nature, 138, 152, 158, 160
Naturphilosophie, 6, 64
Naturwissenschaften, 175
Neighboring group, 243
Nelson, John Maurice, 124
Neptunium, 161

Nernst lamp, 12
Nernst, Walther Hermann, 11, 12, 15, 34, 35, 69, 122, 124, 260
Neumann (von), John, 38
Neutron, 91, 92, 100, 101, 132, 156–162
Newcomen, Thomas, 3
Newton equations, 173
Newton, Isaac, 72, 173, 187, 213
Nicholson, John William, 120, 172

O
Octet, 123, 126, 127, 231
Octet rule, 125, 126
Oettingen (von), Arthur, 63
Of irreversible processes, 21, 22
Olah, George Andrew, 240, 244–248
Olsen, Jeppe, 207
Onnes, Heike Kamerlingh, 7, 20, 188
Onsager, Lars, 17–22, 79–81
Onsager limiting law, 17, 80
Onsager reciprocal relations, 19–21
Operator, 95, 128, 175–177, 182, 185, 208
Oppenheimer, Robert, 95, 186, 194, 201
Orbit, 117–121, 128, 131, 157, 172, 173, 178, 179, 191, 202, 203
Orbital, 179, 180, 193, 195, 196, 198–209, 211, 248, 249, 251–253
Orbital quantum number, 121, 129
Ørsted, Hans Christian, 6
Ortho-hydrogen, 46, 47
Osmometer, 76
Osmosis, 76
Osmotic pressure, 33, 69, 75–77, 125
Ostwald's dilution law, 65
Ostwald, Wilhelm, 11, 14, 28–30, 33, 43, 44, 62–69, 77, 124, 125
Overtone, 53, 176
Owens, Robert Bowie, 144, 146
Ozone, 42, 51

P
Packard, Martin, 98
Painleve´, Paul, 142
Pais, Abraham, 157, 173
Paneth, Friedrich Adolf, 155
Papin, Denis, 3
Paracelsus, 70
Para-hydrogen, 46
Paramagnetic relaxation, 100
Pariser–Parr–Pople method, 211
Parmentier, Antoine Augustin, 40
Parrinello, Michele, 216
Parr, Robert, 204, 212
Parson, Alfred Lauck, 125
Paschen series, 119
Pasta, John, 213
Paterno‘, Emanuele, 256
Patterson, Arthur Lindo, 90
Pauli exclusion principle, 130, 131, 185, 201
Pauling, Linus, 190–198, 202, 234, 251, 263
Pauli, Wolfgang, 101, 129–132, 177, 182, 190–198, 202, 234, 251, 263
Paul, T., 239
Payen, Anselme, 41
Pechblende, 140
Peerdeman, Antonius Frans, 91
Peierls, Rudolf, 21
Peltier effect, 16
Penrose, Roger, 1042
Perey, Marguerite, 156
Pericyclic reaction, 250, 252
Periodic system, 115, 121, 123, 131
Perkin, William Henry, 229
Perpetuum mobile, 8
Perrier, Carlo, 154
Perrin, Jean-Baptiste, 111
Persoz, Jean-Francois, 41
Perutz, Max Ferdinand, 93
Pettersson, Otto, 67
Pettit, Rowland, 250
Pfeffer, Wilhelm, 32, 76, 77
Pfund series, 119
PH, 66, 234
Philosophical Magazine, 5, 148
Phonon modes, 92
Phosphorescence, 136, 255
Photochemistry, 11, 254–264
Photoelectric effect, 88, 171
Photon, 98, 100, 171, 178, 252, 254
Physical Review, 195, 204
Physisorption, 46
Pickering, Spencer Umfreville, 69
Picosecond, 95, 262, 263
Pietsch, Erich, 49
Piezoelectric effect, 138
Pimentel, George Claude, 93
Pines, Herman, 47
Pisarzhevsky, Lev Vladimirovich, 48
Placzek, George, 94
Planck, Max, 12, 65, 89, 118, 119, 160, 168–170, 173
Planck's hypothesis, 170
Platinum sponge, 42
Plesset, Milton Spinoza, 207

Plucker, Julius, 88, 110
Plumpton, Betty, 104
Poincare´, Henri, 136, 137, 169
Poisson equation, 79
Polanyi, John, 38
Polanyi, Michael, 36–38, 47, 48
Polarography, 81, 82
Polonium, 140, 141, 152
Pontifical Academy, 96
Pople, John Anthony, 209, 211–213
Popper, Carl, 155
Porta, (della) Gian Battista, 73, 93
Porter, George, 52, 258–260
Porto, da Silva Sergio Pereira, 94
Positive–negative rule, 225
Potential energy, 8, 37, 48, 173, 213, 215
Potential energy surface, 36, 37
Pound, Robert Vivian, 98, 101
Poynting, John, 150
Pre-exponential factor, 33
Price, Charles C., 204
Prigogine, Ilya, 21, 22
Principal quantum number, 121, 129
Proceedings of the National Academy of Sciences (PNAS), 192
Procopio dei Coltelli, Francesco, 73
Proctor, Warren, 102
Prokhorov, Aleksandr, 97, 98
Promethium, 153, 154
Protactinium, 144, 149, 160
Proust's theory, 27
Pryce, Maurice Henry Lecorney, 104
Pullman, Alberte, 210
Pullman, Bernard, 210
Puluj, Ivan, 88
Purcell, Edward Mills, 98, 101, 103

Q

Quadrupole, 186
Quadrupole interactions, 187
Quantum number, 97, 99, 119, 127–130, 175, 199, 203
Quantum operators, 173
Quantum transition, 130, 174

R

Rabi, Isidor Isaac, 100, 101, 130
Racemization, 235
Radioactivity, 115, 135–163
Radium, 140–145, 147, 148, 152, 155, 156, 159
Radon, 145, 152, 155, 156
Rahman, Aneesur, 215, 216
Raman, Chandrasekhara Venkata, 94, 96, 203
Raman effect, 94
Ramsay, William, 15, 82, 145, 148, 159
Ramsey, Norman Foster, 102, 141
Ramsperger, Herman Carl, 39, 264
Randall, Harrison McAllister, 15, 93
Randall, John Turton, 97
Randal, Merle, 15
Raoult, Franc¸ois-Marie, 74, 75, 77
Rasetti, Franco, 91
Rate constant, 31, 50, 241
Rayleigh diffusion, 94
Rayleigh (Lord). *See* Strutt, John William
Reaction mechanisms, 19, 42, 46, 50, 124, 128, 197, 223–229, 236, 237, 242, 248, 250, 254, 258, 260, 262, 263
Reaction rates, 28, 29, 31, 37–39, 55, 63, 64, 68, 235, 238, 241, 262, 264
Recurrence paradox, 169
Redeal, Eric, 258
Relaxation times, 102, 104, 260
Residual valences, 46, 47, 224, 225
Resistivity, 62
Resonance, 101–104, 195–199, 225, 249
Reversible reactions, 11, 28, 31–34
Rice, Oskar Knefler, 39, 40
Rice, Ramsperger, Kassel, Marcus (RRKM) theory, 39, 264
Rice–Ramsperger–Kassel theory, 39
Richards, Theodore William, 124, 149
Rideal, Eric Keightley, 47, 52
Righi, Augusto, 97, 111
Riiber, Claus Nissen, 19
Rohrer, Heinrich, 50
Rolla, Luigi, 153
Röntgen, Wilhelm Conrad, 88, 135, 136
Roothaan, Clemens C.J., 204
Rosenbluth, Marshall N., 214
Rossetti, Francesco, 74
Rotational spectroscopy, 96–100
Roughton, John Worsley, 258
Royal Institution, 6, 52
Royds, Thomas, 116, 146
Rubens, Heinrich, 96
Ruedenberg, Klaus, 204
Rumford, 2
Rutherford, Ernest, 112, 115–118, 132, 140, 141, 144–149, 152, 156–159, 186
Ruzicka, Leopold, 257
Rydberg, Johannes Robert, 119, 120, 131, 152

S

Sabatier, Paul, 44, 45
Sabatier-Senderens catalytic process, 44–45
Sackur, Otto, 15
Salvetti, Oriano, 210
Sartorius, Balthasar, 93
Saunders, Frederick A., 196, 200
Savery, Thomas, 3
Sayvetz, Aaron, 187
Scanning tunneling microscope (STM), 50
Scattering experiments, 92
Scerri, Eric R., 155
Schachtschneider, Jerry H., 215
Schenzinger, Karl Aloys, 43
Scheraga, Harold A., 214, 215
Scherrer, Paul, 90
Schlenk, Wilhelm Johann, 239
Schlesinger, Hermann Irving, 245
Schmidt, Carl, 63
Schmidt, Gerhard Carl, 143
Schneider, Abraham, 242
Schoenflies, Arthur Moritz, 12, 95
Schönbein, Christian Friedrich, 42
Schoolery, James, 102
Schrödinger equation, 130, 177, 180, 181, 183, 186, 192, 212
Schrödinger, Erwin, 175–178, 183, 184, 191, 192, 202
Schultes, Hermann, 49
Schuster, Peter, 260, 261
Schwab, Georg-Maria, 49
Schweidler (von), Egon, 140
Scrocco, Eolo, 210
Seaborg, Glenn T., 161–163
Searles, J. S., 257
Second principle of thermodynamics, 6–8, 13, 14, 16, 33, 77
Seebeck effect, 16
Seel, Fritz, 245
Segre‘, Emilio, 154
Self-consistent field, 181
Semipermeable membranes, 32, 75, 76, 82
Semyonov, Nikolay, 35
Senderens, Jean Baptiste, 44, 45
Shells, 90, 123, 126–128, 202
Shen, Yuen-Ron, 51
Sheppard, Norman, 93
Sherman, Jack, 198
Shikata, Masuzo, 82
Shimanouchi, Takehiko, 96
Shingu, H., 251
Shull, Clifford, 92
Sidgwick, Nevil Vincent, 197
Silbermann, Johann Theobald, 10
Silber, Paul, 257
Sinanoglu, Oktay, 207
Sklodowska, Bronisława, 138
Sklodowska, Manya, 135
Sklodowska, Marie, 138
Slater, John Clark, 172, 173, 180–182, 189, 195, 196, 201, 202
Slater orbitals, 196
Sleeping energy, 42
Smekal, Adolf, 94
Smith, Hopkins B. James, 154
Smithsonian Institution, 93
S_N1 mechanism, 235, 236, 242
S_N2 mechanism, 235, 236
Snyder, Robert G., 215
Soddy, Frederick, 144, 145, 147–149
Solid state lasers, 262
Solvay conference, 141
Sommerfeld, Arnold, 22, 36, 89, 120, 121, 123, 128, 130, 183, 189, 200, 204
Somorjai, Arpad Gabor, 51–53
Sorensen, Soren, 66
Spin, 21, 100–105, 129–132, 175, 185, 189, 191–193, 195–197, 201, 206, 207, 212, 255
Spin-lattice relaxation, 104
Spin magnetic moment, 100, 101
Spin-orbitals, 196
Spin quantum number, 101, 129, 130, 132
Stark, Johannes, 254
Statistical thermodynamics, 13–15, 17, 169
Staudinger, Hermann, 240
Steam engine, 3
Stefan–Boltzmann law, 167
Stefan, Josef, 17
Stengers, Isabelle, 22
Stern, Otto, 100
St. Gilles, Pean, 28
Stieglitz, Julius, 237
Stillinger, Frank, 215
Stohmann, Friedrich, 42
Stoicheff, Boris Peter, 94
Stokes law, 79
Stoner, Edmund Clifton, 128, 129, 131
Stoney, George Johnstone, 111
Stopped flow method, 259
Strassmann, Fritz, 159, 160
Strömholm, Daniel, 144
Strutt, John William, 16, 94, 115, 117, 170, 187, 199

T
Tackle, Ida, 154
Tait, Peter Guthrie, 112
Tammann, Gustav, 63, 75
Tamm, Konrad, 260
Tantalum, 153
Tautomeric effect, 234
Tautomerism, 226, 227, 232, 249
Technetium, 154
Teller, Edward, 45, 203, 204
Temperature absolute scale, 8
Tension force, 8
Téry, Gustave, 142
Tetrode, Hugo, 15
Thenard, Louis Jacques, 41
Thermal engines, 3–8
Thermodynamic forces, 18, 20
Thermodynamics, 1–2, 5, 6, 8, 13–22, 66, 77, 78, 125, 139, 167, 170, 174
Thermomultiplier, 93
Thiele, Friedrich Karl Johannes, 224, 225, 228–231, 239
Thomas, John Meurig, 52
Thomas, Llewellyn Hilleth, 212
Thompson, Benjamin, 1, 2
Thomsen–Berthelot principle, 11
Thomsen, Hans Peter Jorgen Julius, 11, 64
Thomson, George Paget, 49, 172
Thomson, Joseph John, 49, 111–118, 123–125, 145, 146, 150, 151, 172
Thomson's model, 111, 113–115
Thomson Thomas, 75
Thomson, William, 8, 16, 71, 112
Thorium, 139, 143–147, 150, 160
Thorpe, Jocelyn, 229, 232
Tilden, William Augustus, 150
Tildesley, Dominic, 216
Tiselius, Arne Wilhelm Kaurin, 83
Tolman, Richard C., 19
Tomasi, Jacopo, 210
Toronto arc, 94
Torrey, Henry C., 101
Townes, Charles Hard, 98, 99
Trajectory, 111, 171, 173, 178, 181, 213, 215
Transactions of the Connecticut Academy, 14
Transition state theory, 36–40, 48, 262
Transport number, 61
Traube, Moritz, 76
Traube, Wilhelm, 160
Tripsine, 41
Trommsdorff, Johann Bartholom€aus, 254
Truhlar, Donald, 216
Turro, Nicholas J., 257

U
Uhlenbeck, George Eugene, 129, 131, 181, 191
Ulam, Stanislaw, 213
Ultrasound, 260
Ultraviolet catastrophe, 168
Ultraviolet photoelectron spectroscopy (UPS), 49
Unstable molecules, 95, 237
Uranium, 116, 135, 137–140, 143, 144, 146–150, 154, 157–162
U rays, 140, 145
Urey, Harold Clayton, 151, 201
Usaybi'a, Ibn Abu, 73

V
Vagner, Egor Egorovich. *See* Wagner, Georg J.
Valence, 122–127, 195, 196, 200, 203, 206, 209, 223–225, 228–232, 251
Valence bond theory, 190–199, 202, 209
Valence electrons, 127, 200
van Bommel, Adriaan J., 91
van den Broek, Antonius Johannes, 152
van der Waals forces, 44, 183, 187, 188
van der Waals, Johannes Diderik, 187
van Helmont, Jan Baptiste, 70
van't Hoff, Jacobus, 11, 29–34, 36, 62–65, 69, 78, 79
van Vleck, John Hasbrouck, 95, 201
Variational principle, 185
Vibrational assignments, 94
Vicinal group, 243, 244
Villard, Paul, 146
Vincent-Geisse, Josette, 94
Virkow, Rudolph Carl, 41
Vis vitalis, 10
Vis-viva, 5
Vol'kenshtein, Fyodor F., 48
Vorländer, Daniel, 228
Vortex theory, 112
Vortices 111–113
Vries (de), Hugo, 32, 76, 77

W
Waage, Peter, 29, 64
Wagner, Georg J., 238, 239, 242
Wagner–Meerwein molecular rearrangement, 239, 242
Wahl, Arthur C., 161
Wainwright, Thomas Everett, 214
Wallenberg Raoul, 51
Wallerius, Johan Gottschalk, 70

Wang, Shou Chin, 188, 189
Wannier, Gregory, 21
Warshel, Arieh, 215, 216
Watt, James, 3
Wavefunction, 131, 176, 178–186, 189, 192, 194, 196, 197, 205, 206, 209, 212
Wave packet, 178
Weber, Joseph, 98
Weizsäcker (von), Carl Friedrich, 160
Welsbach, Carl Auer von, 141
Welsch, Harry Lambert, 94
Wenzel, Karl Friedrich, 28
Werner, Adolph, 254
Weyl, Hermann, 95, 175
Wheatstone bridge, 62
Wheatstone, Charles, 62
Wheland, George, 197, 251
Whewell, William, 59
Whitney, Willis Rodney, 63
Whytlaw-Gray, Robert, 145
Wichelhaus, Carl Hermann, 122, 197
Wicke, Ewald, 259
Wien, Wilhelm, 112, 167–170
Wigner, Eugene Paul, 46, 95
Wilhelmy, Ludwig Ferdinand, 28, 40, 41
Williamson, A.W., 31, 61, 62
Wilson, Edgar Bright Jr., 38, 95, 97, 198
Wilson's GF matrix method, 95
Winkler, Carl A., 39
Winstein, Saul, 243–245
Wislicenus, Johannes, 30, 63, 75
Wittig, Georg, 245, 254
Wöhler, Friedrich, 63
Wollan, Ernest Omar, 91
Wood, Robert William, 138, 170
Woodward–Hoffmann rules, 252, 253
Woodward, Robert, 206, 250, 252–254
Wüllner, Adolph, 75

X

X rays scattering, 171
X ray's spectrograph, 90

Y

Yang, Weitao, 212, 216
Yntema, Leonard, 154
Yonezawa, T., 251

Z

Žáček, August, 97
Zaitsev, Aleksandr Mikhailovich, 238
Zavoisky, Yevgeny Konstantinovich, 103, 104
Zeeman effect, 130
Zeeman effect anomalous, 130, 131, 138
Zeiger, Herbert, 99
Zeitschrift für physikalische Chemie, 34, 43, 64, 259
Zemplen, Geza, 245
Zermelo, Ernst, 169
Zero differential overlap, 211
Zewail, Ahmed, 263
Zinin, Nikolai Nikolaevich, 249
Zinin reaction, 250
Zischka, Anton, 43

MIX
Papier aus verantwortungsvollen Quellen
Paper from responsible sources
FSC® C105338
FSC
www.fsc.org

If you have any concerns about our products,
you can contact us on
ProductSafety@springernature.com

In case Publisher is established outside the EU,
the EU authorized representative is:
Springer Nature Customer Service Center GmbH
Europaplatz 3, 69115 Heidelberg, Germany

Printed by Libri Plureos GmbH
in Hamburg, Germany